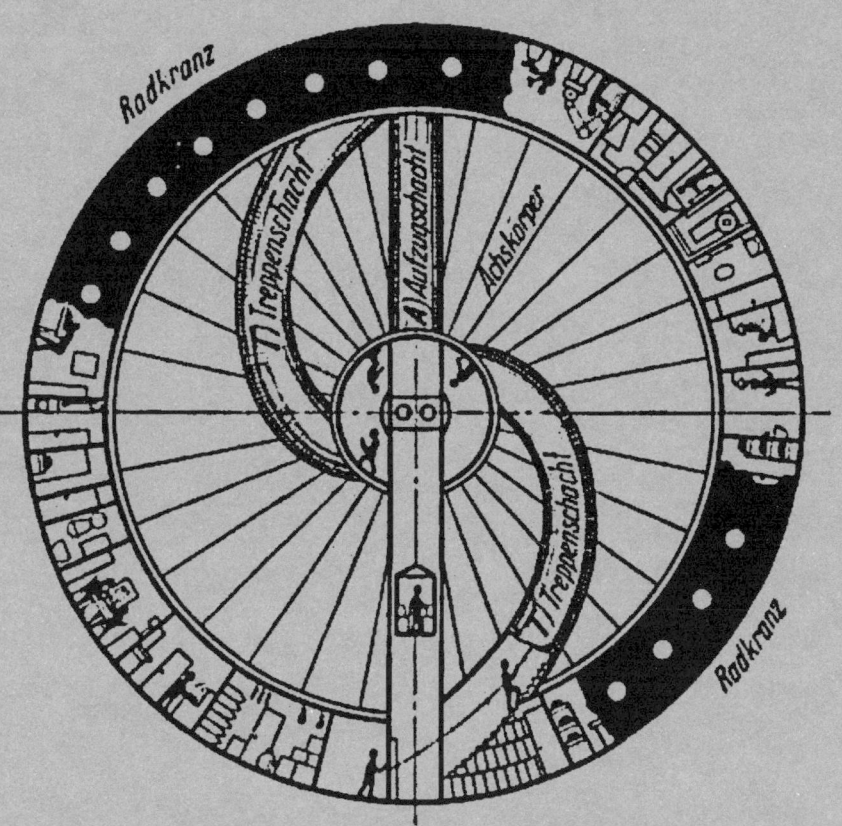

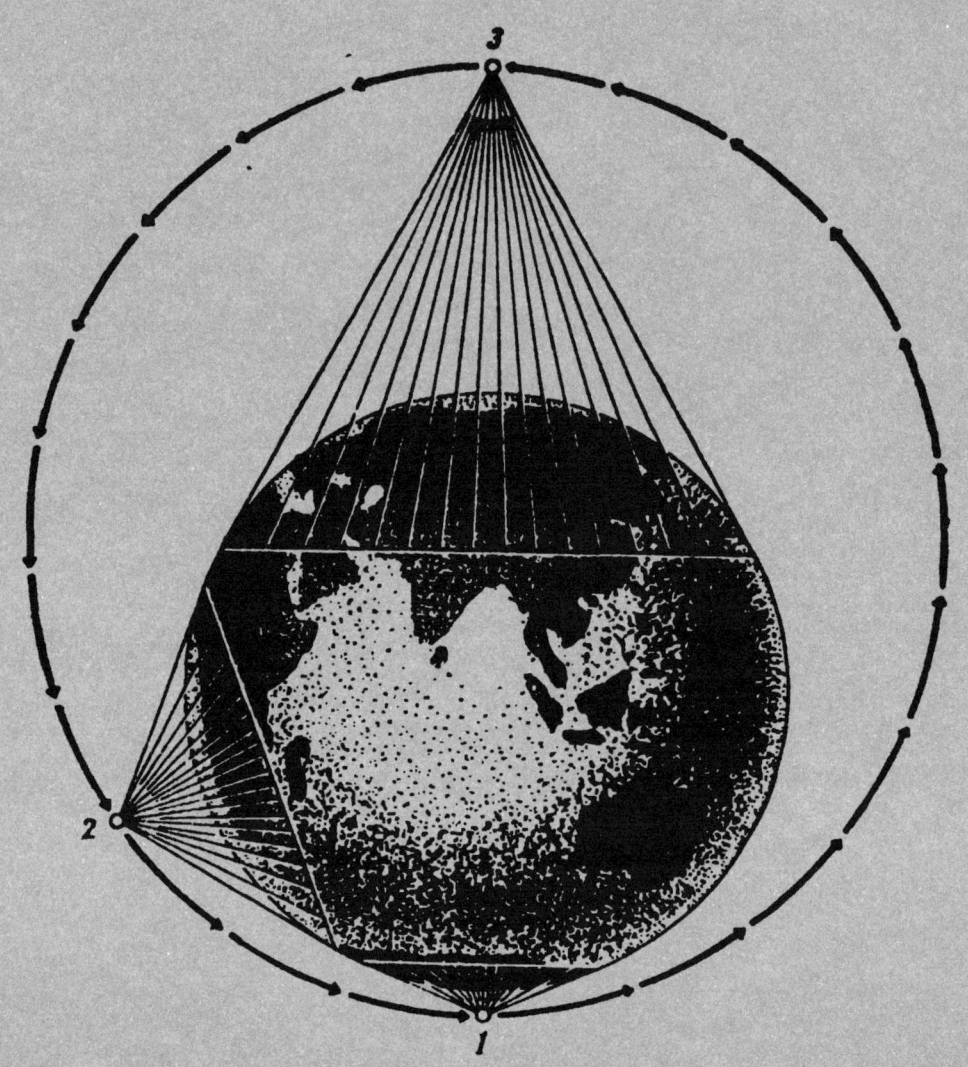

SPACE JOURNAL

DALLAS CAMPBELL

SPACE JOURNAL

ART, SCIENCE AND COSMIC EXPLORATION

WITH OVER 500 ILLUSTRATIONS

Introduction 8

ACT 1.

14

'DREAM'

SPACE DREAMERS AND FLIGHTS OF FANCY

1. Johannes Kepler and Co. 16
Early space visionaries

2. Jules Verne 26
Voyages Extraordinaires!

3. Chesley Bonestell 34
Destinations. The space artist who painted the new frontier

4. Arthur C. Clarke 44
Science fiction writer and Futurist who lived forever in tomorrow

ACT 2.

CAELUM SPECTANTES 52

'STARING INTO SPACE'
THE SCIENTISTS AND THINKERS WHO MADE
SENSE OF THE UNIVERSE

5. Galileo Galilee 54
The astronomer who
brought the stars
within reach

6. Isaac Newton 60
For every action ...

7. Percival Lowell 66
Believing is seeing.
The astronomer who
mapped the Martian canals

8. Mary Ward 74
Ireland's first lady
of the lens

9. Sara Seager 84
Second home.
Exoplanet explorer

ACT 3.

IGNITIO 90

'IGNITION'
LEAVING THE EARTH: THE ROCKET ENGINEER

10. Konstantin Tsiolkovsky 92
Leaving the cradle.
The spiritual father
of Soviet cosmonautics

11. Robert H. Goddard 102
From a cherry tree
to the Moon. America's
first rocket scientist

12. Hermann Oberth 112
The true founding father
of German rocketry

13. Wernher von Braun 120
From Wunderwaffe to
Apollo. The dark legacy
of Dr Space

14. Gerhard Zucker 128
Par fusée. The German
who delivered rocket
mail to Scotland

15. The Chief Designer 134
The secret architect
of the Soviet space
programme

ACT 4.

ITER STELLARUM

140

'STARRY JOURNEY'
SPACE TRAVELLERS

16. Soviet Firsts 142
Let's Go! Gagarin, Tereshkova and Leonov

17. Laika 152
The first of the Soviet space dogs

18. John F. Kennedy and Theodore C. Sorensen 158
Rhetoric and politics in space travel

19. NASA 166
Government organization turned graphic design icon

20. Neil Armstrong and the Legacy of Apollo 174
One small step ...

21. Ryan Nagata 186
How to make a spacesuit

22. 'Little Old Ladies' 196
A stitch in space-time. The women whose needlework took us to the Moon

23. Trevor Beattie 204
Buying a ticket to space

ACT 5.

VITAM QUAERITUR

210

'THE SEARCH FOR LIFE'
WHERE IS EVERYBODY?

24. Oliver Postgate 212
Animator and writer

25. Carl Sagan 218
Talking to ET

26. Jocelyn Bell Burnell 226
Transmission unknown. The PhD student who discovered pulsars

27. Jill Tarter and the Origins of SETI 230
The Search for Extraterrestrial Intelligence

28. Tomohiro Nishikado 236
Insert coin. The original Space Invader

ACT 6.

TEMPORIS PEREGRINI 244

'WANDERERS IN TIME'
RETRO-FUTURISTS, ARCHAEOLOGISTS
AND THOSE PUSHING TO NEW HORIZONS

29. Ron Jones 246
What would it take to reach the stars?

30. Freeman Dyson 252
The interstellar scientist ahead of his time

31. Paul Van Hoeydonck 260
The legacy of the *Fallen Astronaut*

32. Bezos, Musk and the High Frontier 264
The billionaire space race

33. Alice Gorman 272
Dr Space Junk - space archaeologist

Acknowledgments 282

Picture Credits 283

Index 285

INTRODUCTION

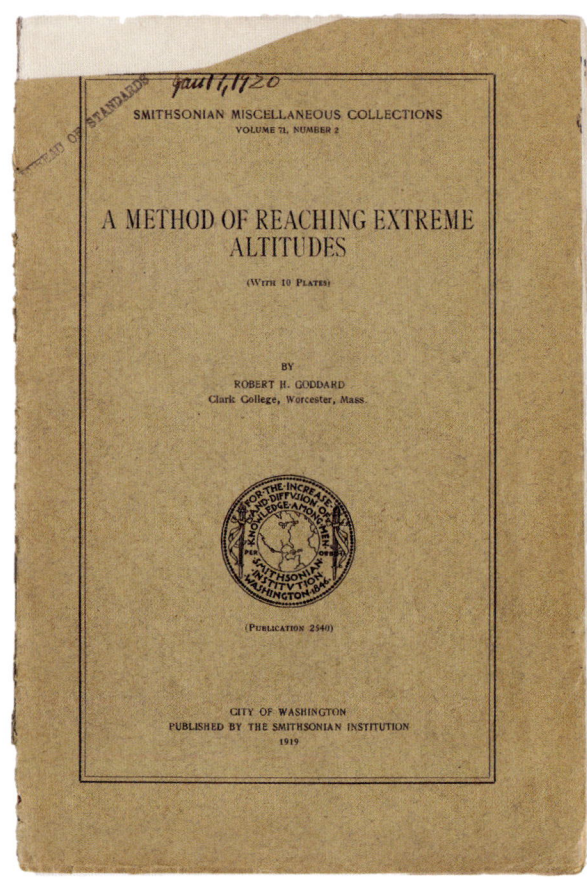

There was a time not so long ago when space travel was impossible. Then, in 1919, the American rocket scientist Robert H. Goddard wrote *A Method of Reaching Extreme Altitudes*, a scientific paper which contained a footnote explaining that as far as rockets were concerned, the sky was no longer the limit. It was this footnote rather than the paper itself that raised eyebrows. Its implication was that rockets could in theory reach an *infinite* altitude. This amused the *New York Times*, who reminded its readers what every high school student already knew: a rocket couldn't work in a vacuum because (obviously) there was 'nothing to push against'. Fifty years later in July 1969, as Neil Armstrong and Buzz Aldrin stepped onto the lunar surface, the newspaper printed a belated retraction:

Further investigation and experimentation have confirmed the findings of Isaac Newton in the 17th century, and it is now definitely established that a rocket can function in a vacuum as well as in an atmosphere. The Times *regrets the error.*

Fast-forward another fifty years and we find ourselves today witnessing Arthur C. Clarke's 'third law' ('any sufficiently advanced technology is indistinguishable from magic') in real time. Hardly a day seems to pass without news of some miraculous example of space technology blurring the line between fantasy and reality. Recently we've seen Katy Perry and William Shatner join a legion of other notables who've touched the edge of space aboard Blue Origin's suborbital rocket. And then there's the engineering miracle that is the James Webb Space Telescope, which in early 2022 unfolded like an origami flower a million miles away from Earth and now silently watches the universe, recording it in unimaginable detail. What secrets will it reveal? We must mention the exponential growth of satellites orbiting the Earth that provide everything from climate monitoring and high-speed internet to navigation, and 1,001 other things that form the bedrock of modern civilization. SpaceX is pioneering rockets that will be rapidly reusable. The booster section of the behemoth rocket Starship, the biggest object ever flown, can now perform a handbrake

---

0.1 Cover of Robert H. Goddard, *A Method of Reaching Extreme Altitudes* (1919).

0.2 Jules Verne, *From the Earth to the Moon* (1874).

0.3 *Popular Mechanics* (March 1930).

0.4 Life imitating art. June 2024, test flight four of SpaceX's reusable Starship. Note its visual similarities with the spaceships from fiction. Imagination is the greatest design tool.

FROM THE EARTH TO THE MOON
JULES VERNE

MARCH
25 CENTS
POPULAR
SEE PAGE
386

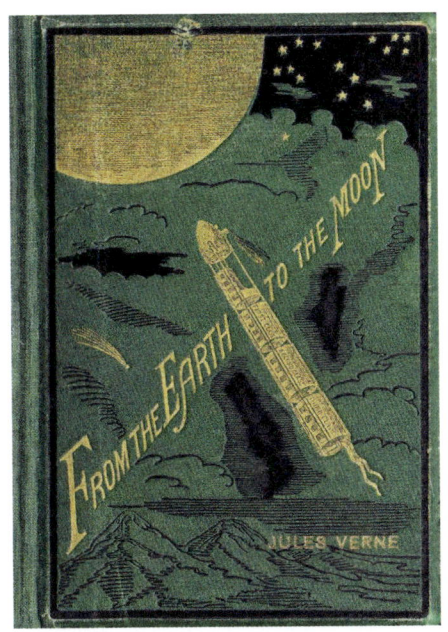

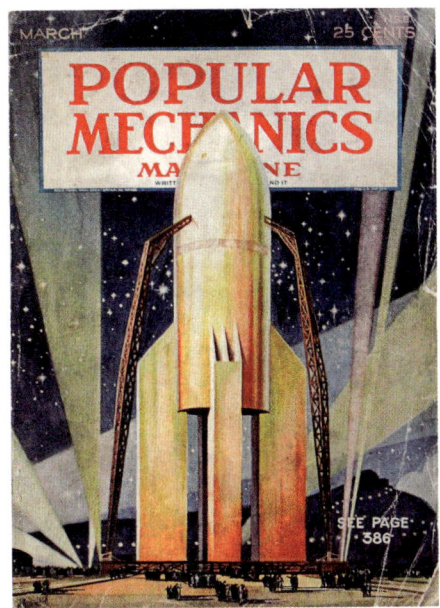

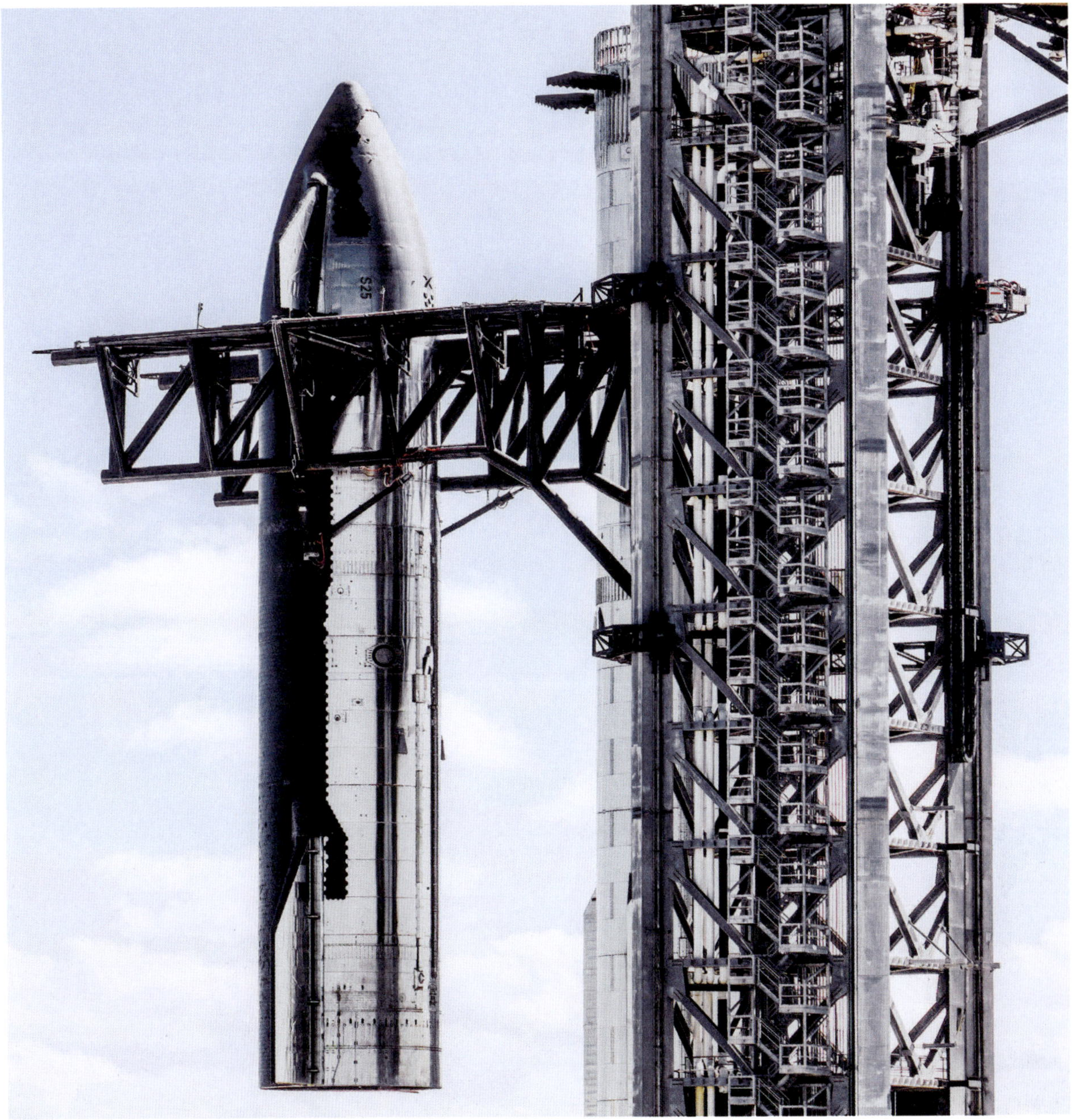

0.2 / 0.3

0.4

0.5 The sunshield on NASA's James Webb Space Telescope is the largest part of the observatory: five layers of thin membrane that was designed to unfurl reliably in space to precise tolerances. The sunshield is about the length of a tennis court and was folded up like an umbrella around the Webb telescope's mirrors and instruments during launch.

0.6 Centre L-R: Lady Bird Johnson, former President Lyndon Johnson and Vice President Spiro Agnew watch the lift-off of Apollo 11, the first crewed lunar landing mission, from the Kennedy Space Center at 9.32 a.m. local time on 16 July 1969.

turn in the upper atmosphere before returning to Earth to be caught by the chopstick arms of the launch tower (*Karate Kid* style), ready to fly again. Why does this matter? Well, imagine the cost of your plane ticket if every flight taken resulted in the brand-new aircraft being scrapped after a single use. That's pretty much what the space industry has been until now. Starship and Blue Origin's heavy-lift rocket New Glenn will form part of the architecture behind a revolution that is making access to space cheaper and faster. These new rockets have profound implications for our future down here on Earth, as well as getting us back to the Moon and onwards to Mars. They might even facilitate Elon Musk's often-cited vision of making humanity a multiplanetary species – a long-imagined fantasy which seems to be crystallizing in a way that would have been incomprehensible a decade ago. There is so much more to come.* Technology, politics and the imagination all travel into the future at very different speeds. When they meet, extraordinary things happen.

This book is about where some of this magic comes from. Or rather who it comes from. While space exploration relies on science and

---

* I'm writing this stuck in the slow lane of the recent past. You're ahead of me. What other technological or political developments have transpired in the gap between the writing (me) and reading (you) of this book? Please feel free to jot your own additions or corrections in the margins. I always enjoy finding people's pencilled thoughts in second-hand books.

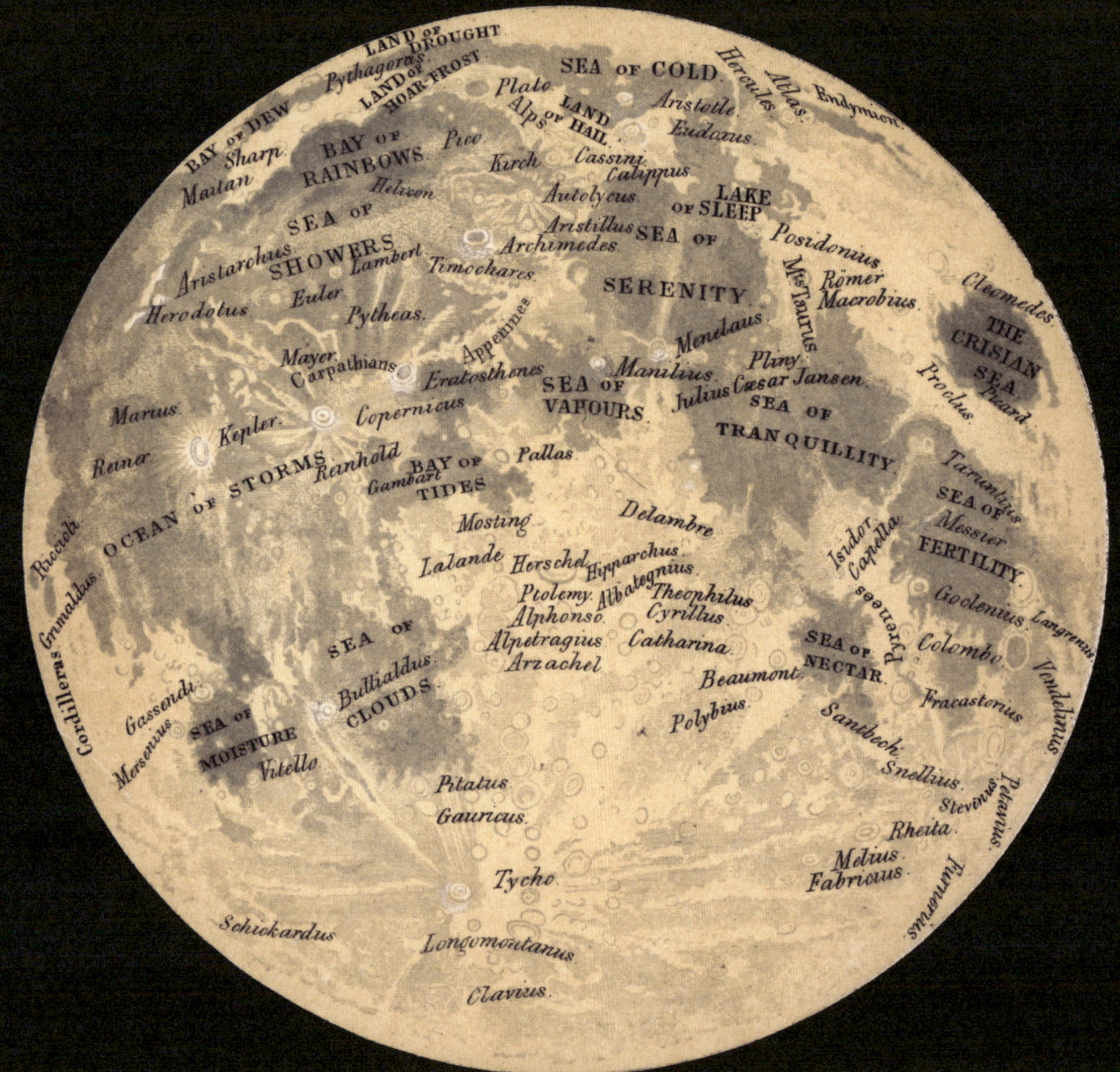

engineering, the fuse is often lit by the artists and dreamers. Fantasy and fiction are vital to our story; they can be as potent as rocket fuel, although I'm reminded of T. E. Lawrence's maxim: 'All men dream; but not equally.' If you're looking for a representative collection of names who might sum up the story of space exploration, you'd do well to look at the Moon. Not at its geology, but at the thousands of craters named after people who throughout time have turned their minds and talents towards advancing human presence in space. The Moon's craters are an A to Z of explorers, scientists, engineers and other notables, from Ernst Abbe (German physicist) to Fritz Zwicky (Swiss astronomer, and presumably always last on any roll-call). The size of crater seems to be directly related to the perceived importance of the individual. Tycho Crater, named after Tycho Brahe, the great Danish astronomer of the pre-Galilean age, is perhaps the most famous, and certainly the brightest and most visible to the naked eye. The aviator Amelia Earhart has one, as does Ernest Shackleton, the polar explorer whose ice-filled crater at the lunar South Pole is the destination for a new political and commercial space race. In the pages that follow, there are many names you'll recognize. Some are less well known – a reminder that history is never definitive. It is messy, full of footnotes, riddled with biases and coloured by the politics of time.

If you've ever been 'mudlarking' in London, or even simply heard of the practice, you'll know that the River Thames washes up all manner of stuff, and at low tide presents it on the foreshore like a historic show-and-tell: Roman pottery sherds next to Victorian clay smoking-pipe stems, next to a Nokia 3310, some Edwardian jacket buttons, unidentified animal bones and an important-looking key. This book is a collection of the stuff that our story has washed up. Treasures which allow a glimpse into the creative process; the scribbles in the margins; the material ephemera of minds that have been left to wander. To tell this story completely is an impossible and pointless task, akin to counting every impact crater on the Moon or tracing the particular cave-dwelling ancestor who first looked to the stars and imagined going to space. I've kept the brief as wide as possible, avoiding too strict a chronological order, much as the Thames does. Instead I've divided our story into rough thematic acts. There'll no doubt be names that you can't believe I've omitted. Or names that I've included that may make you scratch your head. Such is the challenge, and fun, of writing a book like this.

I hope you enjoy this stroll along the foreshore of this most exciting of human endeavours. Perhaps something will glint in the light and catch your eye. You never know where it might take you.

'Too great a burden of knowledge can clog the wheels of imagination.'
ARTHUR C. CLARK, PROFILES OF THE FUTURE (1962)

0.7 Mary Ward, 'Map of the Moon with the names given by Riccioli in the 18th century, and others added by succeeding observers', in Telescope Teachings (1859).

ACT 1.

SOMNIUM

'DREAM'
SPACE DREAMERS AND FLIGHTS OF FANCY

JOHANNES KEPLER AND CO.

1.

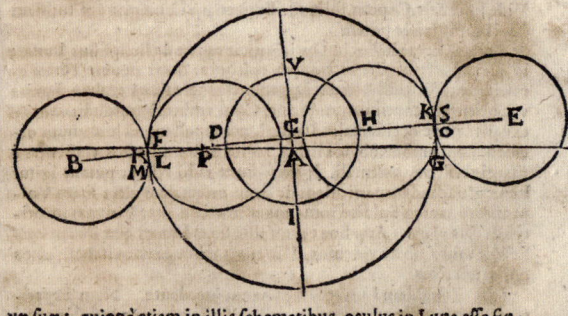

Iceland is the perfect place for lucid dreaming about space travel. An alien landscape of geology and mythology sculpted by ice and fire; a borderland where old worlds and new worlds touch. Here we meet Duracotus ('science'), a young student, and his mother Fiolxhilde ('ignorance'), who makes her living by selling magical herbal concoctions in goat-skin bags. One day, Duracotus is whisked away on a ship bound for Denmark. Upon landing, he tracks down the great Danish astronomer Tycho Brahe (1546–1601) and with him studies the new cosmology of the age. After a few years the prodigal son returns to share his knowledge with his now-dying mother. He is curious about her occultism, and she tells him about the spirit beings who travel between the Moon and the Earth. Duracotus is eager to learn more, so they travel to a crossroads where Fiolxhilde summons a Moon-dwelling Daemon ('knowledge'). The Daemon appears and in a rasping voice explains how these lunar entities travel along the brief shadow that forms during eclipses, bridging the two worlds. He also states that humans can make the perilous journey too, but they have to have *the right stuff*. Their bodies are violently thrown up into the air by a legion of spirits, 'as if by gunpowder'. They are anaesthetized by opiates to protect them from the shock. Their limbs are bound together to stop their bodies being ripped apart by the force. Damp sponges are inserted in the noses of the 'astronauts' to prevent asphyxiation. After the violence of the launch, the natural pull of the Moon takes over, gently guiding them down the shadow corridor to their destination. For the rest of the story, the Daemon describes at length what earthly travellers will find. He details the lunar topography, its geography, calendar, climate and astronomy. The Moon is an inhabited world of stark contrasts; it is a world playing by the same earthly laws of physics. The same, but different.

The seventeenth century was a golden era for scientists, storytellers and space travellers. It revealed a radical intellectual shift in our understanding of the natural world, and in particular marked the move away from the Earth-centred cosmology of Ptolemy and Aristotle to the reality of the Sun-centred model of the solar system, as described by Nicolaus Copernicus in 1543 and confirmed by Galileo Galilei and others in the decades that followed. One of the architects of this scientific revolution was Johannes Kepler (1571–1630), the German mathematician and

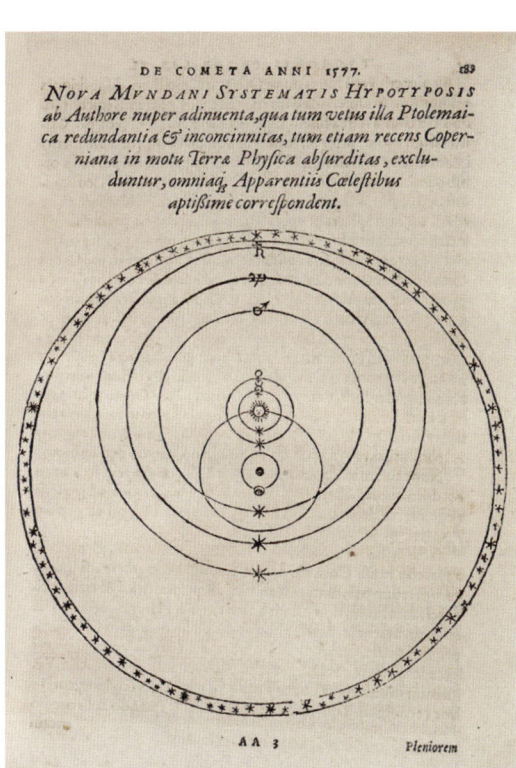

1.1 Page from Johannes Kepler's *Somnium* (1634), with a diagram showing an eclipse.

1.2, 1.3 Audio/visual artist Joshua Ellingson's KEPLER/SOMNIUM murals, created during a 2013 residency in Anger, Austria: 'I found that while in Austria, Kepler wrote *Somnium*, considered by many to be the first published work of science fiction. So, for this residency I thought it would be fitting to pay tribute to Kepler and his story.'

1.4 Page from Tycho Brahe's *De mundi aetherei recentioribus phaenomenis liber secundus* (1603) detailing the results of his research on a comet of 1577. Attempting to determine its distance from Earth, he reached the conclusion that it orbited around the Sun.

1.5 (OVERLEAF) Details from Frederick de Wit, *Planisphærium coeleste* (1680). This beautiful double hemisphere celestial map is surrounded by six supplementary diagrams illustrating the cosmologies of Descartes, Ptolemy, Brahe (left, top to bottom) and Copernicus (right, middle), as well Pierre Gassendi's phases of the Moon (right, top) and the annual motion of the Earth around the Sun by Philippe van Lansberge (right, bottom).

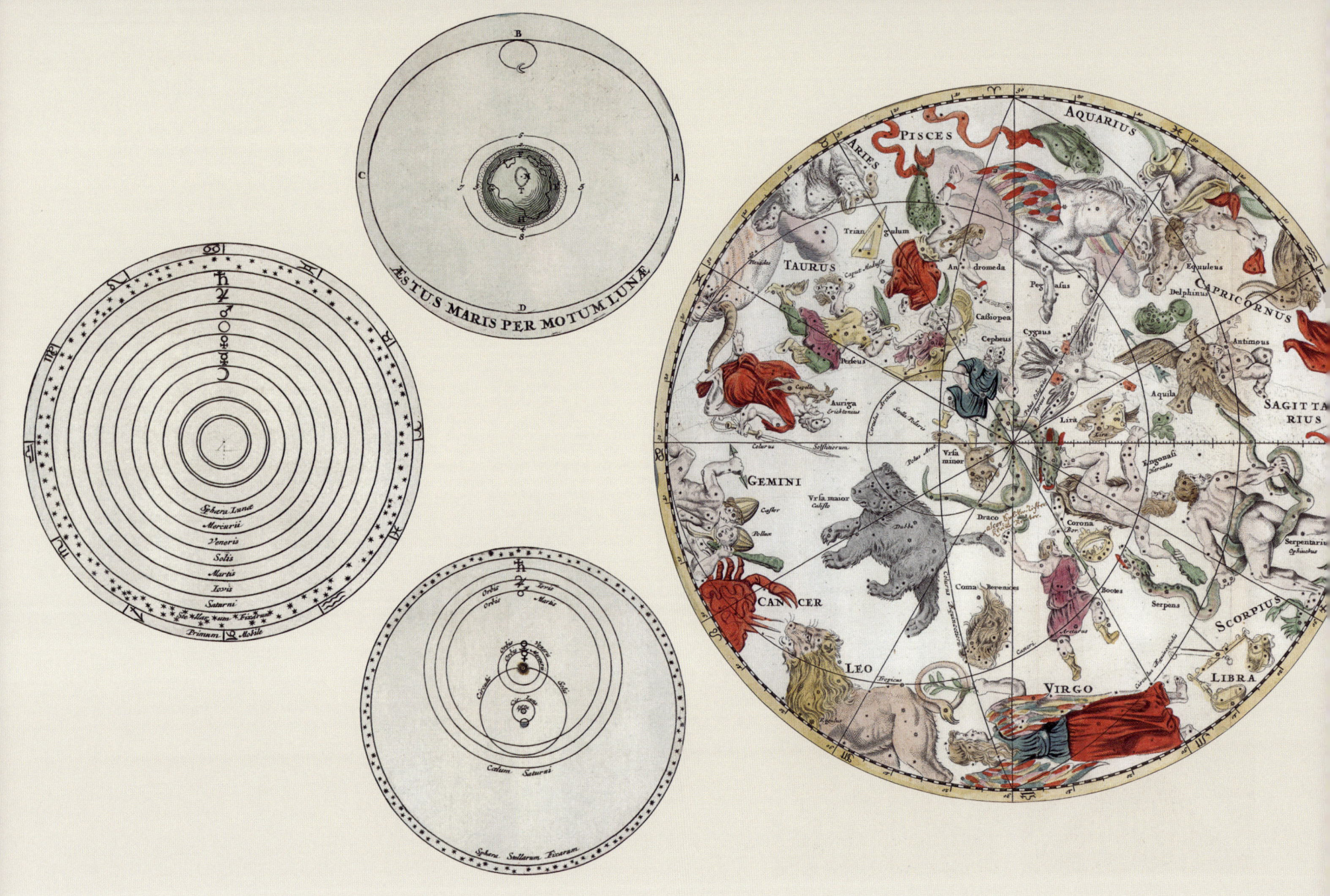

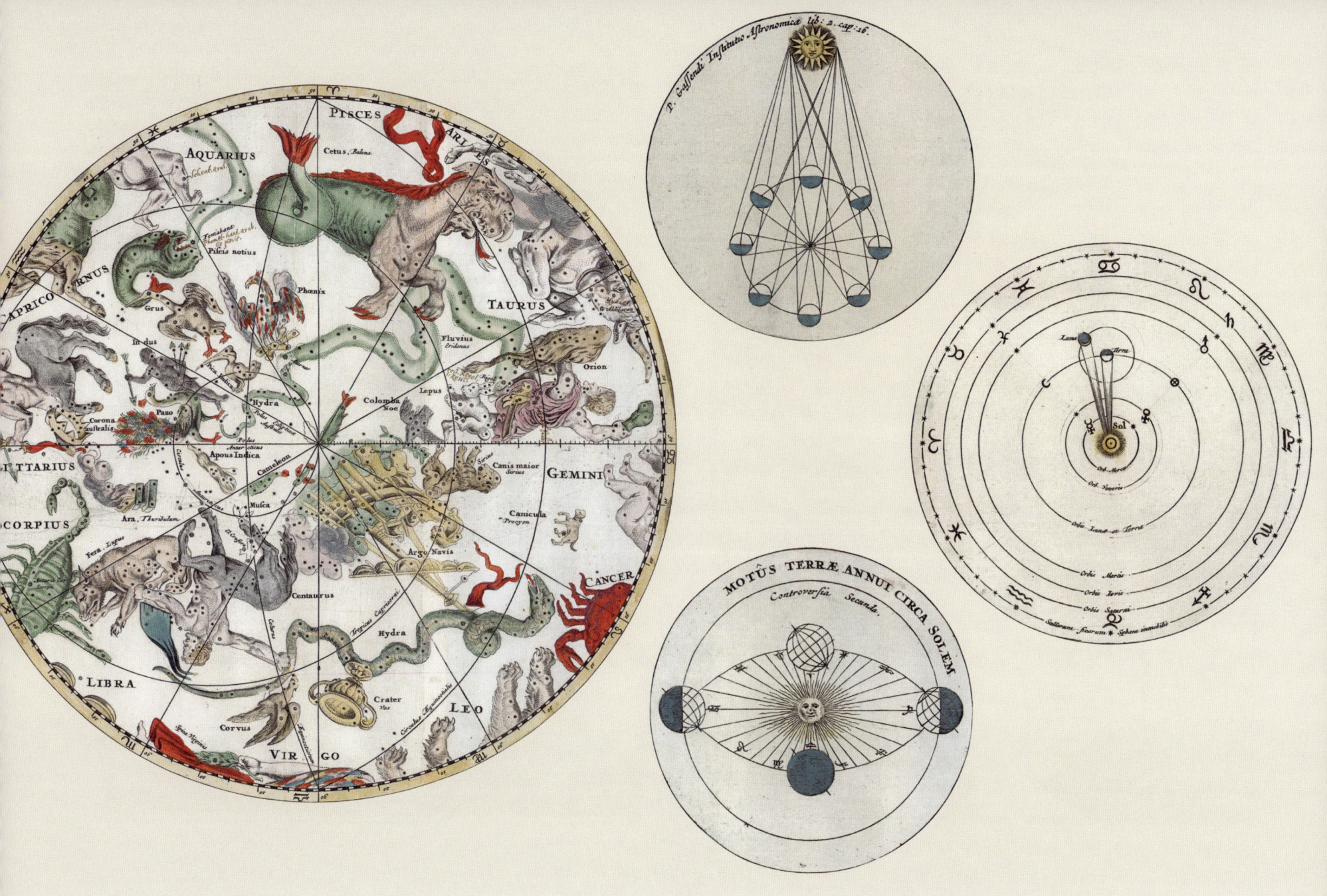

1.6

1.7

astronomer whose laws of planetary motion changed our understanding of the cosmos. His peculiar short story *Somnium*, in which we meet Duracotus, Fiolxhilde and the Daemon, was published posthumously in 1634, but the idea had germinated as a thought experiment posed when he was at university sometime in the 1590s. How would astronomy differ if we were viewing the heavens from the Moon? Over the next thirty years, that idea was shaped and moulded into what became, arguably, the first modern science fiction story. *Somnium* was really a device that allowed Kepler a broad and speculative defence of the new Copernican model of the solar system, exploring the conflict between the old ways of seeing and the new science *du jour*. Kepler was influenced by writers from antiquity who were exploring even earlier ideas of space travel, notably a wild space adventure called *True History* by Lucian of Samosata, a Syrian-born satirist from the second century CE. Despite the story's name, Lucian goes to great lengths to assure us that none of it is actually true, in case his readers were in any doubt. In the story a sailing ship is lifted into space by a giant whirlwind, and so begins an eight-day voyage towards a glistening island orb in the sky. Here the Moon become the stage for a saga between warring factions in a solar system populated by nightmarish alien lifeforms.

The imagination was built for space travel. But 'magic' as a solution for getting to and from the Moon won't get you beyond the page. If you actually want to fly to the Moon, you'll need the mind of an engineer. The Bishop of Hereford Francis Godwin in his 1638 adventure *The Man in the Moone* tells of the diminutive Spaniard Domingo Gonzales. Stranded on the island of St Helena in the Mediterranean, Gonzales trains a flock of geese called 'Ganza' to carry him away by leashing the birds to a chair. But after a few island-hopping test flights, the restless geese begin to feel the irresistible migratory pull of the Moon, 'as the lodestone draweth iron'. Gonzales and his goose-powered contraption speed through circumlunar space before landing on the lunar surface, where he spends several months living with the native Lunarians, communicating with them through a musical language, rather like a Jacobean *Close Encounters of the Third Kind*.* Like Kepler's story, *The Man in the Moone* is an elegant philosophical dance between fact, fantasy, politics and science, and the implications these have on important matters of faith.

The French writer Cyrano de Bergerac (1619–1655) also dabbled in some ingenuous seventeenth-century aerospace engineering in his *Comical History of the States and Empires of the Moon*. The astro-protagonist Cyrano collects morning dew in glass bottles and ties them to his body. The evaporation of the dew in the heat of the Parisian sun is sufficient to carry him aloft – but only as far as Quebec. His next attempt at

* Or indeed *The Clangers*, whom we'll meet later.

1.6, 1.7 *The Snare of Vintage* and *Lucian's Strange Creatures*, 1894. Engravings by Aubrey Beardsley, from Lucian's *True History*.

space travel is in a 'machine' which propels him into space via tiers of fireworks. Perhaps Cyrano had heard of the legend of Wan Hu, a Chinese official who on deciding to visit the Moon ordered his assistants to attach forty-seven fireworks to a chair and set them alight. There was a huge explosion, and Wan Hu vanished in a cloud of smoke, nowhere to be seen. Had he been blown to smithereens? Or had he indeed become the first astronaut? You're free to speculate. There is a crater on the Moon named in his honour just in case.

Perhaps no one of this period took space travel more practically than science popularizer and co-founder of the Royal Society John Wilkins. He explored in great detail the possibilities of interplanetary travel and extraterrestrial life in his *Discovery of the World in the Moon*, published in 1638, the same year as Godwin's goose fantasy. The cover-page illustration (fig. 1.11) reveals much of what we can expect to come: Copernicus stands opposite Galileo, clutching his telescope; above them, a diagram of the new solar system with the Sun firmly in the centre, orbited by the planets and a universe of stars stretching out beyond. Inside the book, Wilkins sets out a series of considered propositions – that the Moon is solid, it is a real world, and that real worlds require real beings to inhabit them.

THE MAN IN THE MOONE:

OR

A DISCOVRSE OF A
Voyage thither
BY
D*r* Francis Godwin B*p* of Landaff
DOMINGO GONSALES
1st Edition
The speedy Messenger.

LONDON,
Printed by JOHN NORTON, and are to be
sold by Ioshua Kirton, and Thomas Warren. 1638.

1.8 Domingo Gonzales being transported to the Moon by a flock of geese. Francis Godwin, *The Man in the Moone; or a discourse of a voyage thither* (1638).

1.9, 1.10 Agnes Meyer-Brandis, *The Moon Goose Analogue* (2011-), part of the Lunar Migration Bird Facility department of Meyer-Brandis's Art and Subjective Science Institute, FFUR. Inspired by Francis Godwin's famous story, in 2011 Meyer-Brandis created a narrative art project in which she hatches, raises and imprints colonies of 'Moon geese' and prepares them for lunar adventure. Her short documentary film, *The Moon Goose Colony* (2012), is well worth searching out.

1.11 Frontispiece to John Wilkins, *The Discovery of a World in the Moone* (1638).

1.12 Detail of Taurus from Johann Bayer, *Uranometria* (Measuring the Heavens, 1603), which aimed to compressively record the positions and magnitudes of the stars across the entire night sky. Much of the information in his fifty-one-plate star atlas came from Tycho Brahe's extensive astronomical records. The stars appear as they would have been seen historically.

'All men dream, but not equally.'
T. E. LAWRENCE, *SEVEN PILLARS OF WISDOM* (1926)

The fourteenth of these propositions states that 'Tis possible for some of our posteritie to find out a conveyance to this other world; and if there be inhabitance there, to have commerce with them'. He was right in the first part. Wilkins reasoned that a 'flying chariot' using mechanical power (clockwork, springs, gears, levers and cogs) would be the solution. Such a flying chariot would indeed make the journey to the Moon 331 years later, carrying three American astronauts, Neil Armstrong, Buzz Aldrin and Michael Collins.

Wilkins, Kepler, Godwin et al. all played with the same fundamental idea as the great maritime explorers of the previous two centuries, Columbus, Drake, da Gama and Magellan – albeit a cosmic extension. The message was simple: the heavens are another ocean to be crossed, and the mysterious points of light within it are inhabited islands, waiting to be explored. Modern space flight is built on the power of story. If the scientific revolution of the seventeenth century marked a fertile period for staring into space, the Industrial Revolution of the nineteenth and twentieth centuries, with its raw energy and inventive spirit, would take us a step closer to making those dreams a reality.

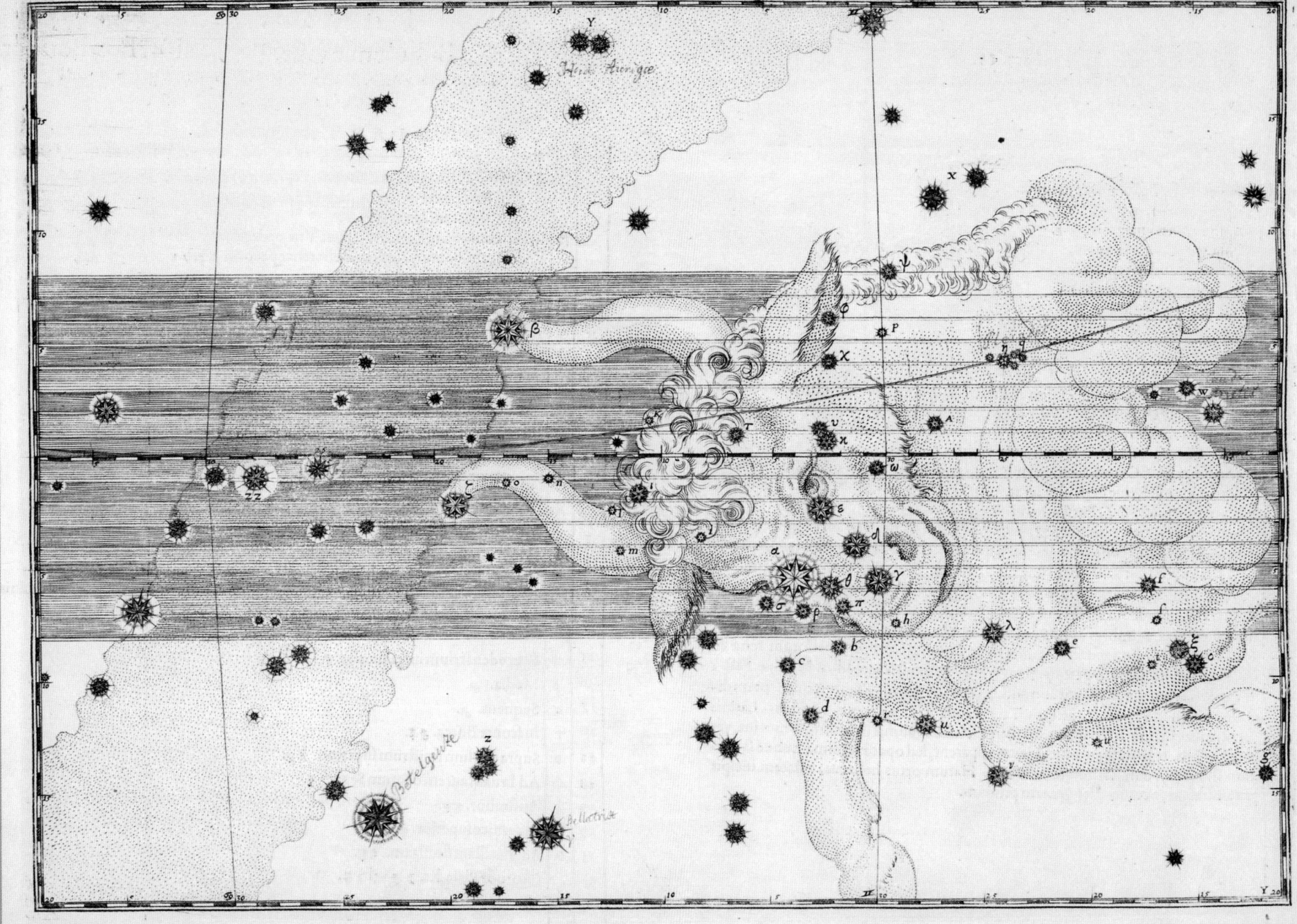

JULES VERNE

There's a pleasing list of similarities between Jules Verne's novel *From the Earth to the Moon* (1865) and the Apollo Program a hundred years later. The Apollo 11 Command Module was named Columbia as a nod to Verne's fictional space capsule Columbiad. Both used similarly shaped and sized capsules made from aluminium. Both were sent into orbit from Florida after debates about the suitability of southern Texas as a launch location. Both transported three astronauts into space; both ended up in the Pacific Ocean; and both were eye-wateringly expensive. You can even argue that the motivations were drawn from similar political wells: Apollo was born from President John F. Kennedy's rousing 'We choose to go to the moon' speech, motivated in large part by Cold War politics. Verne's post-American Civil War social satire also starts with a rousing call to arms, 'Three cheers for the Moon!', in response to a speech by central protagonist Impey Barbicane. A brief reminder of the plot: the Baltimore Gun Club in Maryland is lamenting the close of the American Civil War, which had been so important for the science of gunnery. They needed a grand project to put an end to the 'deplorable inactivity' facing them. To test their energies and skills, a giant cannon would be built to fire a projectile to the Moon. The story is akin to reading a scientific paper. Questions are posed. Calculations are calculated. Budgets are budgeted. Wagers are waged. The Moon's viability for habitation pondered. As word spreads internationally of this great endeavour, a Frenchman, Ardan, announces he will fly in it. He is joined on this grand adventure by Barbicane and Captain Nicholl. We are left with a cliff-hanger ending.

It's nineteenth-century technical progress that is the real protagonist in Verne's saga. Despite the impossibility of a giant space gun reaching the Moon, Verne's meticulous historical knowledge, technical detail and language lend a credibility to the fantastic; a believability to the inconceivable,

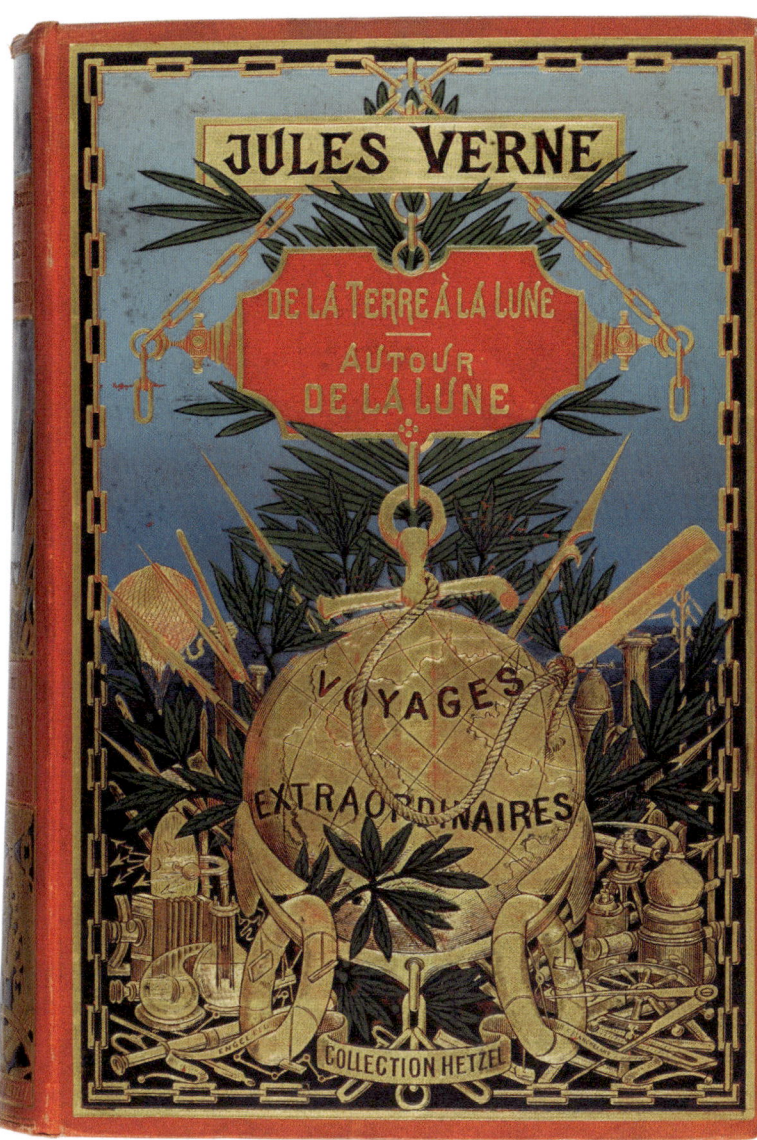

2.1 Poster advertising Jules Verne's complete works, 1889 (detail), featuring the Columbiad and its cannon launcher, among other inventions.

2.2 Émile Bayard, 'White all, Barbicane', for Jules Verne, *From the Earth to the Moon* (1874). The splashdown is in the Pacific Ocean, not far from the point where the Apollo travellers landed.

2.3 The Apollo 11 Command Module Columbia with astronauts Neil Armstrong, Michael Collins and Buzz Aldrin aboard splashed down at 11:49 a.m. CDT, 24 July 1969, about 812 nautical miles southwest of Hawaii.

2.4 Jules Verne *De la Terre à la Lune/Autour de la Lune* (From the Earth to the Moon/ Round the Moon, 1865).

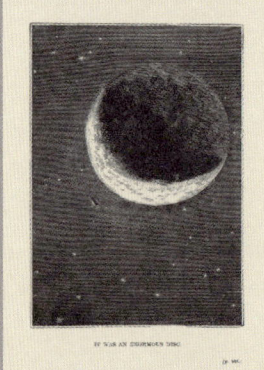

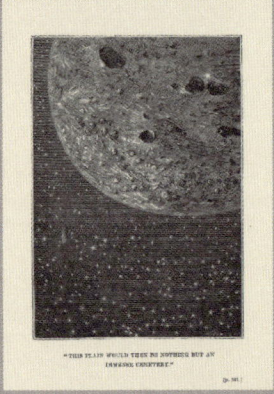

2.5–2.22 Illustrations from Jules Verne, *From the Earth to the Moon* (1874).

2.23 Manuscript page of *From the Earth to the Moon* (1865). The page is split in two: Verne writes only in the left part and leaves free the right for corrections, remarks, annotations or diagrams.

2.24 Manuscript page of *Round the Moon* (1869), showing a diagram of a lunar eclipse and accompanying calculations.

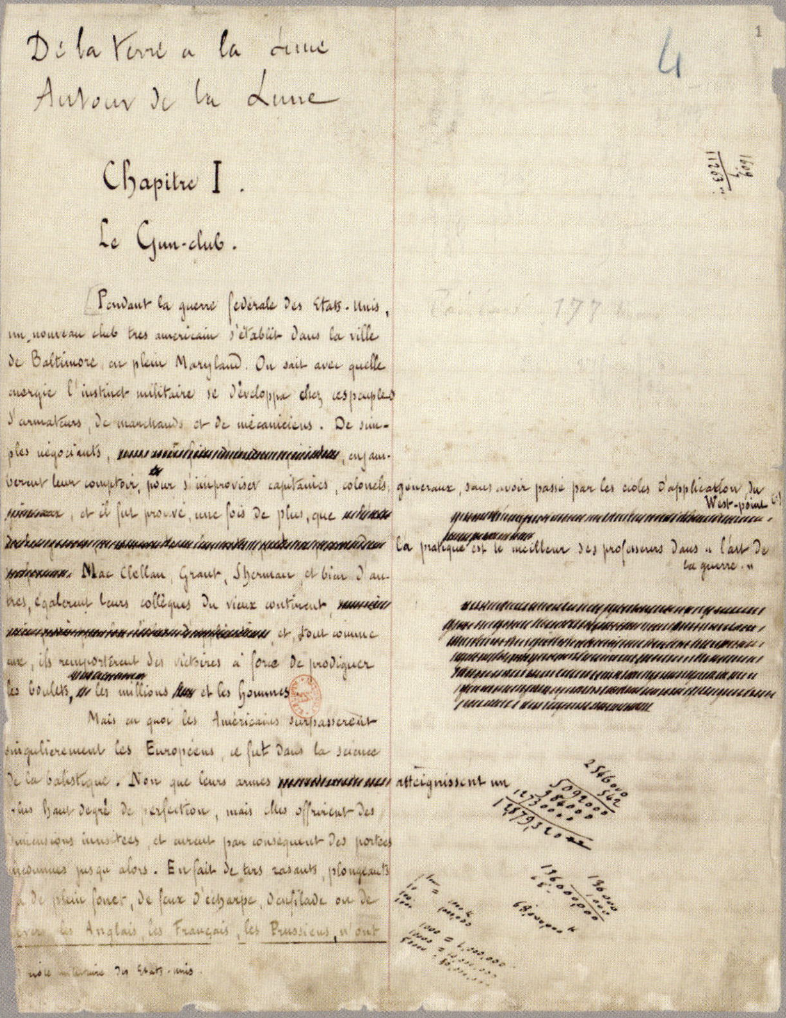

2.23

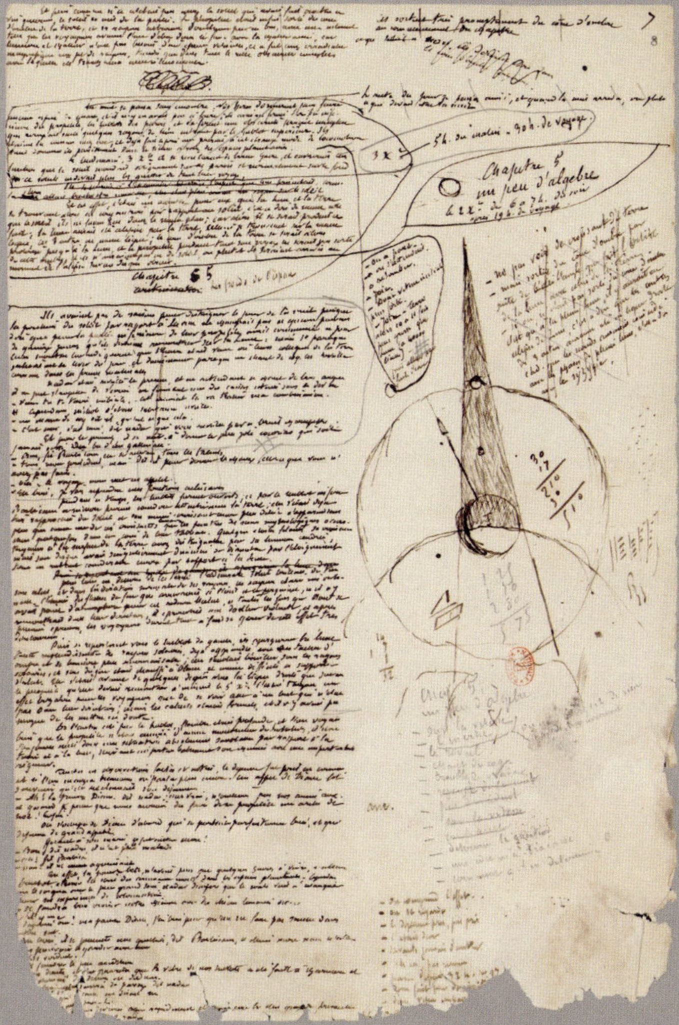

2.24

and as such Victorian readers were permitted to travel vicariously with a higher fidelity than ever before. Context is important: The Future was all the rage in the later nineteenth century. A second Industrial Revolution was injecting energy and innovation into America and the West through electrification, manufacturing power and steel production; the new railways were getting us around quicker than ever before; and the world was shrinking even further with the invention of the telegraph, and later radio. The American Civil War acted as a catalyst, speeding up the need for technological progress in communication, medicine, transport and weaponry. The message was clear: solid engineering rather than *deus ex machina* was the only way to fly to the Moon.

Verne wasn't writing in a vacuum. This was a particularly fertile time for space fiction writers. In 1835 the journalist Richard Adams Locke famously satirized the prevailing scientific vogue for speculating about life on the Moon in an effort known as *The Great Moon Hoax*. He published a series of fabulous tabloid articles calling upon the work of the reputable English astronomer Sir John Herschel, purporting to detail exotic life that Herschel had observed on the Moon via telescope, complete with lush forests, mini zebras, blue goats and flying man-bats. Verne's lunar novel is a direct homage to those whose shoulders he was standing on, namely the seventeenth-century writers Francis Godwin, Cyrano de Bergerac, Bernard de Fontenelle and in particular his contemporary the American writer Edgar Allan Poe ('Cheers for Edgar Poe!'), who had written his own lunar adventure. In 'The Unparalleled Adventures of One Hans Pfaall' (1835), Poe's

2.25 Richard Adams Locke, *The Great Moon Hoax: Lunar Animals*, 1835. Locke's articles described fantastic animals, including unicorns and bat-like winged humanoids (Vespertilio-homo) who built temples.

2.26 'A view of the inhabitants of the Moon, as seen through the telescope of Sir John Herschel', woodcut centrefold in *The History of the Moon; or an Account of the Wonderful Discoveries of Sir John Herschel* [by Richard Adams Locke] (1835).

hero, in a desperate quest to escape his earthly financial woes, builds a balloon fit for space travel using a machine that condenses the rarefied atmosphere into breathable air. Some years later Edward Everett Hale's novella *The Brick Moon* (1869) was published. It's the first story to feature an artificial satellite that could be used as a navigation aid. The story takes a turn, however, as the brick moon is accidentally launched via its giant rotating flywheels, taking its hapless crew with it, thus becoming the first crewed space station in fiction.

If science and technological change are a muse for speculative fiction, then the reverse is even more profound. Jules Verne directly influenced a new generation of explorers and scientists – Simon Lake, who designed the first generation of submarines, did so after reading Verne's *20,000 Leagues Under the Sea* (1870). Igor Sikorsky was inspired by him to design aircraft and helicopters. Most importantly for us, the three fathers of rocket science, the Russian Konstantin Tsiolkovsky, the Hungarian/German Hermann Oberth and the American Robert H. Goddard, were all devoted fans of Verne. Their work in independently inventing the space rocket was a direct answer to the questions he and other writers posed. Verne was neither a scientist nor an engineer, but as a master storyteller he is in contention to be one of the most important names in the history of space flight. It's no surprise that the new medium of cinema in the twentieth century began with Georges Méliès's version of Verne's story, *Le Voyage dans la Lune*. As the world changes radically once again, with technologies indistinguishable from magic, the power of storytelling in all its forms continues to function beyond the scope of distracting entertainment, as a source of ideas, of hope and of action. The stories we share, while infinite in their hues, all rely on the same simple formula – the quest for a solution. Where shall we go? How will we get there? What will we find?

2.27 Lithograph from Leopoldo Galluzzo, *Altre scoverte fatte nella luna dal Sigr. Herschel* (1836).

2.28 Edward Everett Hale, *The Brick Moon* (1869).

2.29 Georges Méliès, *Le Voyage dans la Lune* (1902), hand-coloured by Élisabeth Thuillier's colouring lab the same year.

CHESLEY BONESTELL

The May 1944 edition of *Life* magazine makes for sober reading. The fatigue of war laid out in black and white. The bruised politics of a world coming to terms with the reality of the past few years. From the pages stern men in uniform meet your gaze, punctuated with moments of frivolity in the adverts for toothpaste, cigarettes and car tyres. Then, like Dorothy waking up in Oz, we arrive at page seventy-eight: an astronomy feature about the solar system. It's the only feature in colour, introduced by six spectacular images of the view of Saturn as seen from the surface of its moons. Pin-sharp physical worlds with panoramic vistas. Dramatic landscapes of smoothly rendered colour offering a tantalizing view outwards instead of backwards, and a moment to leave earthly concerns behind. The pictures were painted by Chesley Bonestell. With them, he invites us to engage not as passive observers of the heavens, or as scientists staring down the end of a telescope, but as explorers gazing out across a universe of infinite possibility and adventure.

Bonestell was born in San Francisco in 1888, growing up in the wealthy district of Nob Hill. At seventeen his first commercial illustrations were featured in *Sunset* magazine, a publication which celebrated the American West with glorious images of Yosemite, the Californian coast and the railroads that made it all possible. His own horizons were stretched further when he visited the Lick Observatory in 1905 and saw Saturn and its rings for the first time, which inspired his first celestial-themed painting. He moved to New York to study architecture, honing his skills as a draughtsman before returning to San Francisco to work for the architect Willis Polk on rebuilding the city of his childhood. During a spell in London after the First World War, he met Scriven Bolton, an amateur astronomer and artist from Leeds who had developed a technique for making plaster lunar landscapes, which he would then photograph and paint. It was a technique Bonestell would adopt for himself. He also became heavily influenced by the French artist Lucien Rudaux, whose astronomical artwork in his book *Sur les Autres Mondes* (1937) would prove an important reference for Bonestell's style. He returned to America in 1926 to take advantage of New York City's skyscraper building boom, where he worked on the Chrysler Building, before heading West once again to produce artwork for the newly designed Golden Gate Bridge, fleshing the utilitarian engineering schematics with glossy art deco renderings that turned buildings and bridges into symbols of American beauty and confidence. But it was in popular culture rather than architecture that he really made his mark. In 1938 he moved to Los Angeles to work for the new film studios as a matte artist, painting photorealistic backdrops for movies. Matte

---

3.1 Space artist Chesley Bonestell pictured in his studio with some of his many paintings.

3.2 Illustration by Lucien Rudaux for *Sur les Autres Mondes* (1937).

painting was the CGI of its day, and Bonestell's understanding of visual depth and perspective made him one of the most successful scenic artists in Hollywood.

He never lost his interest in astronomy, however, or in painting his interpretation of the heavens. In fact, it was his dramatic views of Saturn from the proximity of its moons, Phoebe, Iapetus, Titan, Dione and Mimas, which he speculatively submitted to the editor of *Life* magazine, that really brought him to the public's attention. His paintbrush lent a resolution and drama that the telescope couldn't provide. He had transformed the solar system into a destination to be explored and settled, much like the Western frontier had been in the nineteenth century. Bonestell's style is much more than an effort to recreate the scientific realism attempted in Rudaux's works. Instead we see similarities with the nineteenth-century romanticism of the Hudson River and Luminist schools of American landscape painters, with their huge skies and vast tracts of wilderness, bathed in pools of ethereal light, often with small human figures in the foreground dwarfed by nature at its most reverential. As viewers, we are invited to stand awestruck in God's creation, humbled by its scale and grandeur. If you look carefully at Bonestell's painting of *Saturn as seen from Mimus* (fig. 3.4), you'll see clusters of tiny figures on the surface, almost invisible in the rugged landscape. It's a recurring motif throughout his work. Bonestell was offering the audience a new, cosmic Manifest Destiny, one of infinite expansion and settlement, unrestricted by political borders or the mere coastlines of a continent. For a jaded post-war American public, it marked the start of a seductive story of American exceptionalism that runs throughout the telling of space history like a stick of rock.

Bonestell's success wasn't just owed to his skill as an artist. It also came from collaborating with the right people – and two people in particular. The first was science writer Willy Ley, the German emigre and popularizer of space who more than anyone else at the time helped tell the story of the new frontier. Bonestell provided the pictures to Ley's words in books such as *The Conquest of Space* (1949), setting out in full colour a grand vision of the coming space age. The second was the father of the V-2 himself, German rocket scientist Wernher von Braun, who despite his Nazi background was now working for the American space programme. In *Collier's* magazine, Bonestell fleshed out von Braun's designs of exotic-looking rocket ships

3.2

3.3 Albert Bierstadt, *Merced River, Yosemite Valley*, 1866. Note the similarities between Bonestell's grand cosmic view and the dramatic, romanticized landscapes of the 19th-century Hudson River school of painting. These two examples both feature tiny groups of figures placed in the foreground, dwarfed by the majestic, illuminated vistas.

3.4 Chesley Bonestell, *Saturn as seen from Mimas* (detail), 1943.

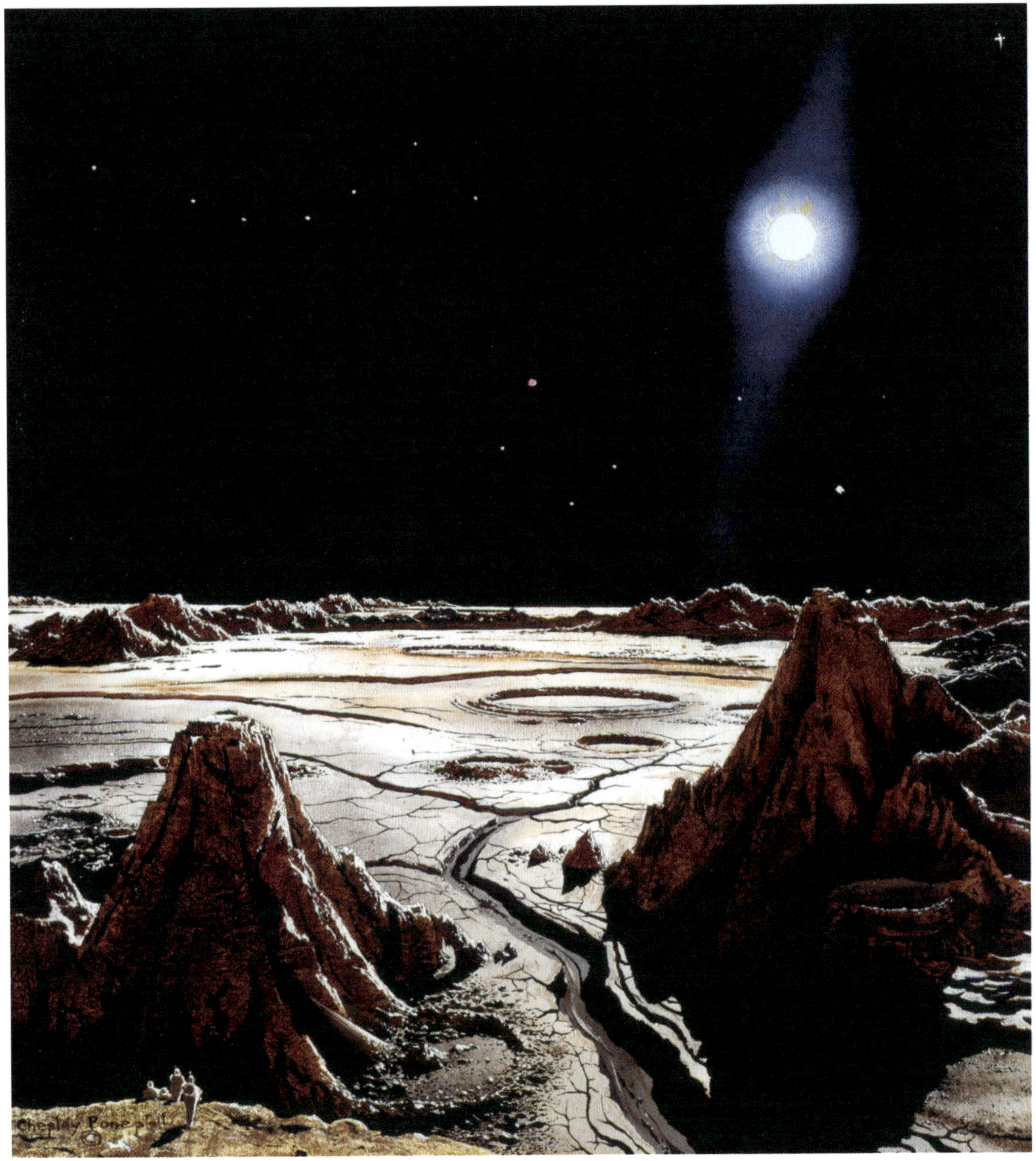

THE CONQUEST OF SPACE
A PREVIEW OF THE GREATEST ADVENTURE AWAITING MANKIND
WITH TEXT AND PICTURES BASED ON THE LATEST SCIENTIFIC RESEARCH
PAINTINGS BY
CHESLEY BONESTELL
TEXT BY
WILLY LEY

DESTINATION
MOON
COLOR BY TECHNICOLOR

3.5

3.6 / 3.7

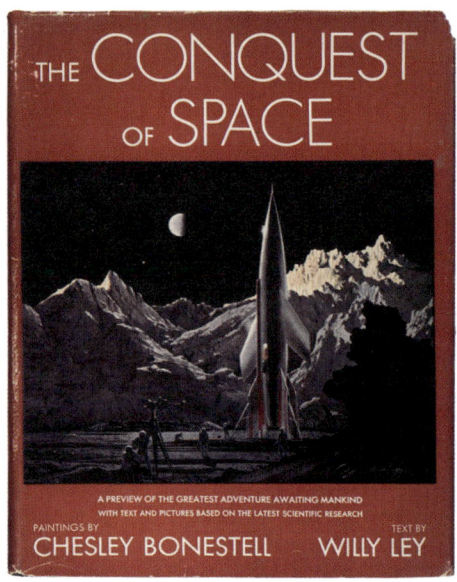

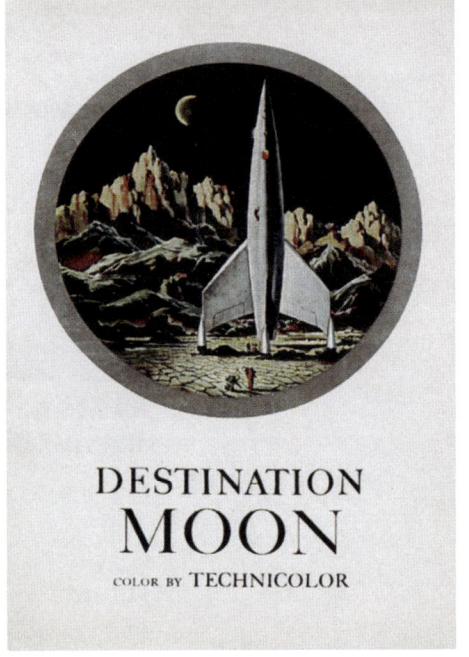

'Do you want to go to lunch? Or do you want to go to the moon?'
DESTINATION MOON (1950)

and futuristic space stations (fig. 13.9), much as he had done for the architects of the Golden Gate Bridge. Their enduring collaboration became known as the 'Collier's Space Program' and was hugely influential in building public support for the emerging space race that was igniting – as well as launching the careers of a thousand space scientists, engineers and would-be astronauts. Hollywood too was cashing in on the new appetite for space adventure. Bonestell did much of the scenic artwork for films such as George Pal's lunar adventure *Destination Moon*, Rudolph Maté's *When Worlds Collide* and the opening sequence of *The War of the Worlds*, and his hand lent them a sense of credibility. His art fuelled a populism, but it also filled in the visual gaps that science had yet to complete.

By the time humans finally reached the Moon at the end of the 1960s, we already knew that Bonestell's pictures didn't match the reality. His jagged lunar mountain ranges had given way to the softer, powdery undulations of the actual lunar surface. The spidery Apollo Lunar Module didn't have the sleek deco sensibilities of Spaceship Luna, the silver vertically landing rocket that he had designed for *Destination Moon*. But reality was not what he was offering. He was offering escapism. The Chesley Bonestell legacy lives on in the mise-en-scène of films such as *2001: A Space Odyssey*, *Star Wars* and countless others. Generations of scientists, explorers and artists owe a nod to the man who made outer space a theatre for our most vivid fantasies. Most importantly, he provided audiences with a sense of thrilling anticipation at a time when humans were still firmly rooted to the Earth.

3.5 Chesley Bonestell, *The Surface of Mercury* (detail), 1948.

3.6 Dust jacket of Willy Ley and Chesley Bonestell, *The Conquest of Space* (1949), featuring the Bonestell painting *Rocket Landed on the Moon*.

3.7 Poster for *Destination Moon* (1950).

3.8 Still from *Destination Moon* (1950).

3.8

Collier's
JUNE 27, 1953 • FIFTEEN CENTS
Are You Afraid of LAWYERS?
I Worked for Three Presidents
A White House Diary on F.D.R.—TRUMAN—IKE
A DOZEN LAUGHS WITH BASEBALL UMPIRES
THE BABY SPACE STATION
First Step in the Conquest of Space

Collier's
March 22, 1952 • Fifteen Cents
Man Will Conquer Space Soon
TOP SCIENTISTS TELL HOW IN 15 STARTLING PAGES

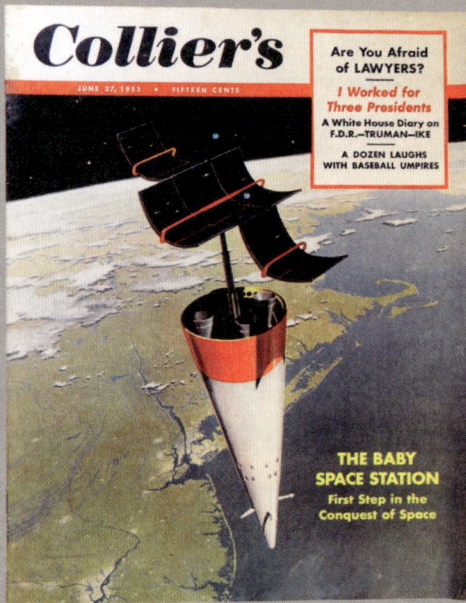

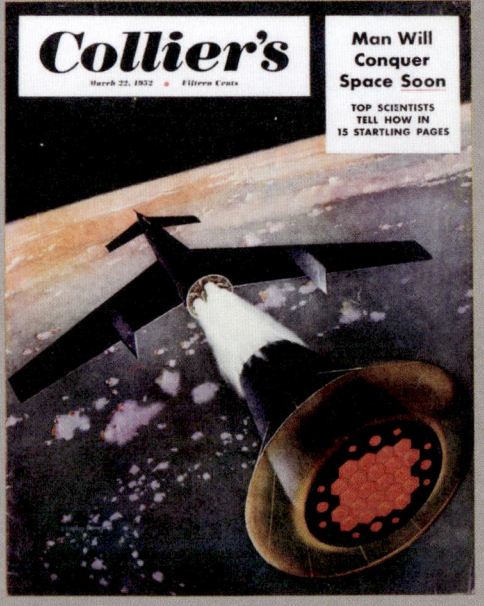

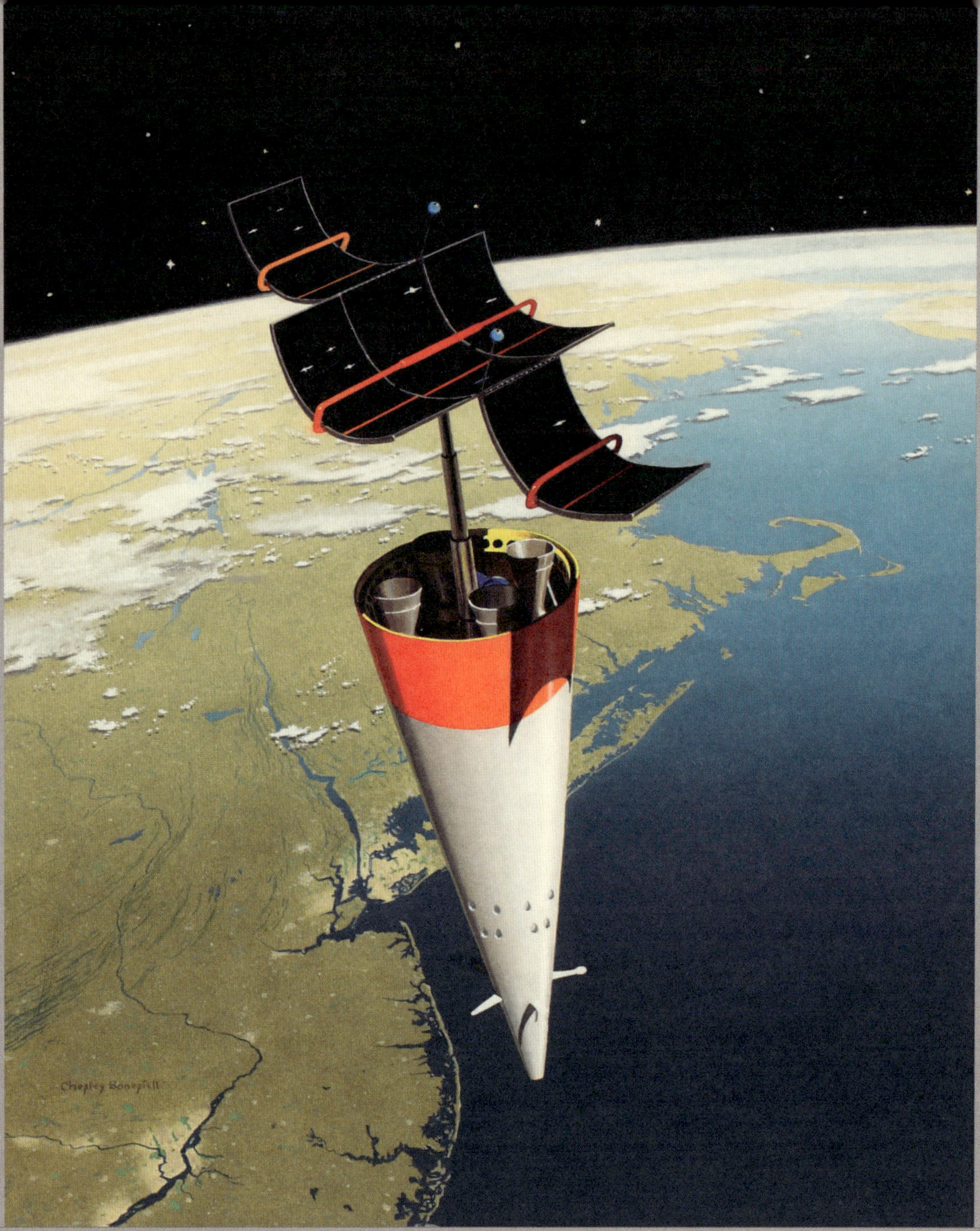
Chesley Bonestell

3.9 / 3.10

3.11

3.9 Cover of the 'Baby Space Station' issue of *Collier's* magazine, June 1953.

3.10 Cover of the 'Man Will Conquer Space Soon' issue of *Collier's* magazine, March 1952. The first of many important collaborations between Chesley Bonestell and the rocket scientist Wernher von Braun.

3.11 Chesley Bonestell, *The Baby Space Station*, 1950.

3.12 Chesley Bonestell, *Man Will Conquer Space Soon: What Are We Waiting For?*, cover artwork for *Collier's* magazine, 22 March 1952. Bonestell imagined this winged space shuttle more than seventy years before SpaceX's giant Starship, which looks remarkably similar.

3.13 (OVERLEAF) Chesley Bonestell, *In Orbit 600 Miles above Mars! Preparing to Land* (detail), 1953. The first catalogued owner of the original work was Wernher von Braun, whose plans to colonize space were outlined in *Collier's* magazine and visualized so vividly by Bonestell.

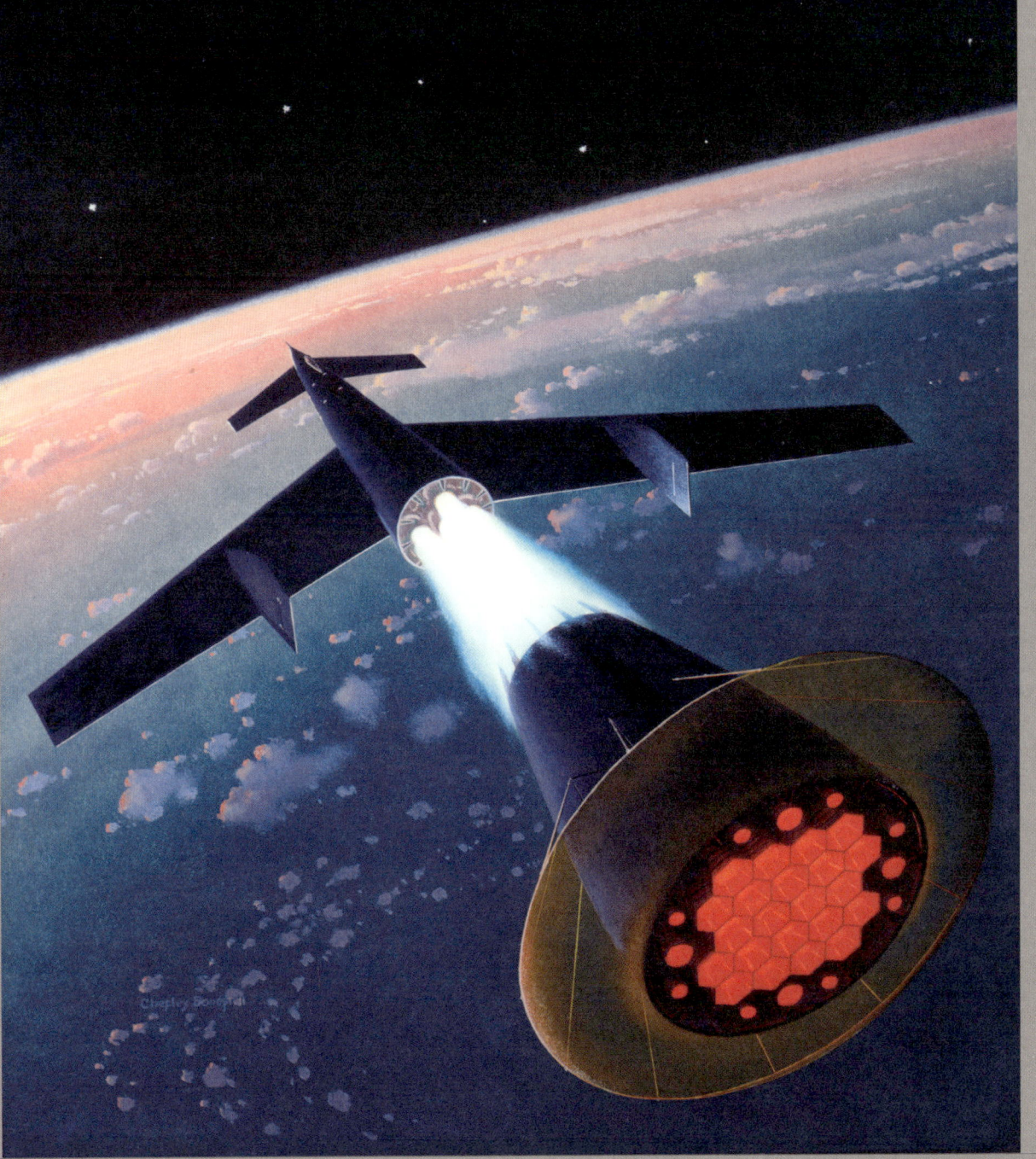

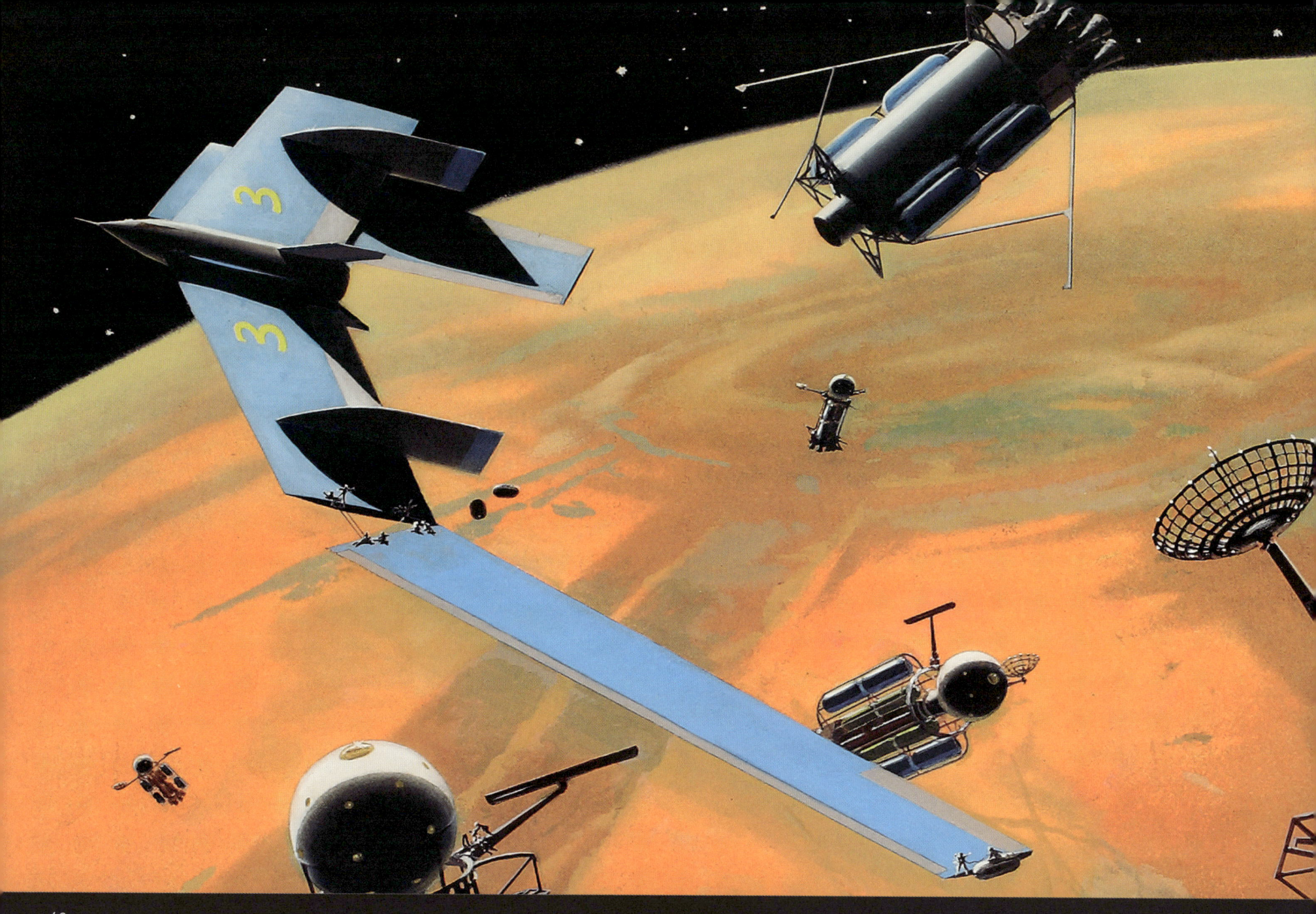

ARTHUR C. CLARKE

You could argue that space exploration began at 7 p.m. on Friday, 13 October 1933, in Room 15 on the second floor of 81 Dale St, Liverpool. This was the inaugural meeting of the British Interplanetary Society, if not the very first then certainly the oldest lay society concerned with the exploration of space that's still going today. A teenage Arthur Charles Clarke (1917–2008), from the small seaside town of Minehead in Somerset, had seen the society's advert in the newspaper calling for members and immediately wrote back: 'Please could you tell me about your particulars as I'd very much like to join it …' He would become its most notable member, and eventually its president – a natural home for the insatiably curious Clarke and other likeminded enthusiasts impelled by the promise of space.

Clarke's career stretches beyond that of a prolific science fiction author. He wrote numerous books and essays on the realities of space exploration grounded in his studies in mathematics and physics at King's College, London. He was also an undersea explorer, as well as a television presenter, hosting his *Mysterious World* series in the 1970s. The name 'Arthur C. Clarke' became a portal through which audiences could peer into the future and marvel at the miraculous innovations that would change our lives. He predicted the digital age, artificial intelligence and the power of satellite communications not just as utilitarian tools but as technologies that would fundamentally alter the course of our entire species.

But where did he get these ideas from? Like others in this book, it was his childhood reading that would come to define his life's course, with three great inspirations in particular. First, the exciting pulp sci-fi magazines that shipped to England from America, edited by the great science fiction impresario Hugo Gernsback.

4.1 Arthur C. Clarke diving between the propeller blades of a wreck off the coast of Sri Lanka, 1955. His underwater adventures in Australia and Ceylon changed the direction of his life and work.

4.2 Frank Phillips, 'The Onslaught from Venus', *Science Wonder Stories*, 1/4 (1929).

The September 1929 edition of *Science Wonder Stories*, for example, features tales such as 'The Human Termites' and 'The Onslaught from Venus', however nestled among its pages is Herman 'Noordung' Potočnik's essay 'The Problems of Space Flying'. This piece brought practical space travel as a concept to the general public via wheel-shaped rotating space stations, orbiting Earth observation satellites, giant space mirrors and rocket propulsion, which was still in its infancy. But it was the cover art as much as the periodicals' content that arrested the imagination. Clarke's particular favourite was the November 1928 edition of *Amazing Stories*, with a vivid picture of Jupiter dominating the scene and in the foreground a group of space explorers pointing in wonder from the lush tropical paradise of one of Jupiter's moons (fig. 4.8). The second inspiration was David Lasser's *Conquest of Space* (1931), the first English-language book to deal with the emerging physics of space flight, albeit wrapped in a fictional lunar voyage. Here was the promise that space travel, although still decades away, was real and tangible. The third was a close

'Come my friends, 'tis not too late
to seek a newer world.
To sail beyond the sunset, and the
baths of all the western stars.'

TENNYSON, 'ULYSSES' (1842)

4.3, 4.4 Pages from Arthur C. Clarke, 'The Space Station: Its Radio Applications' (1945), with text and diagrams describing 'the use of a chain of space-stations' in geostationary orbit. Best known for his science fiction writing, Clarke developed numerous practical (and at the time, wildly speculative) engineering applications for space. This privately circulated paper proposed that a network of satellites could be placed in geostationary orbits and used for global communications. He was a man very much ahead of his time.

4.5 Arthur C. Clarke in his study, c. 1936, aged about eighteen. The shelves behind him are lined with his beloved collection of science fiction magazines.

4.6 Arthur C. Clarke on his first flight, c. 1927. When he was almost ten years old, his mother took him flying at Taunton, Devon, in a British Avro 504 biplane owned by the Cornwall Aviation Company. His mother, Mrs Mary Nora Clarke, is sitting at the rear. The pilot is Captain Percival Phillips. For Arthur, this was to be the first of innumerable worldwide flights.

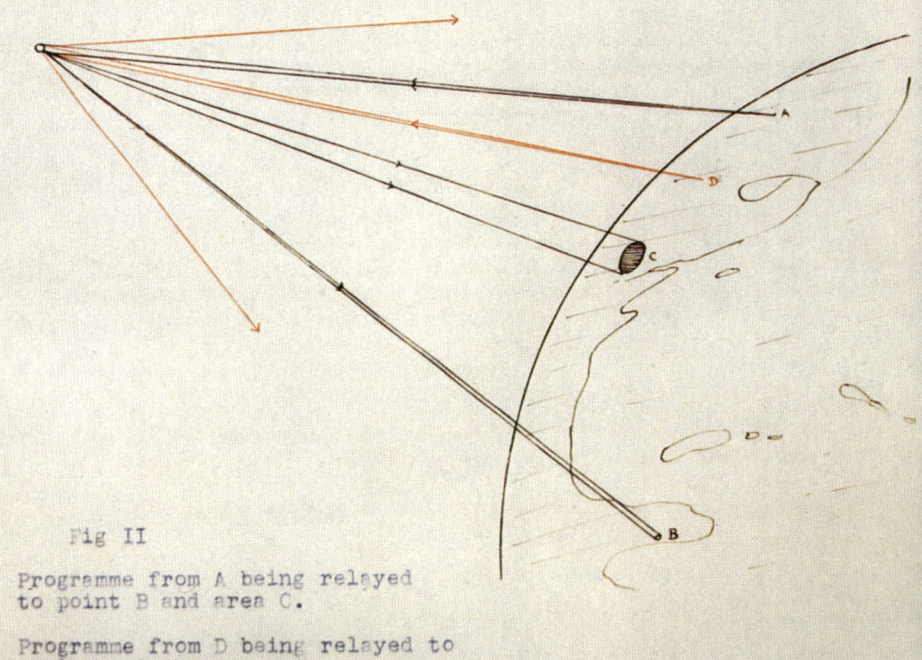

a reflector only a few feet across would give a beam so directive that almost all the power would be concentrated on the earth. Arrays a metre or so in diameter could be used to illuminate single countries if a more restricted service was required.

11.   The stations would be connected with each other by very-narrow-beam, low-power links, probably working in the optical spectrum or near it, so that beams less than a degree wide could be produced.

12.   The system would provide the following services which cannot be realised in any other manner:-

    a) Simultaneous television broadcasts to the entire globe, including services to aircraft.

    b) Relaying of programmes between distant parts of the planet.

13.   In addition the stations would make redundant the network of relay towers covering the main areas of civilisation and representing investments of hundreds of millions of pounds. (Work on the first of these networks has already started.)

14.   Figure II shows diagrammatically some of the specialised services that could be provided by the use of differing radiator systems.

Fig II

Programme from A being relayed to point B and area C.

Programme from D being relayed to whole hemisphere.

manner in which these can be provided.

8. All these problems can be solved by the use of a chain of space-stations with an orbital period of 24 hours, which would require them to be at a distance of 42,000 Km from the centre of the earth. (Fig 1.) There are a number of possible arrangements for such a chain but that shown is the simplest. The stations would lie in the earth's equatorial plane and would thus always remain fixed in the same spots in the sky, from the point of view of terrestrial observers. Unlike all other heavenly bodies they would never rise nor set. This would greatly simplify the use of directive receivers installed on the earth.

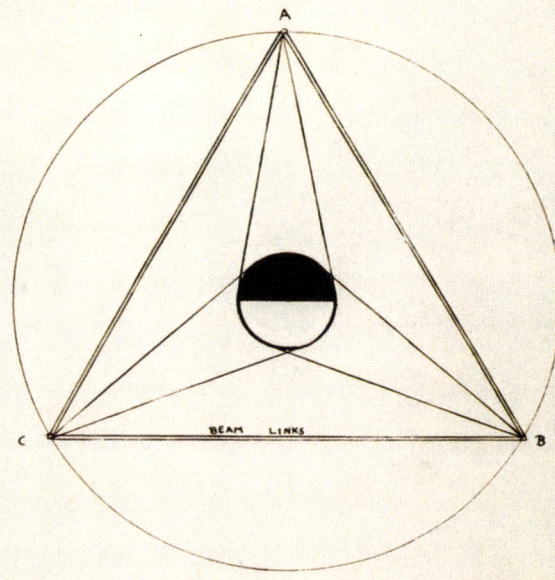

Fig 1.

9. The following longitudes are provisionally suggested for the stations to provide the best service to the inhabited portions of the globe, though all parts of the planet will be covered.

    30 E - Africa and Europe.
    150 E - China and Oceana.
    90 W - The Americas.

10. Each station would broadcast programmes over about a third of the planet. Assuming the use of a frequency of 3,000 megacycles, a

encounter in Minehead Library with the science fiction writer, philosopher and political activist Olaf Stapledon, whose books have been largely overshadowed by the better-known Jules Verne and H. G. Wells. Two of his novels in particular stayed with Clarke: *Last and First Men* (1930), a grand mythic exploration of deep time and the evolution of our species from the twentieth century to billions of years into the future; and *Star Maker* (1937, also illustrated fig. 30.10), which takes its protagonist on a grand cosmic odyssey from an English hillside under a star-filled night sky to the furthest corners of the universe and back. For Stapledon, time and space, however distant, were exotic locations to be explored.

Of all the hundreds of Clarke's science fiction stories, novels, books and television programmes, his most enduring creative work has to be his collaboration with Stanley Kubrick on the masterpiece *2001: A Space Odyssey*. The central story was borrowed from his 1949 short story 'The Sentinel', about the discovery of a mystifying artefact on the Moon. The film's production design chimes with Clarke's future-facing sci-fi tropes: routine space travel, video calls, artificial intelligence, lunar bases and rotating space stations with artificial gravity. It's notable that the film came out in 1968, the year Apollo 8 took us to the Moon and back for the first time. But perhaps even more intriguing is the film's final act, which tears us away from the drama of David Bowman's struggle with the rogue HAL 9000 computer and immerses us into the famous 'Jupiter and Beyond the Infinite' sequence, which continues to perplex and delight audiences. 'What does it mean?!' has been the response for over

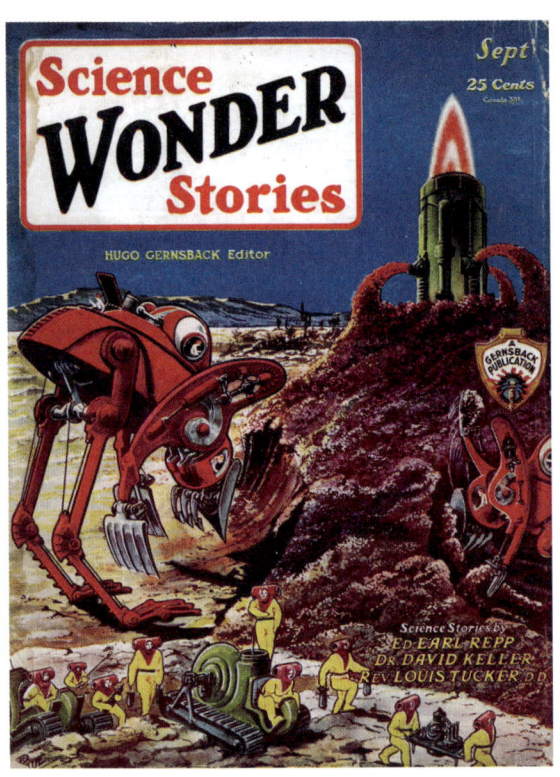

4.7 Cover of *Science Wonder Stories*, 1/4 (1929).

4.8 Cover of *Amazing Stories* (November 1928), which fired the imagination of the young Arthur C. Clarke.

4.9 Olaf Stapledon's time charts for *Star Maker*. Stapledon was a great influence on Arthur C. Clarke. These handwritten charts show his speculative plotting of the deep time scales of the universe - a central theme of his two most famous and ambitious novels, *Last and First Men* (1930) and *Star Maker* (1937).

4.10 Arthur C. Clarke, 'Sentinel of Eternity', *10 Story Fantasy* (Spring 1951).

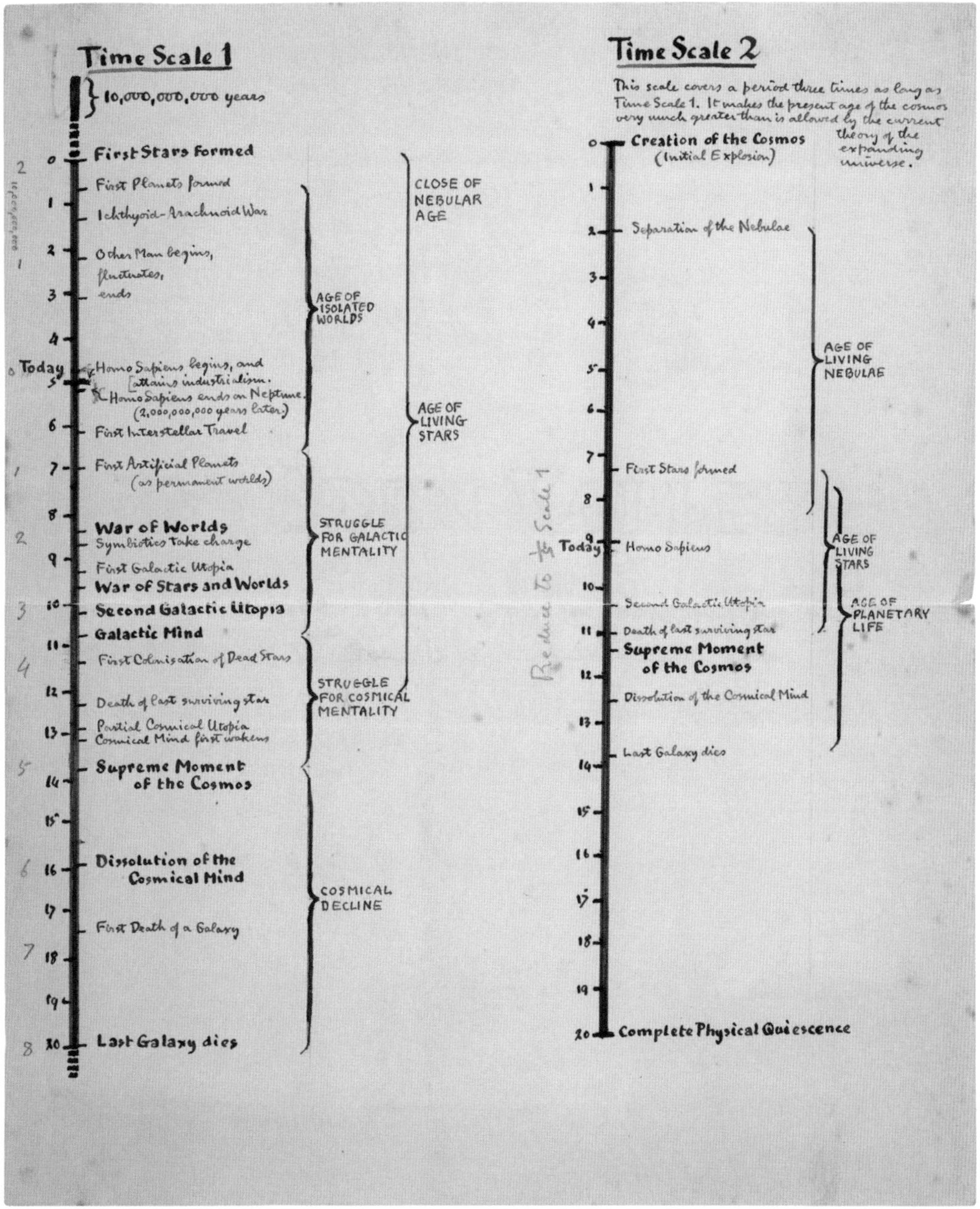

4.9

Sentinel of Eternity

Before there were men on Earth, that signal-sending pyramid had stood alone on a lifeless moon. What would happen now that its alarm was silenced?

By ARTHUR C. CLARKE

THE NEXT TIME you see the full moon high in the South, look carefully at its right-hand edge and let your eye travel upwards along the curve of the disc. Round about 2 o'clock you will notice a small, dark oval: anyone with normal eyesight can find it quite easily. It is the great walled plain, one of the finest on the Moon, known as the Mare Crisium—the Sea of Crises. Three hundred miles in diameter, and almost completely surrounded by a ring of magnificent mountains, it had never been explored until we entered it in the late summer of 1996.

Our expedition was a large one. We had two heavy freighters which had flown our supplies and equipment from the main lunar base in the Mare Serenitatis, five hundred miles away. There were also three small rockets

4.10

fifty years: ape touches mysterious black monolith and becomes man; man touches monolith and transforms into Nietzschean Übermensch, or some long-dreamt post-human entity. One thing is certain, Clarke and Kubrick's film loudly echoes the cosmic visions of Stapledon that touched Clarke so deeply as a youth.

Without Stapledon, Gernsback or Lasser – or the British Interplanetary Society – there would be no Arthur C. Clarke. And without Arthur C. Clarke, we wouldn't have travelled into space quite so stylishly. We would have plodded inevitably into the future, of course, but without the same excited anticipation. Clarke made the story of space travel in all its vivid colour infinitely richer. And story and colour are what Clarke understood better than anyone else.

4.11 Arthur C. Clarke on the set of *2001: A Space Odyssey* (1968).

4.12 Production still from *2001: A Space Odyssey* showing the giant centrifuge set of the Discovery One spacecraft.

4.13 British actress Margaret Tyzack in a still from *2001: A Space Odyssey*.

4.13

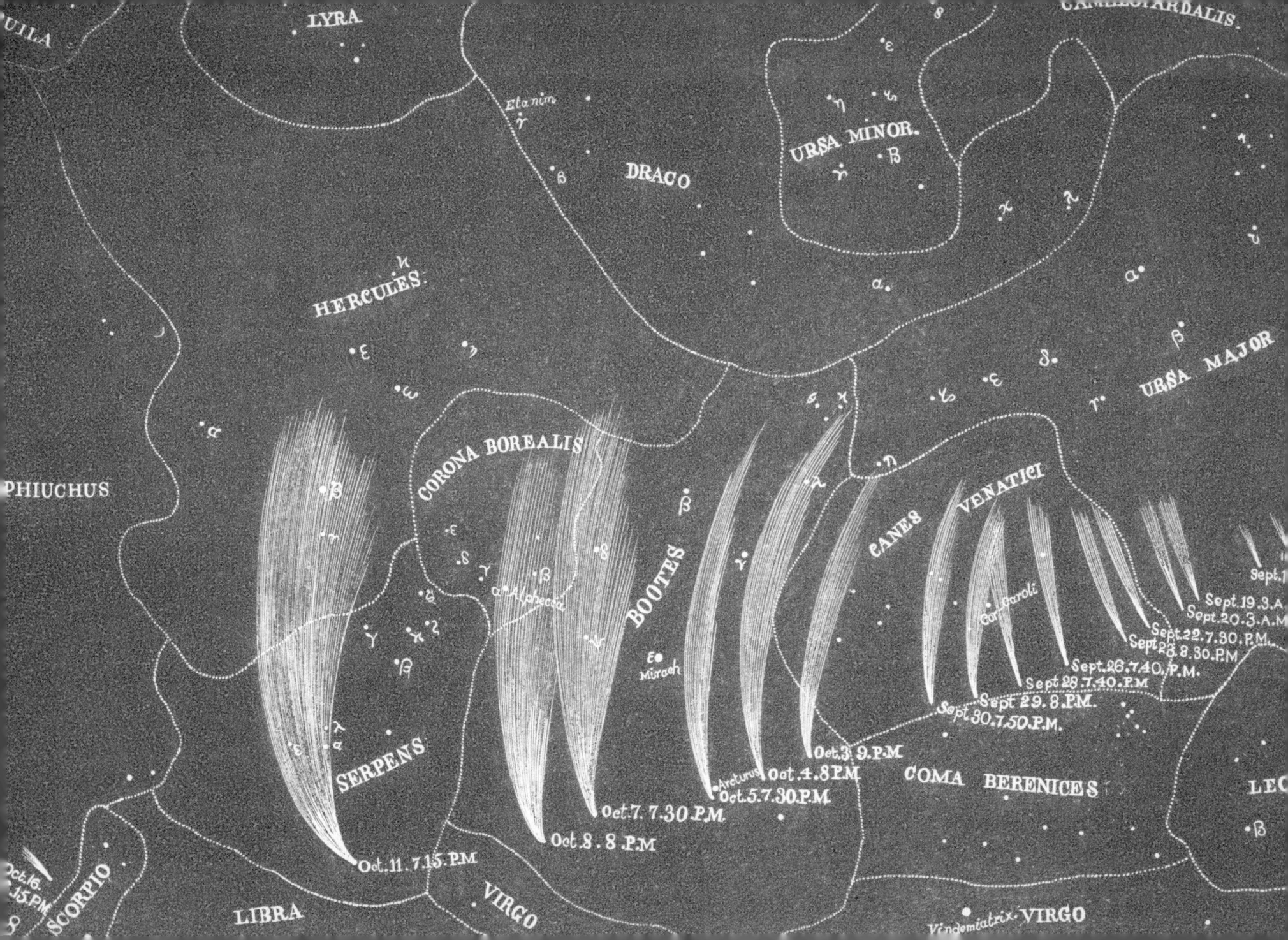

ACT 2.

CAELUM SPECTANTES

'STARING INTO SPACE'
THE SCIENTISTS AND THINKERS WHO MADE SENSE OF THE UNIVERSE

GALILEO GALILEI

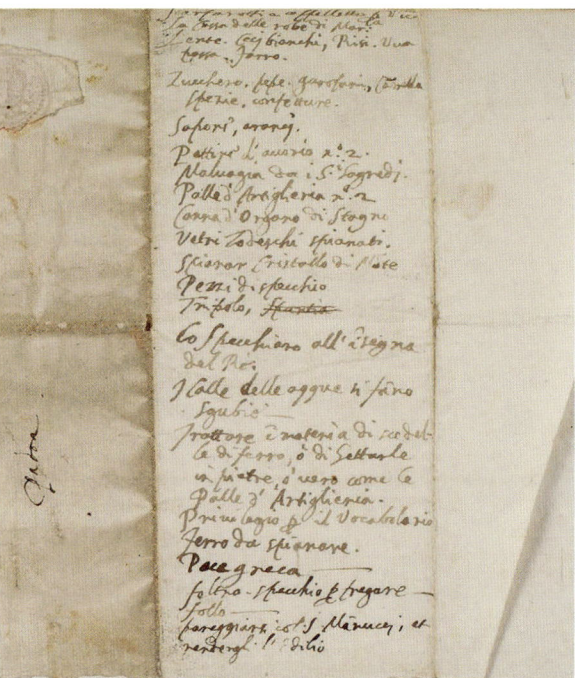

Even history's finest minds have to do the daily chores. On 23 November 1609, on the back of an old letter, the mathematics lecturer from the University of Padua Galileo Galilei (1564–1642) had scribbled down a to-do list. Alongside a reminder to pay a debt to 'Lord Mannucci' were the groceries: chickpeas, lentils, rice, sugar, various spices. He also needed to pick up new shoes and a hat for his son, and some slightly more unusual odds and ends – clues as to what was really on his mind: German spectacle lenses, two artillery balls and an organ pipe made of tin. Luckily he was in and around Venice at a time when artillery balls and organ pipes were presumably easy to find. In the seventeenth century Venice was (and still is) home to the finest glassmakers, particularly on the islands of Murano; the artillery balls were the perfect tool for grinding glass into concave lenses, and an old organ pipe the perfect receptacle in which to hold them. We know Galileo as a scientist and mathematician, but he was also a highly skilled maker of scientific equipment, from compasses to thermometers to pendulum clocks. It's his skill as a craftsman, alongside his talent for self-promotion,* that would prove key to transforming our knowledge of the heavens.

The component parts of the telescope were not new. Since antiquity there had been a fascination with optics and an understanding of the distorting effects of lenses, water or magnifying 'reading stones'. Even a simple tube was known to aid distant vision, concentrating the eye, framing the view and blocking out the surrounding light at the same time. But things came together when news arrived in Venice of an exciting gadget that had appeared in the Netherlands. On 2 October 1608 the spectacle-maker Hans Lipperhey was refused a patent for what he described as a device 'for seeing things far away as if they were nearby'. His invention consisted of two lenses, one convex and one concave, held inside either end of a tube, which would give the viewer 3× magnification – that is, an object 3 miles away would look only 1 mile away. The significance and practical applications of such a device, particularly as a military aid, were not lost on Galileo and his peers.

---

* He would have been all over Instagram.

5.1 Ottavio Leoni, *Portrait of Galileo Galilei*, 1624.

5.2 Galileo's shopping list: *Lista annotate suuna lettera di Ottavio Brenzonidel 23 novembre 1609.*

5.3 Donato Creti, *Moon* (detail), 1711. This image and fig. 5.13 are part of a series of astronomical paintings by Creti, commissioned by Count Luigi Marsili of Bologna to demonstrate the new importance of astronomy.

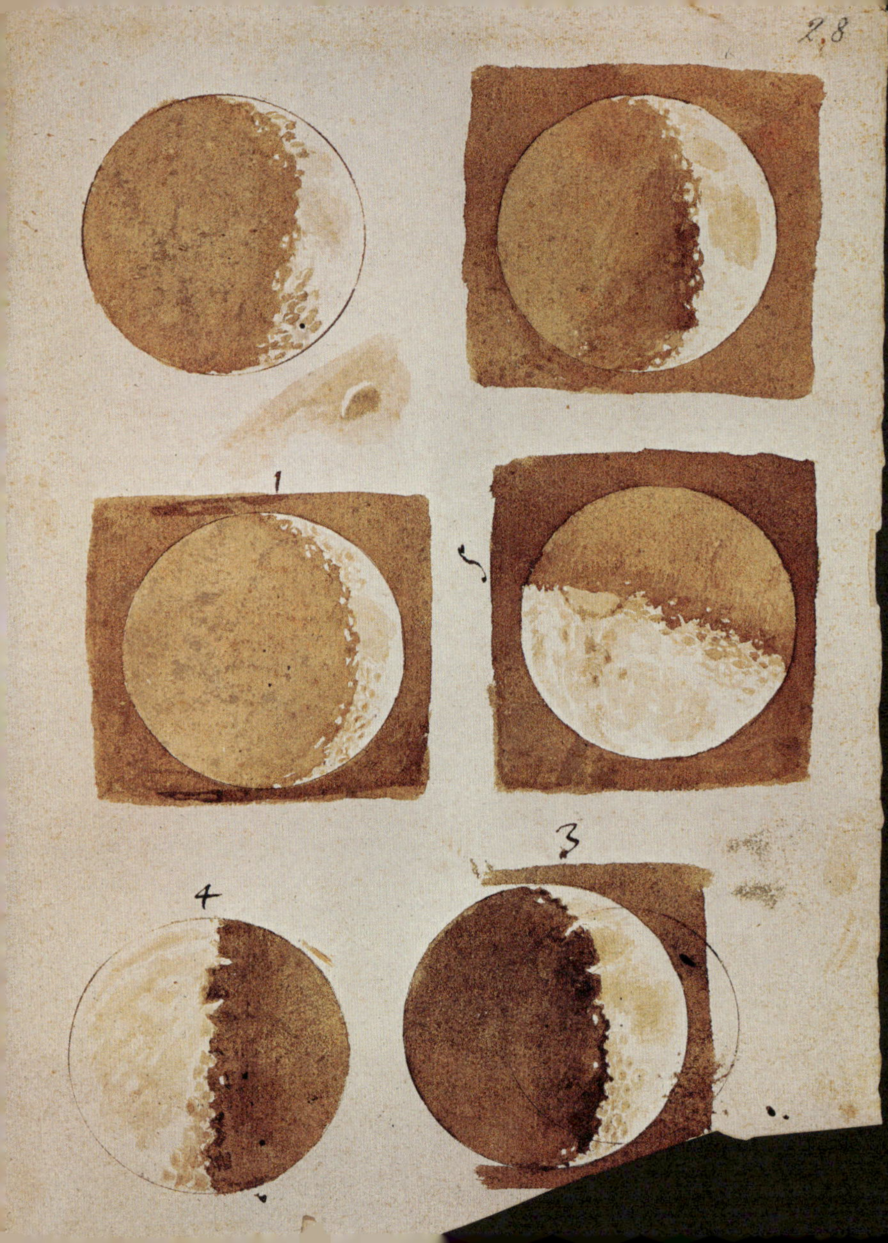

'Forsaking terrestrial observations, I turned to celestial ones, and first I saw the moon from as near at hand as if it were scarcely two terrestrial radii away.'

SIDEREUS NUNCIUS (1610)

5.4 Galileo Galilei, *Drawings of the Moon*, November–December 1609.

5.5 One of Thomas Harriot's drawings, made between 1609 and 1611, detailing features of the Moon's surface such as mountain ranges, craters and 'seas'.

By the summer of 1609, Lipperhey's simple Dutch spyglass had been transformed by the skilful hands of Galileo, who made a version that was three times more powerful. In a political move engineered to elevate his own prominence in Venetian society, he gave the telescope to the doge of Venice, Leonardo Donà, as a gift to the state, and in return gained a pay rise as well as tenure in the University of Padua.

On 1 December 1609, our view of the heavens changed forever when Galileo pointed an even more powerful spyglass of his own design towards the Moon. He was not the first to do this, however. The English mathematician Thomas Harriot had made his first drawings of the Moon using a lesser spyglass in July that year, but crucially Harriot lacked Galileo's visionary interpretation, as well as an understanding of the importance of publishing first. When Harriot looked through his spyglass, he saw the Moon with the mind of a cartographer; when Galileo looked, he saw the Moon as a scientist and explorer. 'I thank God from the bottom of my heart', he wrote in a letter, 'that he has pleased to make me the sole initial observer of so many astounding things, concealed for all these ages.' Galileo saw for the first time that the Moon, which had been fixed, perfect and immutable in the heavens throughout history, was not an unblemished orb as had previously been thought.

The veil had fallen and the Moon was revealed as another Earth-like world, 'uneven, rough and crowded', with 'chains of mountains and depths of valleys'. He noted that the lunar terminator – the line which divides the illuminated part of the Moon from the shadowed – was far from smooth, as it would have been on a perfect sphere. Instead it was jagged and sinuous. He described the cratered complexion of the lunar surface, decorated with spots 'like the blue eyes of a Peacock tail', and the play of light across it. Above all he sketched in beautiful detail what he saw, making precise and accurate mathematical measurements. His observations of the Moon revealed the depth of the Milky Way galaxy itself. Where once there was just a smudge of light, there was now an ocean of stars. Behind the three stars of the belt of Orion the Hunter and the six in his sword were eighty more, now seen for the first time: 'What was observed by us in the third place is the nature or matter of the Milky Way itself, which, with the aid of the Spyglass, maybe observed so well that all the disputes that for so many generations have vexed philosophers are destroyed by visible certainty, and we are liberated from wordy arguments.'

Jupiter too was bright in the Venetian sky in the new year of 1610 and presented Galileo with more revelations, the first being a strange alignment of three hitherto unseen stars alongside the planet. The next night of observation produced a fourth star, and each following night their positions shifted – erratic sidereal bodies performing their revolutions around Jupiter. He had discovered Jupiter's system of moons, which he named 'Medicean Stars' to flatter the Florentine ruler Cosimo II and his three brothers, but which we now know as the Galilean moons of Io, Europa, Ganymede and Callisto.

Within months, Galileo had published his findings in a short and elegantly written booklet, *Sidereus nuncius* (Starry Messenger). The power of the spyglass to bring the universe closer was further enhanced thanks to that other transformative piece of technology: the printing press. The book was a sensation – for both his detractors, who cast doubt on what he had seen, and his supporters, who were captivated by the implications – and it marked the start of the telescope revolution that transformed astronomy.

The real power of the telescope, however, comes not from the quality of the lenses but from the minds of those who are using it. Even with perfect resolution, our understanding of nature is still at the mercy of fallible human reasoning. It is the scientific method itself that remains the best corrective lens we've ever made, and one we have to thank Galileo Galilei for helping to shape and polish so beautifully.

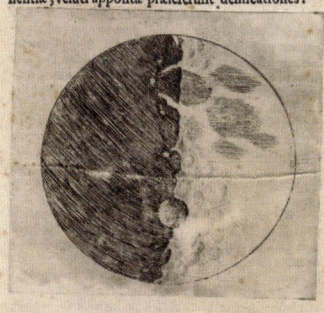
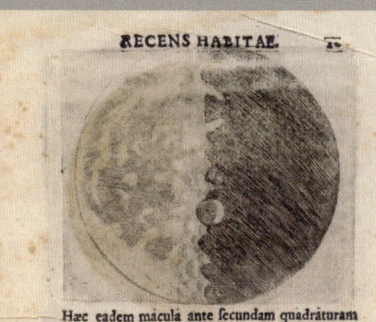
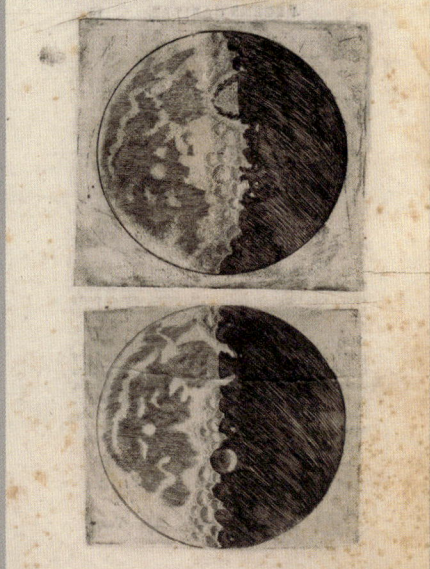
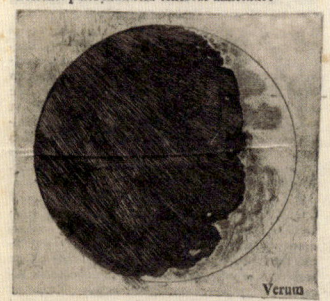

Sidereus nuncius (1610) was the book that brought Galileo's radical observations to the world. From the four Jovian moons to the composition of the Milky Way, it was fundamental in transforming our understanding of both astronomy and our place in the cosmos.

5.6 The Nebulas of Orion and Praesepe (the Beehive Cluster).
5.7 Observations of Jupiter.
5.8–5.11 Detailed drawings of the Moon's mountainous, cratered surface.

5.12 Maria Clara Eimmart, *Aspect of Jupiter*, late 17th century. One of a set of twelve depictions of heavenly phenomena by Eimmart, this work is based on Galileo's discoveries.

5.13 Donato Creti, *Jupiter* (detail), 1711.

IVPITER cum suis Satellitibus.
GRIMALDI et RICCIOLI observationes.
HVGENII.
HOOKII.

5.12

5.13

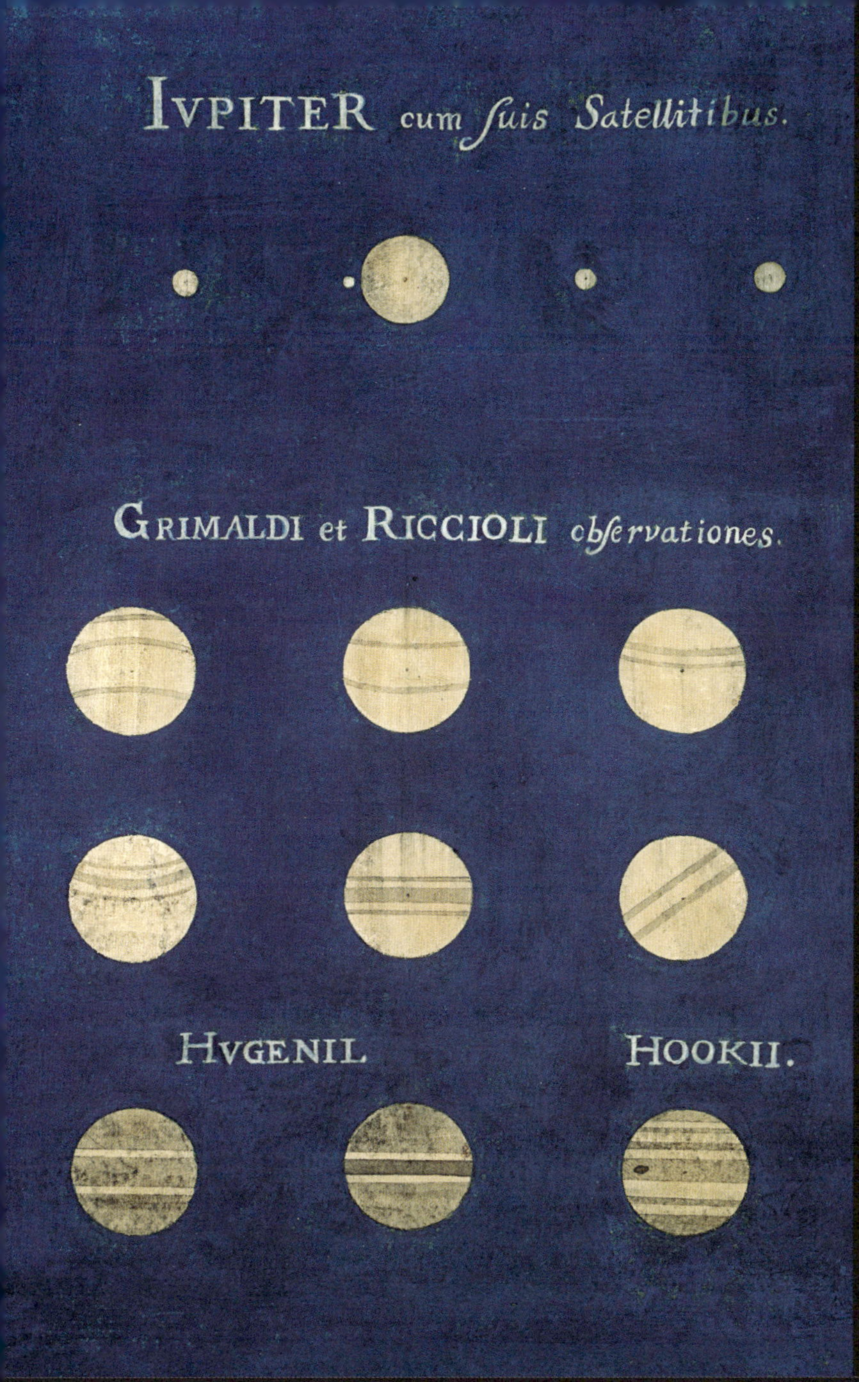

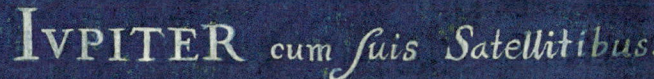

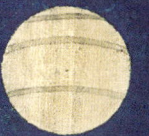

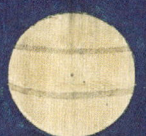

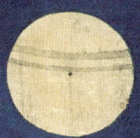

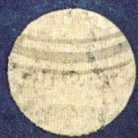

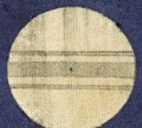

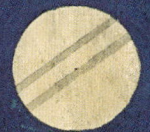

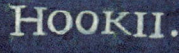

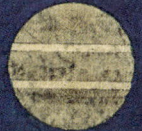

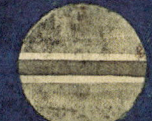

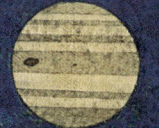

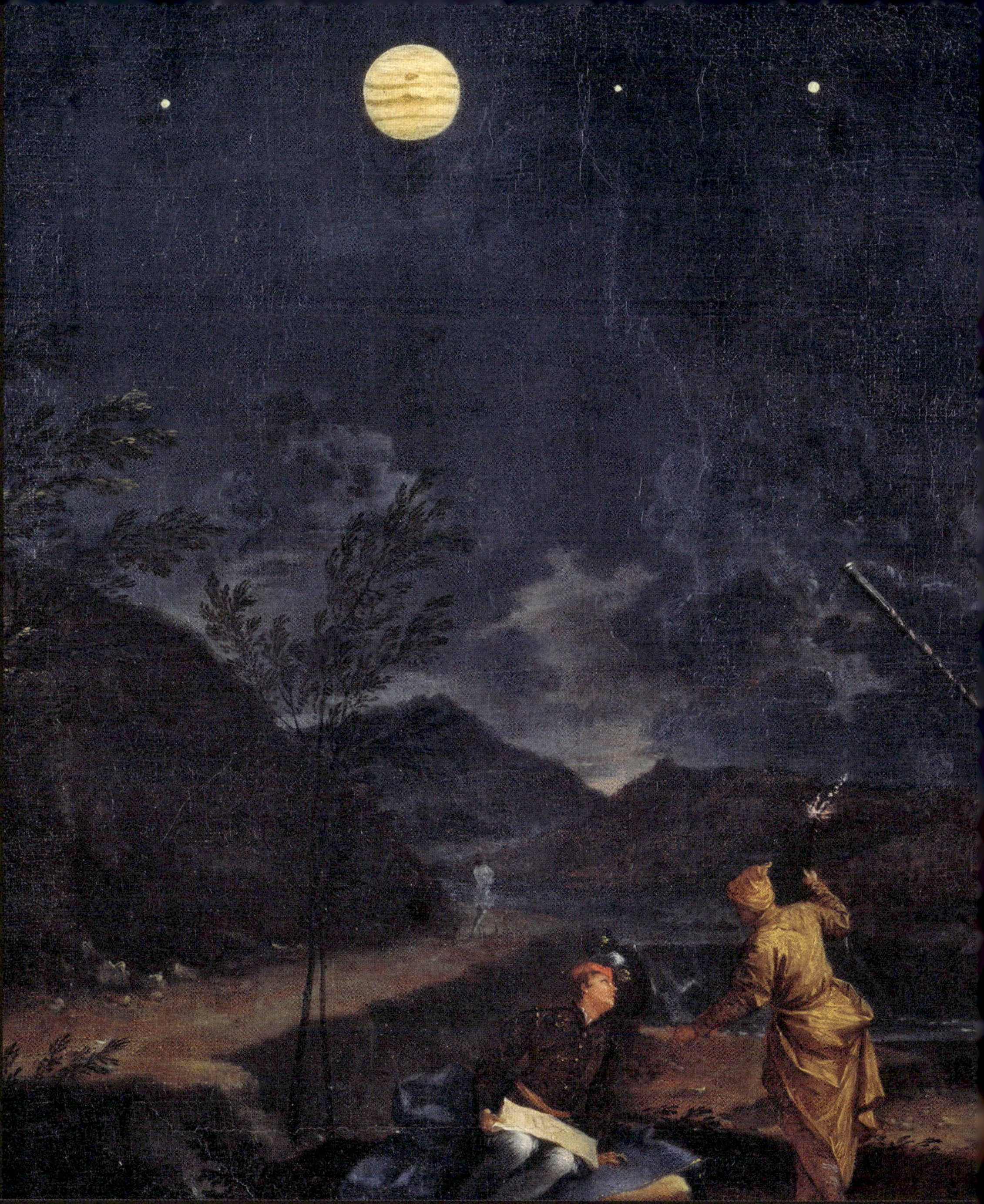

ISAAC NEWTON

Newton kept himself busy during lockdown. During the plague months between 1665 and 1666 he retreated from his student digs at Trinity College Cambridge back to his family home, Woolsthorpe Manor in Lincolnshire. Here, in relative isolation in the quiet rural setting, he managed to invent an entire branch of mathematics (calculus) and revolutionize the field of optics, all while working on his law of universal gravitation. This law would prove that the force of gravity here on Earth – that which causes a tear to roll down a cheek, apples to fall to the ground, my marmalade jar disaster this morning* – is the same force that extends across the universe. It keeps the Moon and the International Space Station in orbit, planets spherical and galaxies knitted together, and is nicely summarized by the poet Samuel Rogers:

> *That very law which moulds a tear*
> *And bids it trickle from its source,*
> *that law preserves the earth a sphere,*
> *And guides the planets in their course.*

Newton's work on gravity and his laws of motion (fundamental to anyone launching a rocket) were published in 1687 in his three-volume *Philosophiæ naturalis principia mathematica*, regarded as the most important scientific book of all time. It unlocked the intricacies of the universe, making moon shots and pinpoint landings on Mars possible.

If *Principia* represents the summit of Newton's achievements, then his unpublished letters, papers and notebooks give us an insight into how he arrived there. It's in these pages that we see the staggering breadth of intellectual inquiry that Newton explored throughout his long life: natural philosophy and mathematics, of course, but also arcane areas of history, language, alchemy and theology. His prolific notes are an extension of his mind, a place where he could explore and organize his thoughts systematically and in private.

As far as we know, Newton's note-taking began with a small vellum-covered pocketbook that he used sometime in his late teens. The book contains clues to his early interests: notes and diagrams on building clocks, sundials and windmills, on drawing technique, on the making of a Copernican model of the solar system, alchemical recipes to fight plagues and ulcers, perpetual motion machines and his own astronomical observations. There's a section on magic tricks which includes a miraculous 'water into wine' routine. He suggests concealing some logwood, a plant that yields a deep red dye, in your mouth and covertly mixing it with a sip of water, before spitting out the claret-coloured liquid to the wonderment of your audience. As well as being an autodidact, Newton was a deeply religious man, and much of his writing is devoted to his radical antitrinitarian ideas, biblical prophecy and prediction. In his Fitzwilliam Notebook he lists at length his various transgressions, as if speaking at a confessional, his tarnished soul laid bare across the page:

---

* See also: Sod's Law.

Caring for worldly things more than God; Peevishness with my mother; Having uncleane thoughts words and actions and dreamese; falling out with the servants; using Wilford's towel to spare my own; and so forth.

In the Trinity College Notebook, purchased at the start of his Cambridge career in 1661, we begin to see the evolution of Newton the independent thinker. Much of the book is devoted to a section titled 'Certain Philosophical Questions', a to-do list of some fifty areas of inquiry intended for study, and one to which he would repeatedly return in the decades that followed. The subjects ranged from the nature of matter and the workings of our senses to the astronomical world of planets and comets, our internal world of the imagination and dreams to the divine world of God and Creation. The book also introduces Newton as the experimental researcher, and in particular the self-experimenter. He records staring at the Sun to better understand its visual effects on the eye,

6.1 Sir Isaac Newton, depicted without his familiar wig. Oil painting by Sir James Thornhill (1675-1734).

6.2 Original manuscript of Newton's *Principia* (1685) that was sent to the printer for publication.

6.3 Newton's own annotated copy of *Principia*, in preparation for the second edition.

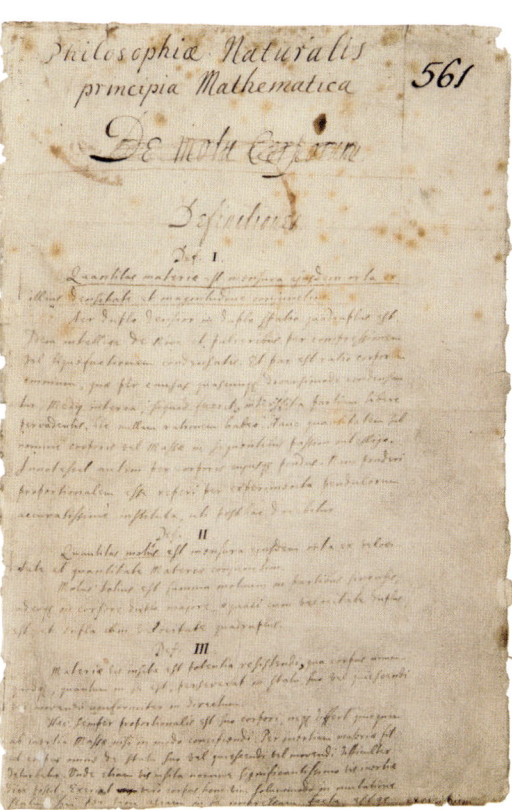

6.4 Title page of Newton's Trinity College Notebook, used c. 1661-65.

6.5 'Changing how we see': Isaac Newton's 'bodkin' experiment, c. 1669.

6.6 The importance of showing your workings: a page of one of Newton's notebooks showing his calculations on universal gravitation. The foundation of modern classical physics in his own hand.

(Repaired by Gray, Cambridge, March, 1963)

Add. 3996

Isaac Newton
Trin: Coll Cant
1661

Isaac Newton

Sep. 25 1727

Not fit to be printed

T. Pellet

Of Colours

56 The powders of Pellucid bodys is white soe is a cluster of small bubles of aire, ye scrapings of black or cleare horne, &c:[because of ye multitude of reflecting surfaces soe are bodys wch are full of flaws, or those whose parts lye not very close together (as metalls, Marble, ye Oculus Mundi Stone &c)[whose pores betwixt their parts admit a grosser Æther into yn yt ye pores in their parts], hence

57 Most Bodys (viz: those into which water will soake as paper, wood, Marble, ye Oculus Mundi Stone, &c) become more darke & transparent by being soaked in water [for ye water fills up ye reflecting pores]

58 I tooke a bodkin gh & put it betwixt my eye & ye bone as neare to ye backside of my eye as I could: & pressing my eye wth ye end of it (soe as to make ye curvature a,bcdef in my eye) there appeared severall white, darke & coloured circles r, s, t, &c. Which circles were plainest when I continued to rub my eye with ye point of ye bodkine, but if I held my eye & ye bodkin still, though I continued to presse my eye wth it yet ye circles would grow faint & often disappeare untill I renewed ym by moving my eye or ye bodkin.

59 If ye experiment were done in a light roome so yt though my eyes were shut some light would get through their lidds There appeared a reddish spot in ye midst at srs, a greater broade blewish darke circle outmost (as ts), & wthin that another light spot srs whose colour was much like yt in ye rest of ye eye as at R. Within wch spot appeared still another blew spot r

The summe of these two summes is equall to ye area befv, supposeing ae=0,9.

And their Difference is equall to ye area bcdv, supposeing ac=1,1. viz:

$befv = 0,10536051566,57826301227,5009,80839392798306,12037,98327,4072...$ If $ae=0,9$. or $eb=-0,1$.
$bcdv = 0,09531,01798,04324,86004,39521,23280,84509,22206,05365,30864,4299...$ If $ac=1,1$. or $bc=0,1$.

In like manner if $x=0,01$. & $a=1$. The calculation is as followeth. $= ax + \frac{x^3}{3a} + \frac{x^5}{5a^3}$
$$\frac{x^7}{7a^5}$$
$$\frac{x^9}{9a^7} + \frac{x^{11}}{11a^9}$$
$$\frac{x^{13}}{13a^{11}}. \&c$$

The summe of these two summes is equall to ye area befv, supposeing ae=0,09. And their Difference to bcdv, if ac=1,01. viz:

$= \frac{xx}{2} + \frac{x^4}{4aa} + \frac{x^6}{6a^4} + \frac{x^8}{8a^6} + \frac{x^{10}}{10a^8} + \frac{x^{12}}{12a^{10}} + \frac{x^{14}}{14a^{12}}$
$+ \frac{x^{16}}{16a^{14}} + \frac{x^{18}}{18a^{16}} \&c.$

$befv = 0,01005,03358,53501,44118,35488,57558,54779,60853,17007,67462,98736...$ If $ae=0,99$.
$bcdv = 0,00995,03308,53168,08284,82153,57544,26074,16887,29609,94005,87984...$ If $ac=1,01$.

For if $x=0,001$. Then The Calculation will be.
$= ax + \frac{x^3}{3a} + \frac{x^5}{5a^3} \&c$

$= \text{summe of } \frac{xx}{2} + \frac{x^4}{4aa}$

Therefore $\begin{cases} befv=0,00100,00003,33583,53350,01429,82... \\ bcdv=0,00099,95003,33083,53...68093,98... \end{cases}$ If $ae=0,999$.

And if $x=0,0001$. Then $0,00010,0000,00333,33347,61904730...$ If $ac=1,001$.

and in one of his laboratory notebooks he details an experiment involving sticking a bodkin needle in his eye socket and deforming the eyeball with it to see what patterns and colours may be produced in the mind (fig. 6.5).

Paper was a valuable resource. Newton's so-called 'Waste Book' was inherited from his stepfather, the minister Barnabas Smith, and contained his theological notes. These were of no interest to Newton, however, who recycled[†] the book's blank pages for his own work. In it we see the seeds of Newton's prowess in mathematics, in particular his evolving work on calculus, and how his laws of motion first began to take shape. All of his notebooks are a record of continuous adaption, complete with his redactions, additions and changes. For Newton they were an archive of ideas that he could return to with new insights and expand upon. There's also a great beauty in the books themselves. We can follow Newton's evolving handwriting, the style of his diagrams and the texture and patina of the paper and ink, the folds, creases and marks, as well as the notations of others who have handled them over the past 300 years.

[†] These days there's big business in the fancy notebooks favoured by poseurs and procrastinators. As a general rule, the quality of the notebook doesn't reflect the quality of ideas. Here's my own law on the subject: 'For every increase in a notebook's fanciness there is an equal and opposite decrease in the quality of the ideas contained within.' Or, for the Classicists reading this (Will and Shomit): *Cuiuslibet libri pulchritudinem augmentum est, eiusdem utilis ideis diminutum est.*

These notebooks are the seedbeds of Newton's genius and are particularly relevant if you're planning to travel in space. In the third book of his *Principia*, *De mundi systemate* (On the System of the World), he provides us with an elegant thought experiment to explain how orbits work. It's particularly useful for understanding why astronauts appear weightless on the International Space Station (ISS). Often you'll hear it's because of the lack of gravity. This is not quite right. Newton's inverse square law of mutual gravitation reminds us that the space station's altitude of 400 kilometres (248 miles) means that there's almost as much gravitational force acting on the astronauts inside as there is at sea level. So what's going on? Newton asks us to imagine a stone being thrown from the top of a mountain (his third law of motion), but let's imagine for fun that there's a cannon, and you're being fired from it. As you fly through the air, with your red cape billowing behind, the force of gravity starts pulling you back to Earth as your velocity decreases. A curved path is the result, just as when you throw a ball for your dog. Let's imagine the cannon has a bit more force. This time, with the greater velocity, you fly even further, but eventually gravity pulls you back down. Hopefully onto something soft. Now imagine you have so much velocity, you fly so far, that the curvature of the Earth matches your curved path of descent. The Earth's curve falls away as you fall to it, and so you never hit the ground. You fall beyond the horizon, around the Earth. This is what's happening with the ISS. And because it's above the atmosphere, there's no air resistance to slow it down (in accordance with Newton's first law of motion). It just keeps going round and round. The astronauts are floating because they are in free fall. Eventually, when it's time for the space station to de-orbit (hopefully without any astronauts on board), they will slow its speed using rockets as brakes and it will fall back to Earth, burning up in the atmosphere as it does. It's such an elegant explanation that the sketch was chosen as one of the images on the Voyager Golden Records (see Chapter 25).

Nothing in this book – the physical or digital one in your hands right now – makes sense without the foundational work of Isaac Newton. The epitaph on his lavish tomb in Westminster Abbey ends with the line, 'Mortals rejoice that there has existed such and so great an ornament of the human race.' If that's a testament to his contribution to science, then his surviving notebooks, if less ornamental, are an equally fitting monument to the power of the creative mind. In them we meet the Newton who lived to the left of the margin: Newton the obsessive, the brilliant, the radical, the ever fascinating.

> '*Amicus Plato, amicus Aristoteles, magis amica veritas.*'
> (Plato is my friend, Aristotle is my friend, but truth is a greater friend.)
>
> ISAAC NEWTON, 1661

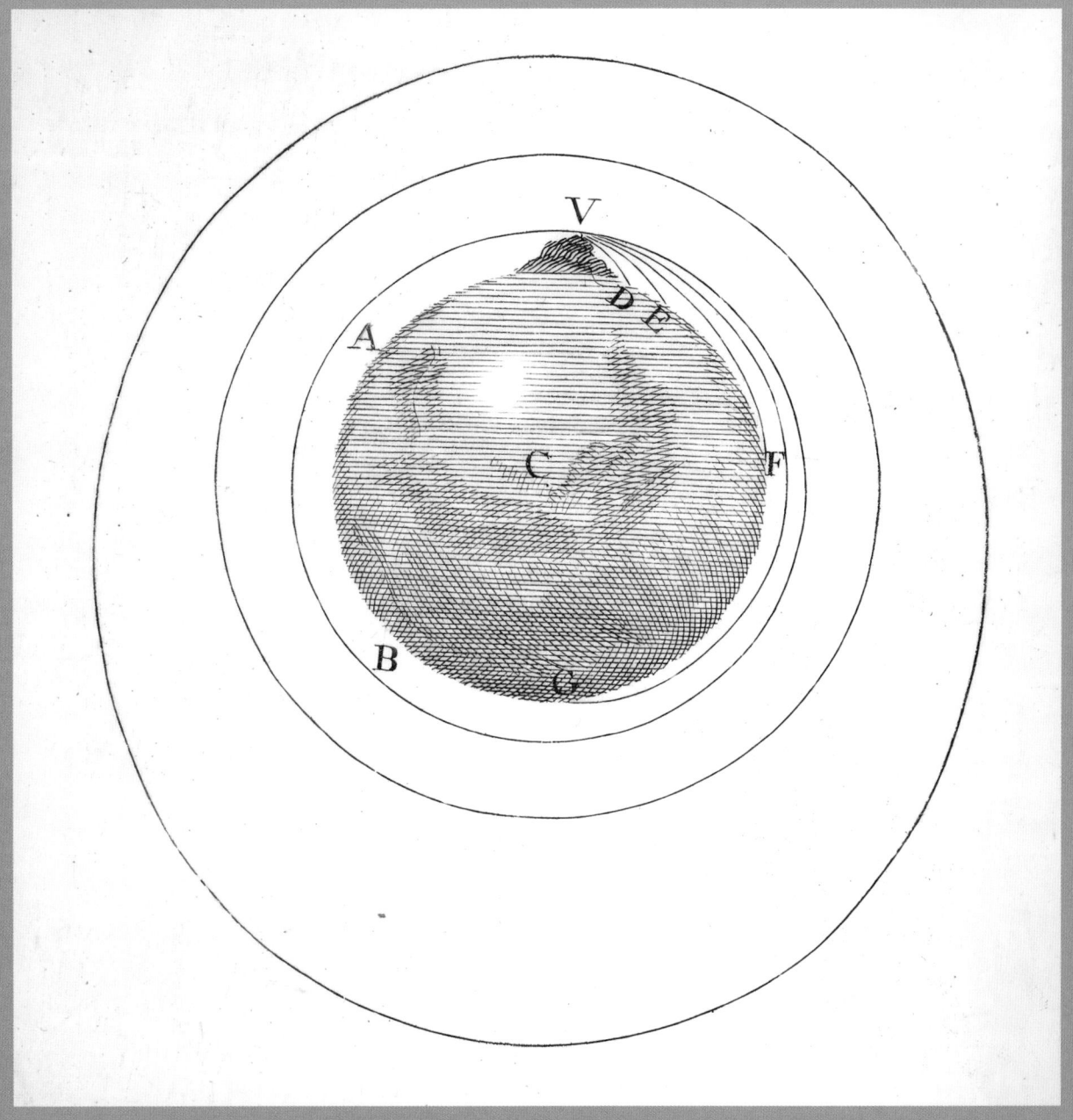

6.7 Isaac Newton, *A Treatise of the System of the World* (1728). The engraving imagines a cannon on a mountain top firing a projectile with increasing amounts of energy until its parabolic flight path arcs beyond the horizon, 'missing' the ground. Without any force acting on it to slow it down it will remain in orbit, falling around the Earth.

PERCIVAL LOWELL

In February 2025 some enthusiastic social media accounts began reposting a historic aerial photograph of the Martian surface taken in 2001 by the orbiting spacecraft Mars Global Surveyor, featuring what looks like the ruined, fragmented perimeter of a mysterious square structure. The podcaster Joe Rogan, an amplifier of fringe voices in archaeology, called it 'fucking WILD'. What could it be? At first glance it does suggest intelligent design. Straight lines and right-angled corners automatically launch our brains into story mode: is it a derelict structure built by a long-lost Martian civilization? Perhaps the same one that built the famous 'face on Mars', as seen in an image taken by the Viking orbiter in 1976?

In that instance, time and better technology revealed that the Sphinx-like alien face was nothing more than the undulating summit of a natural mesa; an optical illusion created by poor camera resolution combined with the play of light and shadow and fallible human interpretation. Where there are gaps in data, our brains fill in the blanks. Where there is visual noise, our brains make pictures. It's what story-seeking brains do. We are all susceptible to motivated reasoning: we seek the evidence which supports a belief we already hold. As the astronomer Carl Sagan noted on the subject of Mars, 'Where we have strong emotions, we are liable to fool ourselves.'

For as long as we've been looking at Mars, we've been seeing strange things on its surface. Astronomers such as Giovanni Cassini (1625–1712) and the Dutch astronomer Christiaan Huygens (1629–1695) sketched mysterious dark patches based on their observations through crude telescopes. William Herschel (1738–1822) looked even closer and noted polar icecaps, making the assumption that 'Its inhabitants enjoy a similar situation to ours.' Then, in 1877, the Italian astronomer Giovanni Schiaparelli suggested that these dark areas were actually linear striations: lines, rather than patches, snaking across the planet's surface like tentacles. He called these features *canali*, meaning 'ditch' or 'channel', but which in English instantly conjured up the idea of canals.

Schiaparelli's exciting observations went on to define the life and work of Percival Lowell (1855–1916), a wealthy American businessman, mathematician and astronomer who had been observing Mars since childhood from his rooftop

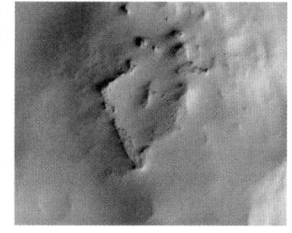
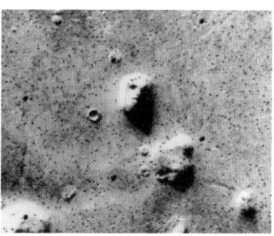

> 'All philosophy, said I, is founded on two things; an inquisitive mind and defective sight …'
>
> BERNARD LE BOVIER DE FONTENELLE, *CONVERSATIONS ON THE PLURALITY OF WORLDS* (1715)

in Boston. In 1894 Lowell built a new observatory in Flagstaff, Arizona, chosen for its elevated position and dry, uniform climate, which would provide a clearer view of the heavens. Through a 60-centimetre (24-inch) refracting telescope, Schiaparelli's mysterious *canali* began to proliferate and straighten, forming a network of fine thread lines which Lowell described in his book *Mars and Its Canals* (1906) as a 'spider's web seen against the grass of a spring morning'. He calculated that the canals varied from a mile to several miles in width and hundreds to thousands of miles in length, some single lines and others double, running parallel. He also suggested that he could see broad bands of vegetation alongside the water channels themselves, like the fertile banks of the Nile, and in the dark nodes where the canals intersected he saw what he called *oases*. For observed phenomena of such complexity, it's notable and revealing that other scientists didn't see them. Lowell attributed this to everything from poor eyesight and a lack of patience to the challenging atmospheric conditions on Earth, which make it impossible to see the Martian surface clearly even with state-of-the-art equipment, as Lowell had. The real reason, of course, was that the canals were simply an optical illusion; products of wishful thinking engineered by the mind rather than Martians. But Lowell's detailed maps legitimized the idea, giving it an authority that helped it spread in the popular imagination. Lowell also drew a conclusion as to their origin and purpose: nothing less than a vast engineering project built by a desperate civilization on a dying planet. The canals were forged to husband water from the icecaps, which were seen to wax and wane, down to the more arid equatorial regions (figs 7.12–7.13). This was picked up by the Victorian social media *du jour* – magazines, newspapers and books – and the

7.1 Percival Lowell observing the heavens through the Clark Telescope, c. 1897.

7.2 Mysterious 'square structure' on Mars, photographed by the Mars Global in 2001.

7.3 The 'Face on Mars', snapped by NASA's Viking 1 on 25 July 1976.

7.4 Four orthographic views of Mars by Giovanni Schiaparelli, according to observations he made at Brera Observatory from September 1877 to March 1878.

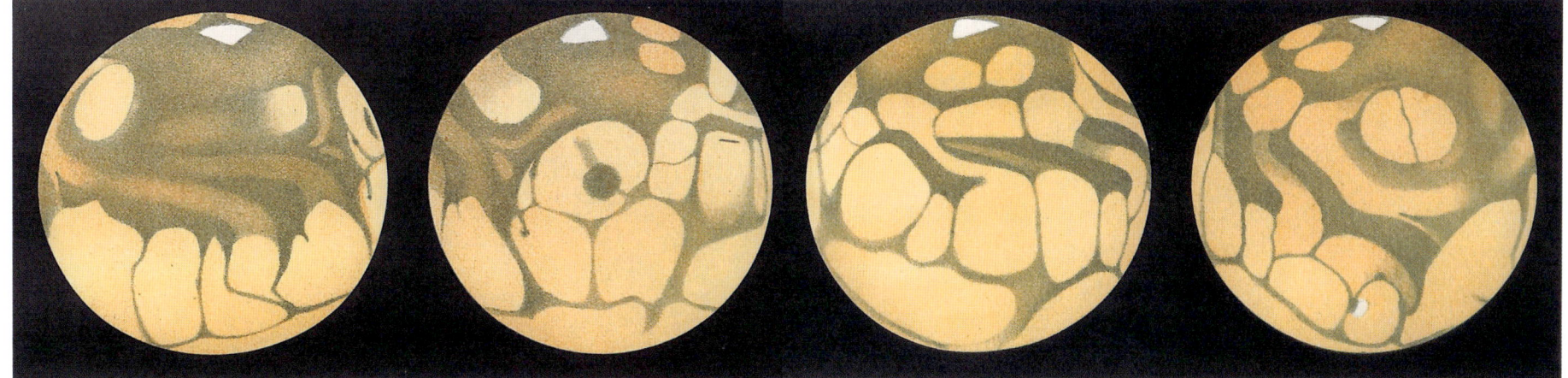

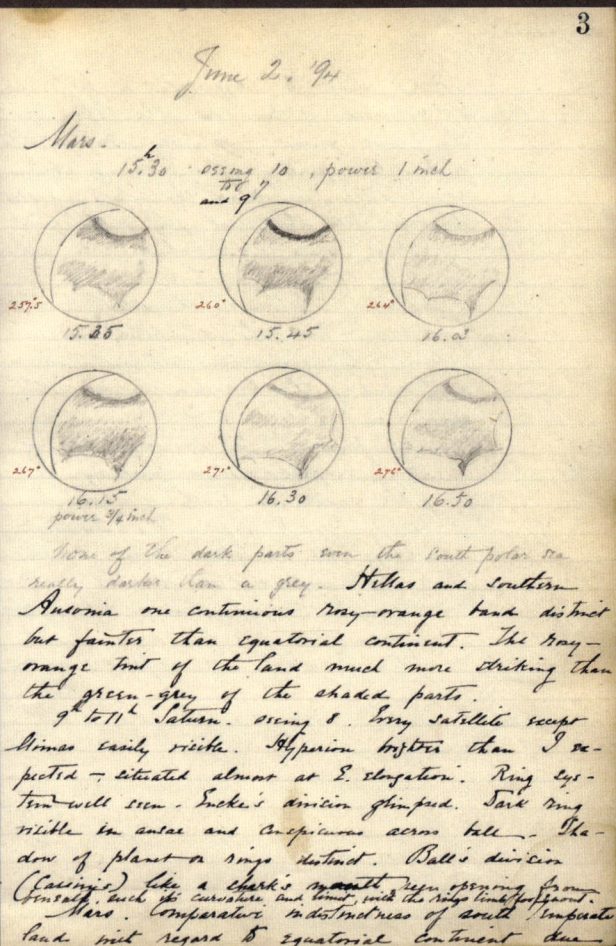

7.5 First observation notes of Mars by Percival Lowell, 1894.

7.6 Observations of Mars by Percival Lowell, from Logbook #18, using the 24-inch Clark refractor telescope, 21-27 March 1905.

7.7, 7.8 A man takes a break at the construction site of the Clark Telescope Dome on Mars Hill in Flagstaff, 5 May 1894.

The writing on the back of the photograph details information about the construction of the telescope dome.

7.9 The Clark Telescope is the oldest telescope at Lowell Observatory. Photographed in 1909 by E. C. Slipher, it has played an important role in research and outreach at the observatory.

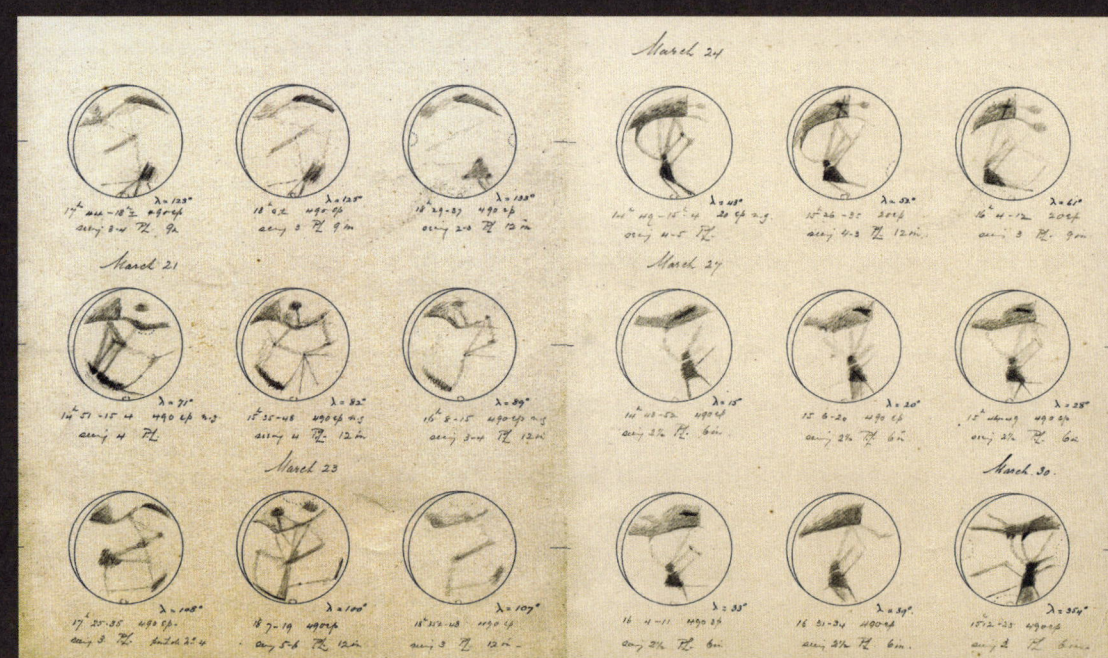

No 20
Site from N. May 5. Saturday
Shows tackle for hoisting
shears, also lower part of pier,
(placed in position on the 7th.) The
two trees near it have been cut
down.

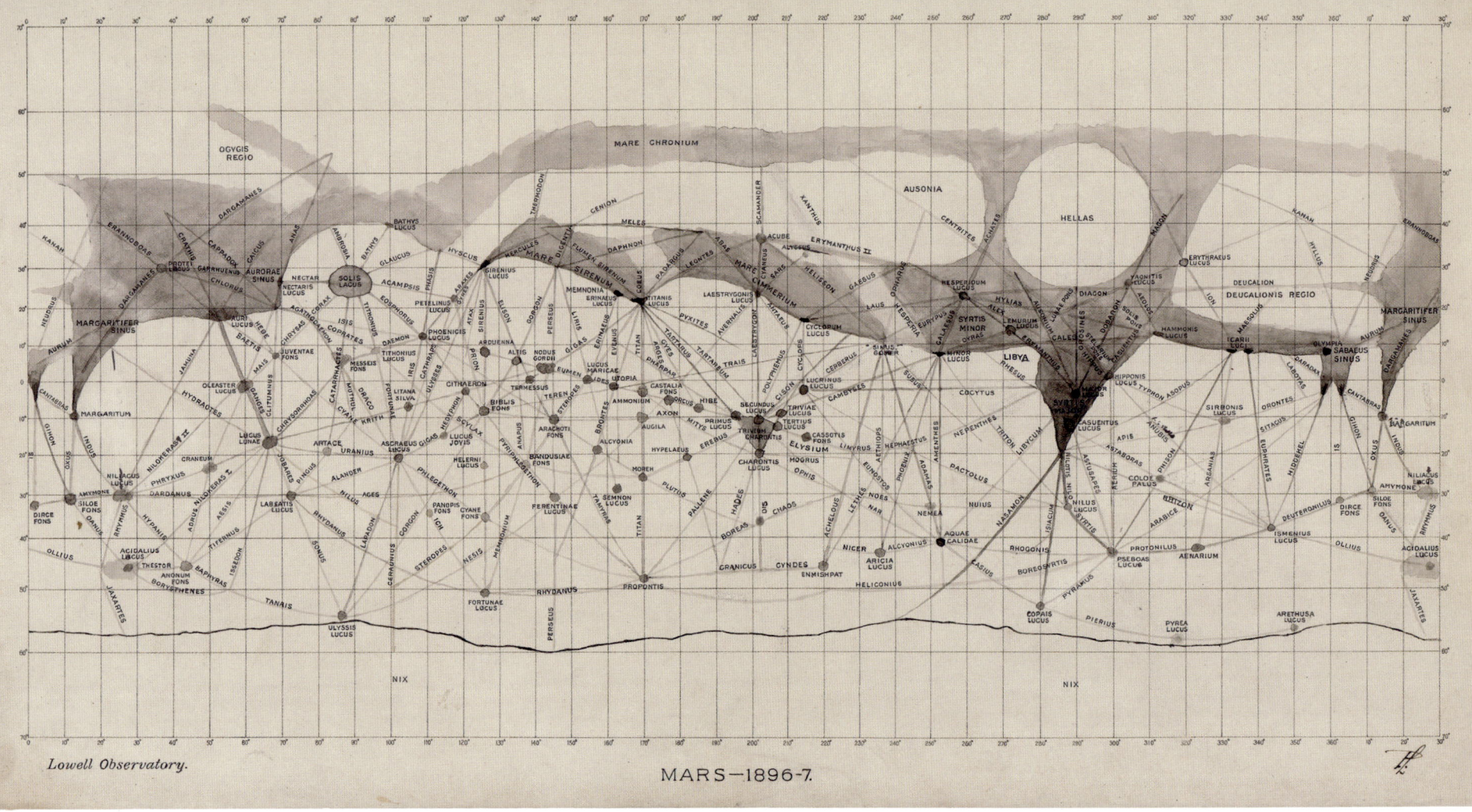

7.10 Percival Lowell's hand-drawn map of Mars, 1896-97.

7.11 Cigarette card from 'Romance of the Heavens' series depicting an imaginary landscape on Mars.

7.12 'Dr. Lowell Tells of Canals, Indicating Irrigation', 1916 article in the *Morning Oregonian* about Lowell's lecture to a Lincoln high school.

7.13 'Life in Mars - Professor Lowell's Defence of His Theories' (excerpt), Lowell's article in the *Daily Mail* defending his various theories, 26 March 1910.

idea lodged deeper into the public consciousness (figs 7.14–7.16). It's no wonder the infamous Martians from H. G. Wells's *War of the Worlds* were so keen to visit Earth.

Lowell's observations didn't go completely unchecked, notably by the great Victorian naturalist Alfred Russel Wallace, credited for independently landing on the theory of evolution at the same time as Charles Darwin. In his book *Is Mars Habitable?* (1907), Wallace elegantly dismantled Lowell's canal hypothesis: 'This idea has coloured or governed all his writings on the subject. The innumerable difficulties which it raises have been either ignored, or brushed aside on the flimsiest evidence.' Apart from the impossibly harsh Martian environment, which would preclude complex life, Wallace pointed out the planet's utterly inadequate water supply. He proceeded to ask some other uncomfortable questions: what evidence of a civilization remains *before* the canals were built? Why didn't the vast network of canals evolve from smaller, local irrigation projects closer to the source of the water? How could the Martians have developed

> 'Yet across the gulf of space, minds that are to our minds as ours are to those of the beasts that perish, intellects vast and cool and unsympathetic, regarded this earth with envious eyes, and slowly and surely drew their plans against us.'
>
> H. G. WELLS, *THE WAR OF THE WORLDS* (1898)

a population large enough and sophisticated enough to construct a planet-wide engineering project? Complexity, whether biological or industrial, proceeds gradually and sequentially, not in a single all-or-nothing leap. Wallace concluded that Mars was not only uninhabited, but uninhabitable.*

The advent of photography at the turn of the century eventually killed Lowell's elegant theory, although it lingers in the public imagination as an entertaining curio and a cautionary tale. Life on other planets is a powerful and persistent idea, but it can also trip us up. However well intentioned our motives, however fine our telescopes and technology, there's no substitute for healthy scepticism – particularly when it comes to our own cherished conclusions.

---

* Wallace would have been a terrific guest on The Joe Rogan Experience.

Magazine Section — **The Salt Lake Tribune** — Sunday, October 13, 1912

Mars Peopled by One Vast Thinking Vegetable!

Interesting Theory of Prof. Campbell, of Lick Observatory, That Explains the "Canals," "Eyes," and Other Puzzling Problems of Our Neighbor Planet

ODD FACTS ABOUT MARS

Mars is the fourth planet from the sun, and the nearest to our earth.

It is called the red planet, and its color is thought to be due to vegetation.

Its size and density are less than ours, and a man weighing 200 pounds here would only weigh seventy-five pounds there.

Mars has atmosphere, seasons, land, water, storms, clouds and mountains.

Mars has two moons, only 3,700 miles away and revolves around it in seven and a half hours — a "fast moving star."

The day on Mars is half an hour longer than ours, and its year contains 687 days.

Professor Lowell has counted 437 "canals" on Mars, and 186 "oases." The canals vary in length from 250 miles to 3,000 miles.

A man on Mars would be able to drive a golf ball fifty miles.

The strength of a man on Mars would be eighty-three times greater than on the earth.

The atmosphere of Mars consists principally of carbonic acid gas.

The water supply of Mars is very slender, and its utilization is the greatest problem of life there.

The Pitcher Plant Devouring a Rat, an Instance of Plant Life Possessing Animal Powers.

Scientists Sure Mars Is Inhabited

NOTED ASTRONOMER SAYS INTELLIGENT BEINGS ARE BUSY WITH THEIR OWN PROBLEMS ON NEIGHBORING WORLD.

Scientists Now Know Positively That There Are
THIRSTY PEOPLE on MARS

- Getting a Water Supply Is the Great Problem on the Dying Planet.
- The Martians Have Mastered the Art of Making Water Run Up Hill.
- The People Are Probably of a Much Higher Type than the Inhabitants of Earth.

7.14 'Mars Peopled by One Vast Thinking Vegetable!', *Salt Lake Tribune*, 13 October 1912.

7.15 'Scientist Sure Mars Is Inhabited', *NEA, Pacific News Service*, 19 October 1916.

7.16 'Thirsty People on Mars', *World Magazine*, 8 July 1908.

7.17 Percival Lowell, drawing of Mars with canals, 1905. The names of Lowell's (and Schiaparelli's) 'canals' and other Martian features were plucked from a variety of real and mythical rivers as well other mythological and classical sources.

MARY WARD

The Hon. Mrs Ward's life was divided between the microscope and the telescope. She wrote and illustrated a handful of elegant, now largely forgotten popular science books; guides for the lay reader of the day on how best to use these wondrous instruments which reveal the opposite ends of nature's scale. She deserves to be remembered as the very best of the Victorian science amateurs: intellectually rigorous and precise in her observations, warm, witty and poetic in her tone. She was Ireland's first lady of the lens, and her own short life was as luminous as the comet or the firefly.

Mary Ward (née King) was born into a well-to-do family in County Offaly, Ireland, in 1827. At that time, formal education for girls of her background was supervised by a governess and restricted to the niceties of reading and writing, studying the Bible and the pleasures of the countryside. Formal scientific education was reserved for men, as were the doors of the great scientific institutions. Her drive for independent inquiry took shape in the gardens and parklands of her home, armed with a butterfly net, sketchbooks and a magnifying glass. She had access to a wealth of periodicals and newspapers and to local libraries, but became frustrated that the books on natural history and astronomy were impenetrable in tone and often a hundred years out of date. She knew she could do better. Her other great influence was her cousin William Parsons, the 3rd Earl of Rosse, who was building the largest telescope in the world at the time in the grounds of his nearby estate in Birr, the construction of which Mary documented in highly skilled drawings. Parsons also brought her into contact with various important men of science, whose respect she won with her extraordinary knowledge. Later in life, her reputation for scientific excellence gained her entry through the doors of the Royal Observatory, Greenwich; she was the only woman permitted at the time.

Her self-confidence as a child led her naturally to share her knowledge, putting on enthusiastic scientific demonstrations for friends, family and staff. One workman looking down the lens of her microscope exclaimed, 'It is beautiful ... but is it true?' A maid exclaimed, 'Oh, Miss M, it makes me want to shout!' The tools of her trade were a 'Ross' microscope (the best available at the time) and her beloved 5-centimetre (2-inch) 'Dollond' telescope. Through these two pieces of technology, she began sharing with the world

'Quicquid nitet notandum' (Whatever shines should be observed)

WILLIAM HERSCHEL (1738-1822)

8.1 Mary Ward, photographed by Nelson & Marshall, 28 January 1861.

8.2 Engraving after Mary's drawing of Lord Rosse's Leviathan Telescope at Birr Castle, fig. 6 from David Brewster's *Life of Newton*, vol. 1 (1860).

8.3 Photos submitted to accompany William Parsons, 3rd Earl of Rosse, 'On the construction of specula of six feet aperture; and a selection of observations of nebulae made therewith', taken by his wife, Lady Mary Rosse, 1861.

8.3

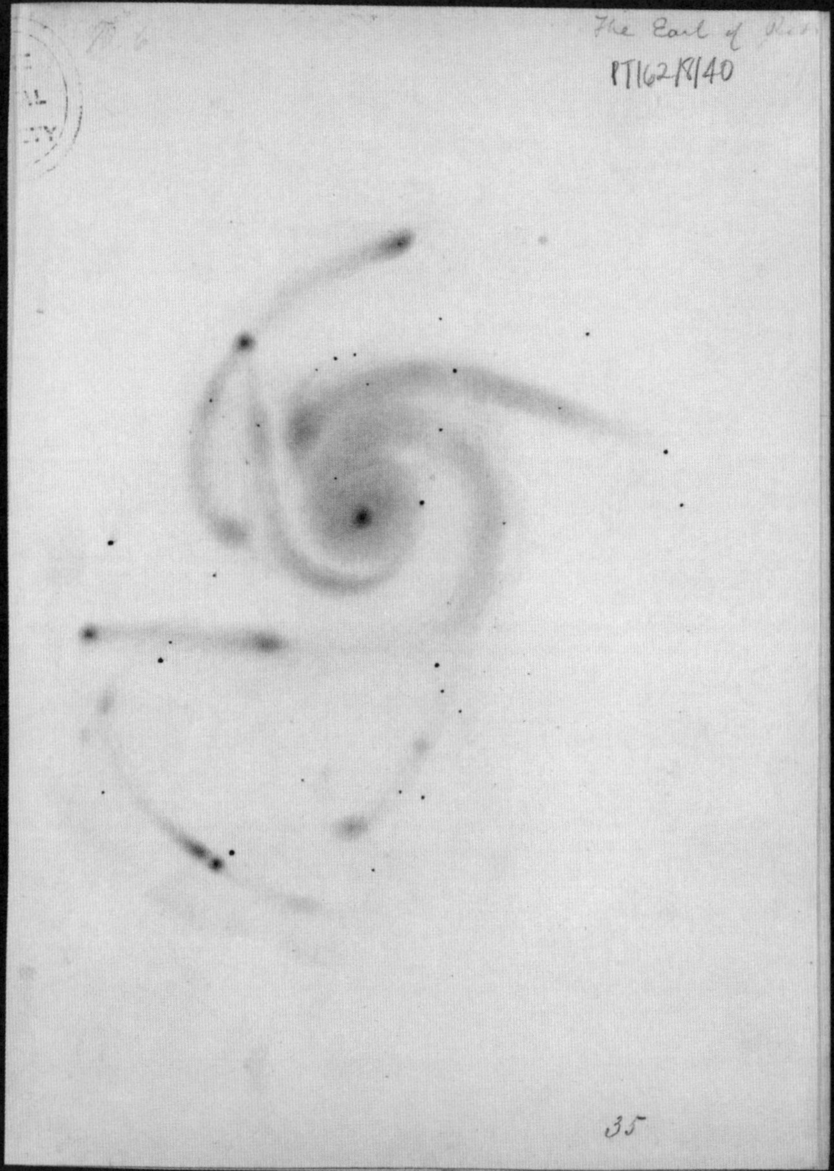
PT162/8/40
35

8.4

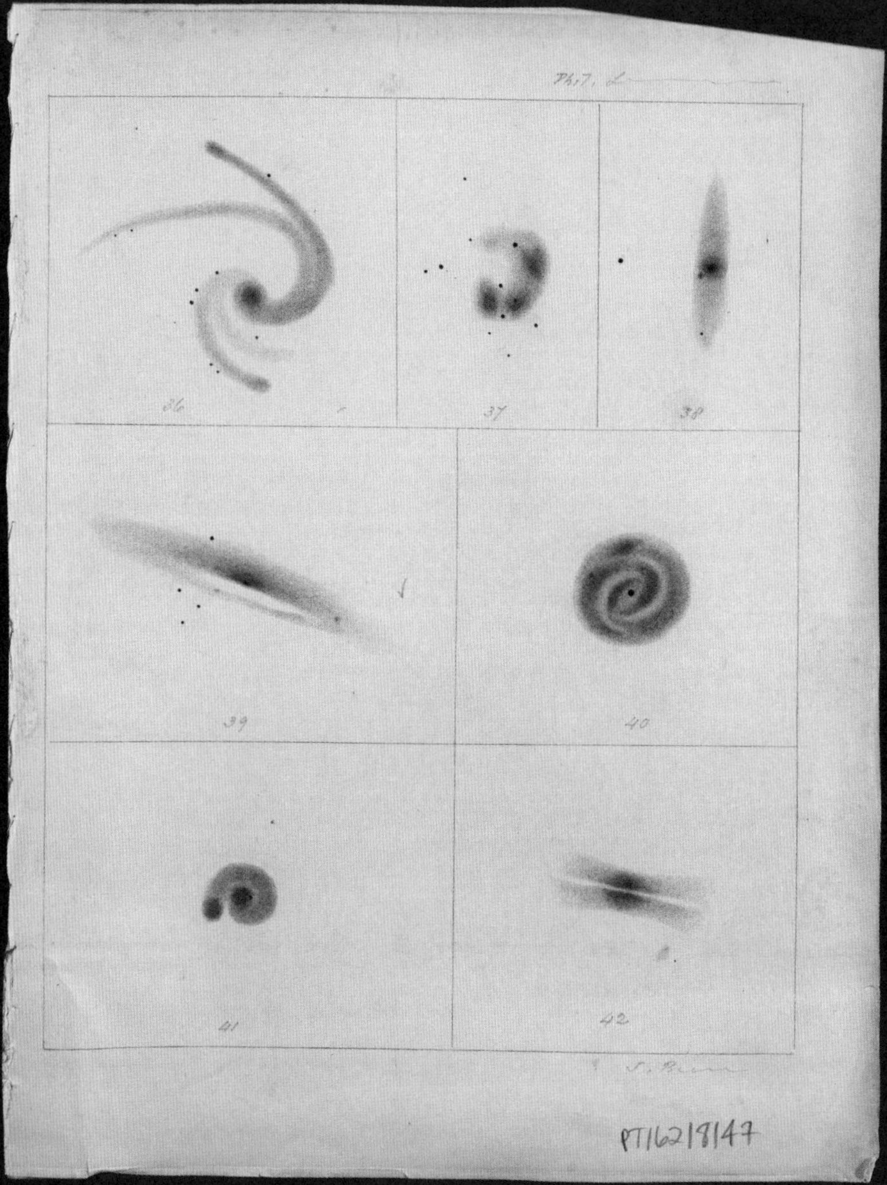
36
37
38
39
40
41
42
PT162/8/47

8.5

'Who shall say what wonders of creation may be contained in those distant worlds? What amazing scenes of grandeur and glory, what triumphs of power unbounded, what monuments of loving kindness no less infinite!
    And let it be remembered that with the greatest telescope we only see one portion of boundless space. Yes: these universes which are shown to us brightly or dimly, are still but a part of the Almighty's works.
    There must be more beyond, for space is infinite.'

TELESCOPE TEACHINGS (1859)

8.4-8.6 Drawings by the Earl of Rosse for the paper 'On the construction of specula of six feet aperture; and a selection of observations of nebulae made therewith', 1861. The power of the telescope allowed the Earl and his team of assistants to observe for the first time that some nebulae are spiral-shaped.

the intricate patterns and structures of nature, from the dragonfly wing to the rings of Saturn.

She would write numerous books over the course of her lifetime, of which the best known are *Microscope Teachings* and *Telescope Teachings*, as well as articles on everything from the transit of Mercury to natterjack toads. Her most productive period came not in the carefree period of her youth, but in the stir and pressure of married life to Henry Ward, 5th Viscount Bangor, and motherhood. Despite marrying into aristocracy, money became a constant battle, as were the domestic pressures of raising eight children. But in the quiet evening or morning, there by the margin, she would gather her thoughts on paper, guiding the reader around the sky. She described the Sun and how to safely look for sunspots; the Moon and the planets; as well as sharing detailed observations of Donati's Comet of 1858, all beautifully illustrated with colour plates. And just behind the observable fixed stars, or the candlelit resolution of the microscope, shone the light of God's realm. Mary was on a mission to share His creation, whose signature in the natural world she could so clearly see.

It's a travesty that her fame came due to the curious nature of her death rather than the quality of her work. She became the first recorded person in history to be killed in a motor vehicle accident. In 1869, aged only forty-two, she slipped off an experimental steam-powered automobile that had been designed and built by her then late cousin; she was crushed by its wheels and killed instantly.

Mary Ward was one of the first popularizers of science and technology. Her legacy, lest we forget, is that the window of science is one through which we're *all* invited to gaze.

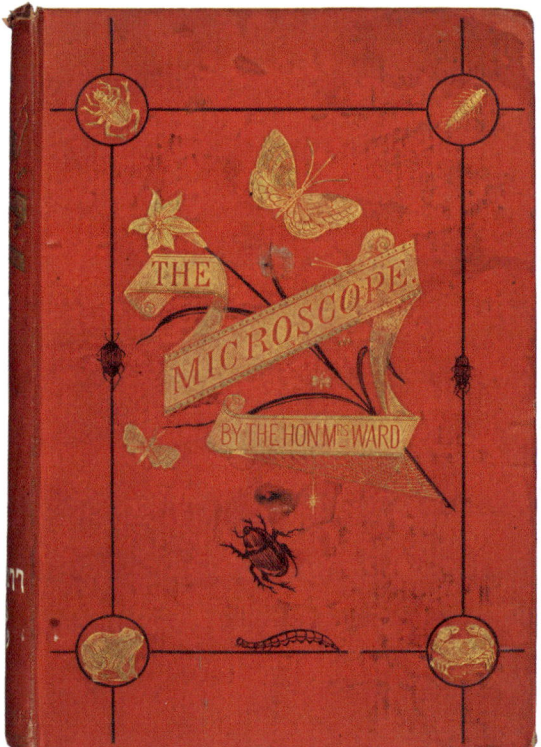

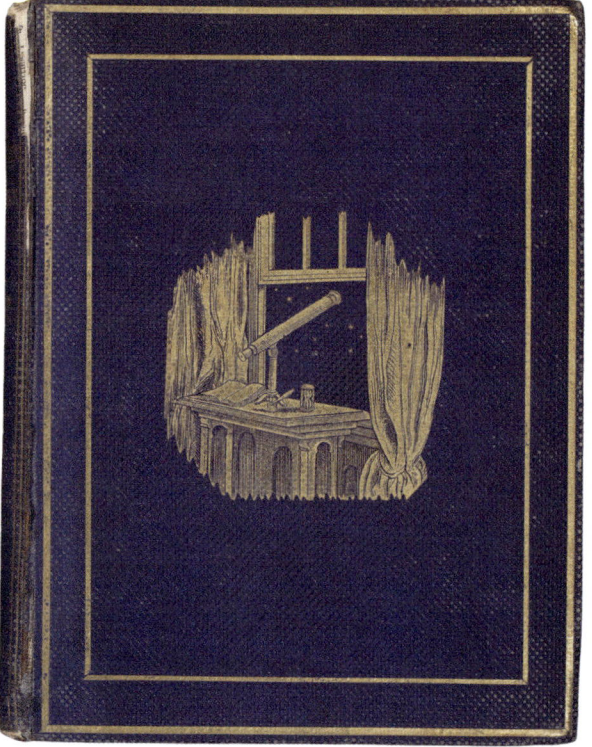

8.7 *The Microscope, or Descriptions of various objects of especial interest and beauty* (1869) by the Hon. Mrs Ward. Illustrated by the author's original drawings.

8.8 Front cover of *Telescope Teachings* (1859).

8.9 'Miniature Chart of the apparent Path of Donati's Comet', plate from *Telescope Teachings* (1859).

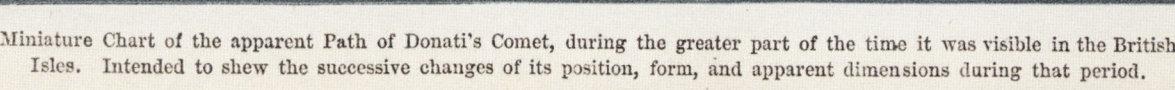

Miniature Chart of the apparent Path of Donati's Comet, during the greater part of the time it was visible in the British Isles. Intended to shew the successive changes of its position, form, and apparent dimensions during that period.

'Can Astronomy be presented as an entertaining study? Has anyone attempted to cull from treatise addressed to the not wholly unlearned in science, facts, anecdotes, the "light literature" of this sublime study, and to tell these things in simple words?'

TELESCOPE TEACHINGS (1859)

Plates from *Telescope Teachings* (1859):

8.10 'The Comet of 1858, as observed on various occasions'.
8.11 'The Comet of 1858 (Donati's)'.
8.12 'A Falling or Shooting Star'.

(OVERLEAF)

8.13 'The Moon'.
8.14 'Portions of the Moon's Surface'.
8.15 'Saturn'.
8.16 'Jupiter and its Satellites'.

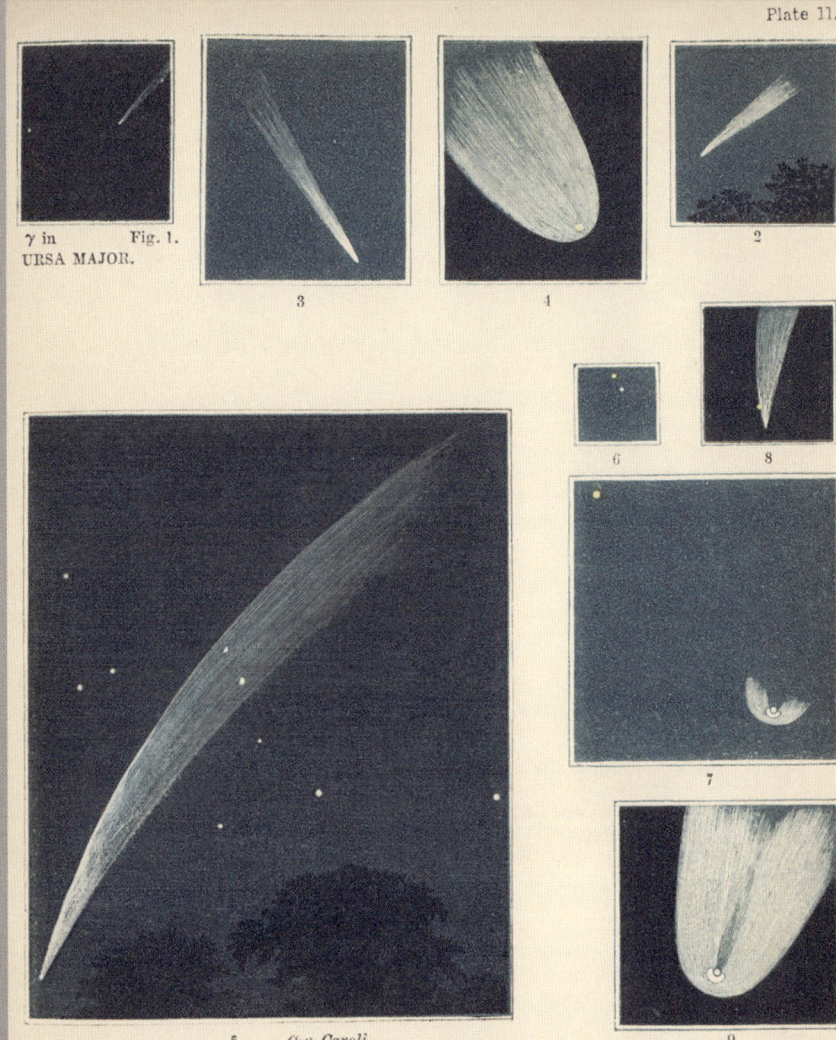

8.10

HERCULES. βγεΛθ SERPENS. δ γβχρς CORONA BOREALIS. ε δ γ Alphecca.

THE COMET OF 1858, (DONATI'S,)
and neighbouring stars, October 11th., 1858, at 7.15 p.m.

A FALLING OR SHOOTING STAR.

8.11   8.12   81

PARTIAL ECLIPSE OF THE MOON,
9.50 p.m., Feb. 27th., 1858.

THE MOON,
Two days and twenty-one hours after New. 8 p.m. April 16th., 1858.

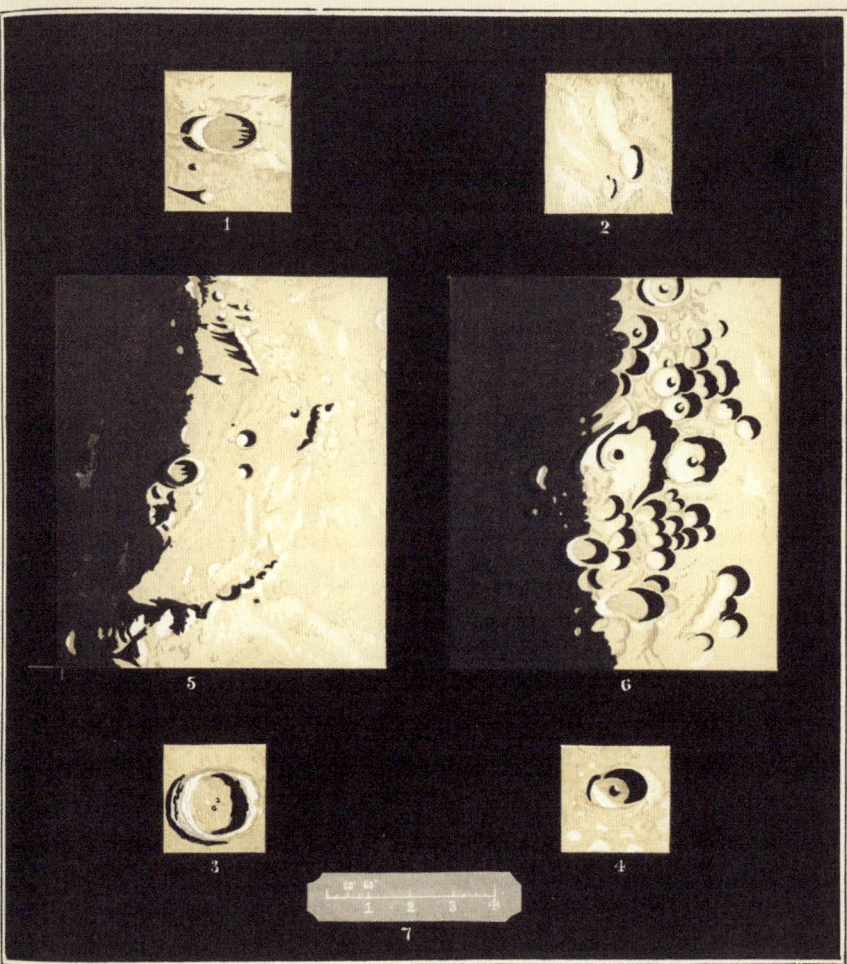

PORTIONS OF THE MOON'S SURFACE,
as seen through the night-glass of a Telescope two inches in diameter.

1. Plato, with the conical mountain of Pico. 2. Aristarchus. 3. Copernicus. 4. Tycho.
5. The craters of Archimedes, Aristillus, and Autolycus, with the Lunar Appenines and Alps, shewing the pointed shadows of the latter. 6. Portion of the mountainous region north of Tycho.
7. Scale in Minutes of a Degree, for figs. 5 and 6. Figs. 1, 2, 3, and 4, are on a slightly larger scale.

Plate 9.

SATURN AND ITS SATELLITE "TITAN,"
as observed Feb. 17th., 1857.

October, 1848. October, 1849. March, 1851.
March, 1853. May, 1855. February, 1858.

SATURN,
as observed from October, 1848, to February, 1858.
These drawings represent the planet as actually placed in the heavens.

Plate 8.

JUPITER AND ITS SATELLITES,
as seen on three successive evenings, in November, 1857.

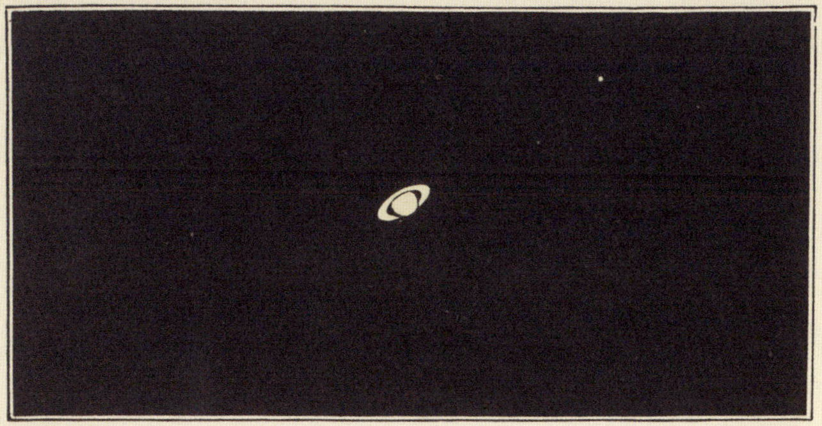

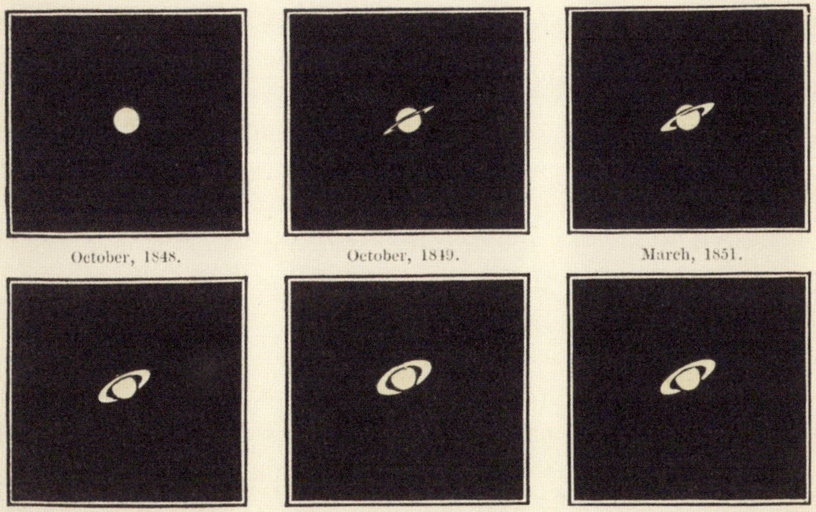

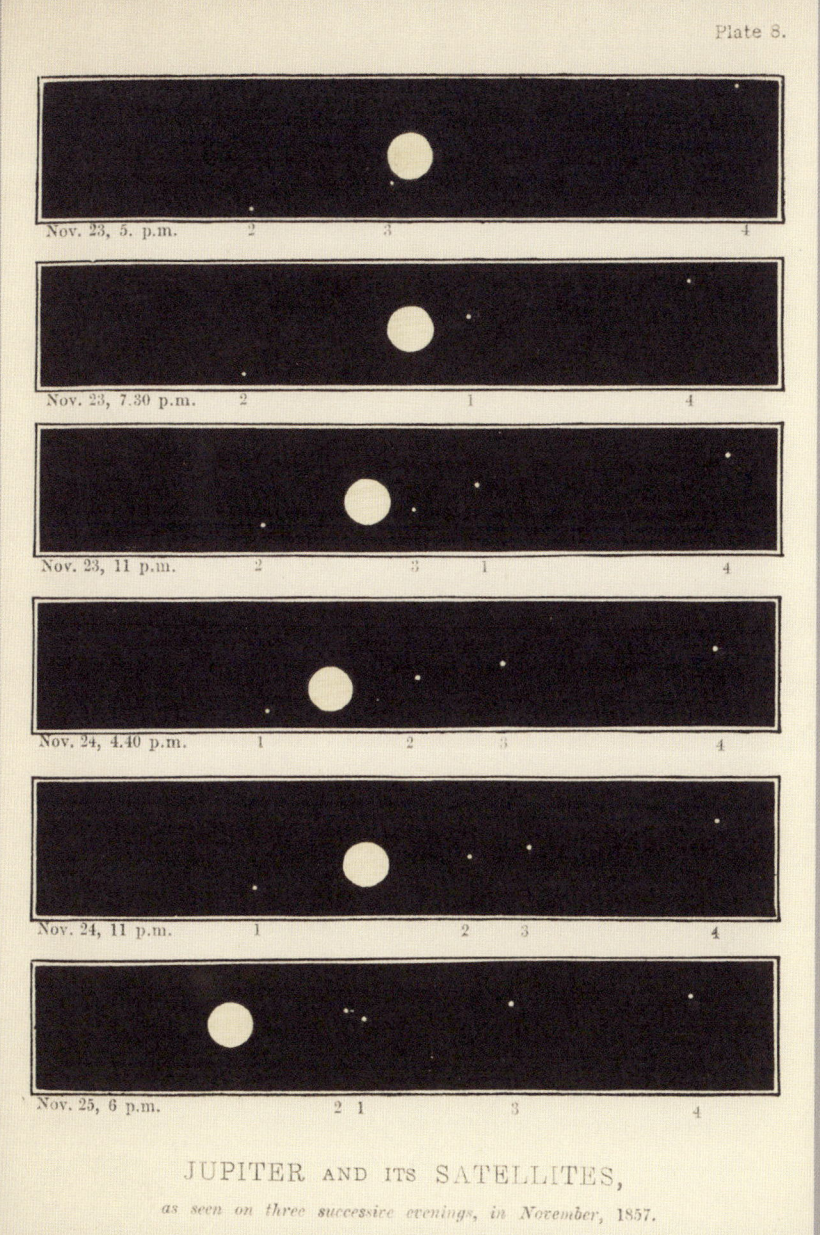

SARA SEAGER

Astronomy is a numbers game. With the naked eye on a clear night, you might be able to see a thousand stars, a tiny fraction of the billions that make up our Milky Way galaxy. And the Milky Way is just one galaxy among billions of galaxies in the observable universe. Scientists (and science fiction writers) have long assumed that most stars have planetary systems just like our own Sun – an inevitable consequence of star formation. If so, the number of *planets* in the observable universe is going to be staggering. Let's call it 1,000,000,000,000,000,000,000, give or take. It's one of those silly numbers that loses all meaning after the first few 0s. It's a number that will keep MIT astronomer and exoplanet scientist Professor Sara Seager (b. 1971) busy for many lifetimes.

Sara grew up in Toronto, Canada. Her scientific career began as a young child, staring out of the car window on long journeys at night with her father, wondering why the Moon always seemed to be following her as they sped along. She remembers the jolting effect that the night sky had during back-country camping trips away from city lights. But Seager didn't want to just be moved by such experiences, she wanted answers, and she's devoted her life to finding them. She studied maths and physics before embarking on postgraduate studies at Harvard University, around the time when the very first exoplanets (planets that orbit other stars) were being discovered in the mid-1990s. This was the Dark Ages of exoplanet research, when it was still teetering on the edge of scientific respectability.

Finding exoplanets is challenging. Studying them even more so. Historically scientists have been able to infer their presence either by detecting the wobble of a star caused by the gravitational tug of a large orbiting planet, or through the 'transit' method, whereby the light of the star dims as the planet passes across its disk. Measuring that slight dimming can tell us something about the size of the planet and its distance from the star. That early trickle of exoplanet discoveries at the turn of the twenty-first century has now become a flood – at the time of writing we now have evidence of around 5,000, with probably a lot more confirmed (or candidates waiting to be confirmed) by the time you read this. Dramatic improvements in technology have seen a revolution in the field: advanced computing power and mathematical models, novel citizen science initiatives and now artificial intelligence are tools that didn't exist thirty years ago. Now, they are changing everything. Most important are the increasingly sophisticated ground-based observatories and the advent of space-based telescopes such as Hubble, the TESS (Transiting Exoplanet Survey Satellite) telescope and in particular the great Kepler space telescope, which was responsible for the discovery of a vast haul of new exoplanets. The next generation of telescopes, led by the James Webb Space Telescope (JWST), is continuing to revolutionize the field, revealing new information about the cosmos in ever finer detail.

9.1 Sara Seager, Canadian-US planetary scientist, with a radio telescope.

9.2 NASA engineer Ernie Wright looks on as the first six flight-ready primary mirror segments are prepped to begin final cryogenic testing at NASA's Marshall Space Flight Center in 2011. This represents the first six of eighteen segments that will form NASA's James Webb Space Telescope's primary mirror for space observations.

Planetary systems around stars seem to be the norm. Our own solar system, with smaller rocky planets closer to the Sun and larger planets further out, seems to be just one arrangement in a vast variety of examples. As you'd expect, bigger planets such as the 'hot Jupiters' that orbit very close to their parent star are much easier to detect. But Seager's interest in exoplanets isn't just a stamp-collecting exercise. She's hunting for smaller, rocky, Earth-sized planets that sit in 'Goldilocks' zones, where liquid water and habitable atmospheres are possible. There are already plenty of candidates – every so often the excited news media announces the discovery of 'Earth 2.0', prompting immediate fantasies of distant foreign travel, such as Kepler 186f (580 light years away) or Kepler 452b (1,800 light years away), or the Trappist-1 system of seven planets that all reside within the equivalent orbital distance of Mercury. Even Proxima Centauri, our nearest star other than the Sun, has a candidate for an Earth-like cousin. Of course we can only imagine what these planets might actually look like. We have no direct images that can resolve an exoplanet's surface detail from such spectacular distances. But luckily we have our imaginations to fill in the blanks – and NASA's 'Eyes on Exoplanets', a 3D visualization platform that allows you to conduct an armchair grand tour beyond our solar system.

Seager's research is really motivated by science's greatest quest: the hunt for life. She and others analyse the chemistry of exoplanet atmospheres by looking at the starlight that passes through them. By analysing the spectra, biosignature gases such as methane, water vapour, ozone and ammonia can be detected; gases which may have been produced by biology rather than geological or other processes. The large volume of oxygen (20 per cent) in our own atmosphere, for example, would be a good indicator of the presence of life for any interested alien astrobiologists watching us from a distance. In early 2025 astronomers excitedly announced the detection of dimethyl disulfide (DMDS), a gas that is produced by simple marine life on Earth, from exoplanet K2-18b, some 124 light years away. A distant water world teeming with alien life? Or something more prosaic? The jury's out, and the search continues.

Seager has also been at the forefront of new and exotic engineering projects that might help our search. One of the problems of direct exoplanet detection is that small planets are very hard to see in the direct glare of a main sequence star like our own Sun. Seager is the scientific lead of the Starshade project – a long-imagined

9.3

9.4

solution to this problem which involves launching a 30-metre (98½-foot) flower-shaped sunshade that can unfurl in space, blocking out glaring starlight. It's like the sunshade in your car that allows you to see the road when sunlight is directly in your eyes. In theory we could then see light from orbiting planet(s) that is billions of times fainter. Science fiction solutions are becoming a reality.

If we do find conclusive evidence of life beyond Earth, it will be because of scientists like Seager who skilfully balance excitement and optimism with scientific caution. We've had many false alarms over the years, but we now have extraordinary technology that suggests a definitive announcement of life elsewhere is inevitable. What happens next? That's a problem for the next generation of astronomers. Knowing life exists elsewhere will make looking into the night sky even more significant. In the meantime, we can enjoy being mesmerized. Stars have always offered us hope. They connect us to something bigger. They give us meaning, however illusionary, in a universe of terrifying scale and indifference. They demand questions from us – does Proxima Centauri's tantalizing Earth-like planet have an atmosphere? Or water? Does it have life? Might that life feel the same joy and grief we do on planet Earth? And of course, can we go there? At 4 light years away, it's still going to take many human lifetimes to reach it. But like the bucket-list destinations on Earth that you'll probably never visit, it's comforting to know they're there.

`9.3-9.6` Stills from NASA's 'Eyes on Exoplanets', detailing the planets discussed in this text.

`9.7` (OVERLEAF) This image shows the bare bones of the first prototype starshade by NASA's Jet Propulsion Laboratory being set up in technology partner Astro Aerospace/Northrop Grumman's facility in Santa Barbara, California, in 2013. The four petals pictured in the image are being measured for this positional accuracy with a laser. As shown by this 20-metre (66-foot) model, starshades can come in many shapes and sizes. This design shows petals that are more extreme in shape, in order to properly diffract starlight for smaller telescopes.

`9.8` (OVERLEAF) Origami folds let the inner disk of NASA's starshade prototype wrap into a cylinder for launch, then unfurl to block starlight reaching a space telescope. Photograph by Craig Cutler.

9.5

9.6

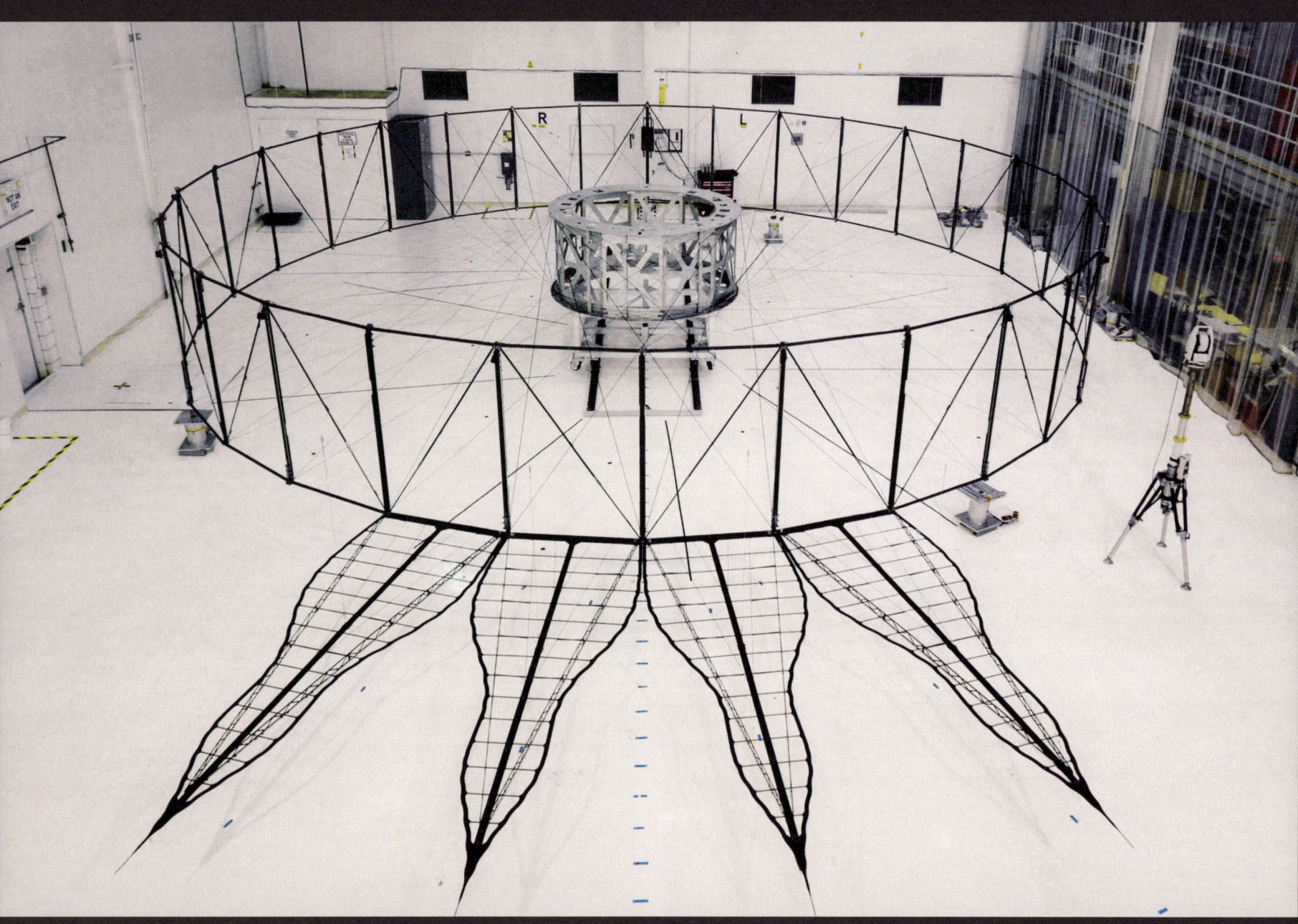

ACT 3.

IGNITIO

'IGNITION'
LEAVING THE EARTH: THE ROCKET ENGINEER

KONSTANTIN TSIOLKOVSKY

All history can be unpicked along a long unbroken line of *what ifs*. Inconsequential events that shape the future. What if nine-year-old Konstantin Eduardovich Tsiolkovsky, the Soviet 'father of astronautics', had stayed at home that cold day in the winter of 1866 instead of going tobogganing? The history of space flight would have had a different cast list.* The chill he caught that day turned into scarlet fever, and in turn he suffered near complete hearing loss. His condition would have a profound effect on the rest of his life. He became withdrawn, self-conscious and isolated, retreating into his inner world of books and his vivid imagination. Perhaps this is why his life's work fixated on a singular idea: escape from the tyranny of gravity. Staring into space would become his legacy.

Tsiolkovsky was the child of a Polish immigrant father, a forester and carpenter, and his mother, a Russian Tatar. It's she who is credited for sparking his interest in science and engineering, and in working with his hands – model-making, kite-making, inventing and tinkering. His physical and social isolation from the world meant that he found himself making all sorts of scientific discoveries, such as developing a theory of gases (although unbeknownst to him, much of what he thought was novel and revolutionary had already been discovered decades earlier). His life played out in a Russia that straddled both sides of the Revolution of 1917, and overlapped with some of the greatest Russian minds who ever lived: Dostoyevsky and Tchaikovsky, Mendeleev, Eisenstein and Stanislavski.

Sometime in his late teens, he went to study in Moscow. He lived a monastic life, spending what stipend he received from his father on things he needed for his experiments. It was in the state library that he met the philosopher Nikolai Fedorov, who was spearheading a new utopian millenarian philosophy known as Cosmism, a broad esoteric school loosely blending mysticism, transhumanism and science. At its heart was the premise that mankind's destiny lay beyond the confines of Earth, and that we should pursue the goal of immortality as we spread throughout the universe.† Science and technology were the tools to achieve this dream, just as the telescope had revealed the infinite cosmos to our ancestors. We can see Cosmist fingerprints all over Tsiolkovsky's writings and drawings, and over the visual symbolism that echoes across Russian

---

* Not unlike the plot of Frank Capra's *It's a Wonderful Life*.

† A recurring theme in our story: Elon Musk is the latest to postulate humans becoming a multiplanetary species.

10.1 Konstantin Tsiolkovsky in his study, 1932.

10.2 Vietnamese stamp printed for the 25th anniversary of the first crewed space flight, c. 1986.

10.3 USSR postage stamp celebrating Cosmonautics Day, 1965.

10.4 North Korean stamp printed c. 1984.

10.5 Pavel Filonov, *Formula of the Cosmos*, 1919. Expounding a new theory of analytical realism in his essay 'The Canon and the Law' (1912), painter and art theorist Pavel Filonov maintained that 'Evolution can be creative, that is, man or any living being observes evolution at work within itself and steers its course toward the desired higher form' - an idea steeped in Cosmist principles.

10.6 The giant cannon used to launch the spacecraft Columbiad, plate from Jules Verne, *From the Earth to the Moon* (1868).

10.7 Tsiolkovsky's designs (1903, 1911 and 1914) for a rocket ship.

10.8 Front cover of Tsiolkovsky's first pamphlet: *Простѣйшій проектъ чисто металлическаго аэроната изъ волнистаго желѣза* (The simplest project of a purely metallic aeronaut made of corrugated iron), published by Tsiolkovsky himself in Kaluga, 1914.

10.9 *Научное Обозрьніе* (Scientific Review), no. 5, May 1903 – the journal in which Tsiolkovsky published his paper 'Exploration of Space Using Reactive Devices'.

space history. It was here, surrounded by books, that Tsiolkovsky read Jules Verne's *From the Earth to the Moon* (fig. 2.4). He realized that Verne's giant cannon wouldn't work: the enormous forces required to launch a huge artillery shell more than 400,000 kilometres (250,000 miles) to the Moon with a single explosive push were impossible, but some kind of Newtonian reaction engine (of which jet engines and rockets are examples) would be the solution. If balloons, aircraft and dirigibles – all the rage in the late nineteenth and early twentieth centuries – had freed us from the Earth's surface, it was the rocket that would free us from the cradle of Earth altogether. As such the rocket became a quasi-religious tool,

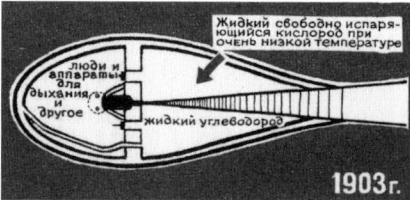
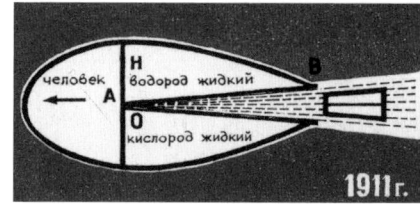
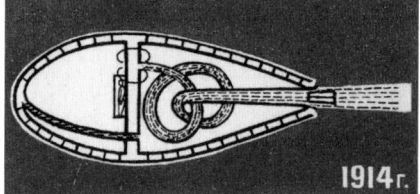

liberating mankind from the two things that had forever imprisoned us: distance and time. For Tsiolkovsky and the Cosmists, space travel, like death, was just an engineering problem waiting to be solved.

In 1871 he took up a mathematics and science teaching post in Borovsk, where he married Varvara Sokolova, before moving in 1882 to a position in Kaluga, the town that would become synonymous with his name. By all accounts he was an excellent teacher and popularizer of science, and his life was to become a balancing act between teaching, his large family‡ and his intensive writing and engineering work. He wrote about all manner of subjects, from the practical (but fascinatingly titled) *How to Protect Delicate Objects Going over Bumps* (1891) to investigations into aerodynamics and fixed-wing flight (a full decade before the Wright Brothers), and of course space travel. In *Free Space* (1883) he first interrogated the movement of bodies in space, free from the resistance of air and gravity. More potent than his writing were his stunningly beautiful drawings and doodles: strange machines, tumbling bodies floating through space and designs for simplistic dirigible-shaped, biological-looking rockets. It's in these scribbles and doodles within the margins that we find the most direct insight into his mind.

In 1903 – the year of the first powered flight, a world away in Kitty Hawk, North Carolina – he published his most famous paper, 'The Exploration of the World Spaces by Reactive Devices'. In it we find his eponymous 'Rocket Equation', the simple recipe that made practical space travel possible. The equation expresses the relationship between a rocket's variable mass (the rocket gets lighter because it burns fuel: most of the mass of a rocket is fuel, and most of that fuel is there to lift all that fuel!) and the velocity of the exiting exhaust gas. Tsiolkovsky also theorized other rocket fundamentals – the need for liquid fuel and an oxidizer; the necessity of staged rockets (that is, rockets on top of rockets) to achieve orbital velocity, which he called 'rocket trains'; and the effects of weightlessness, as well as his more esoteric ideas about the destiny of mankind to evolve into a spacefaring species. It's hard to see the line that divides the science

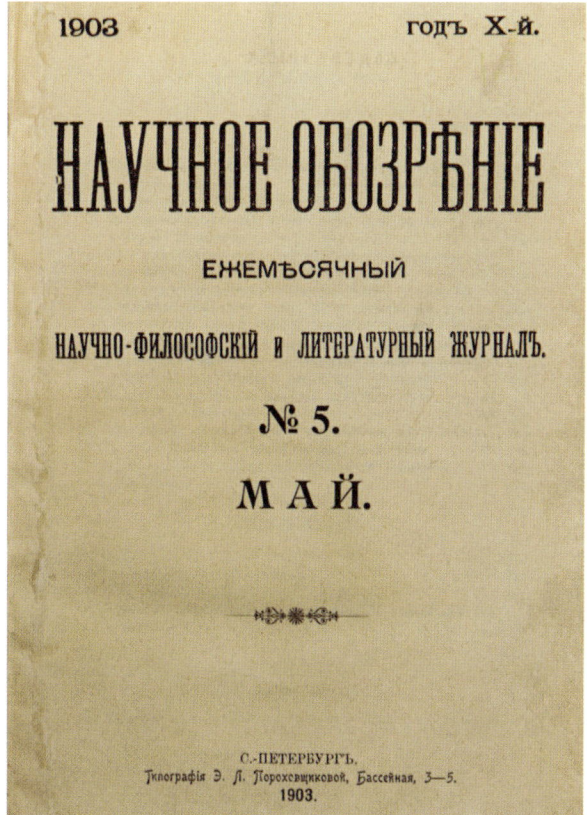

‡ According to a footnote in his autobiography, there were eleven family members living in his small wooden house!

> 'Earth is the cradle of humanity. But one cannot live in the cradle forever.'

KONSTANTIN TSIOLKOVSKY, 1911

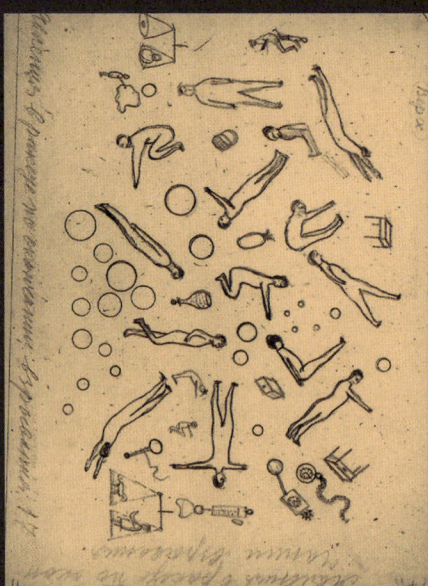 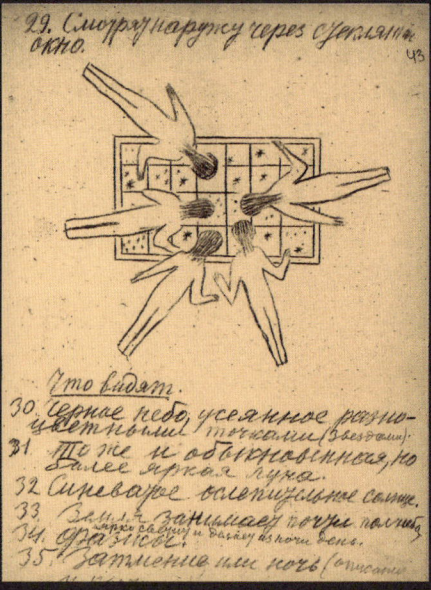

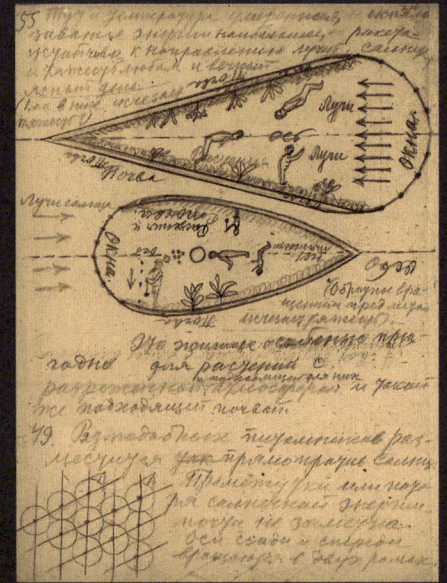 

10.10-10.15 Pages from Tsiolkovsky's *Album of Cosmic Journeys* (1933). Beyond his prose and mathematics, Tsiolkovsky's simple illustrations express ideas of rocket flights and space stations as well as touching on his wider philosophy: a grand vision for humanity living in space, free from earthly constraints - the freedom to move unburdened throughout the cosmos.

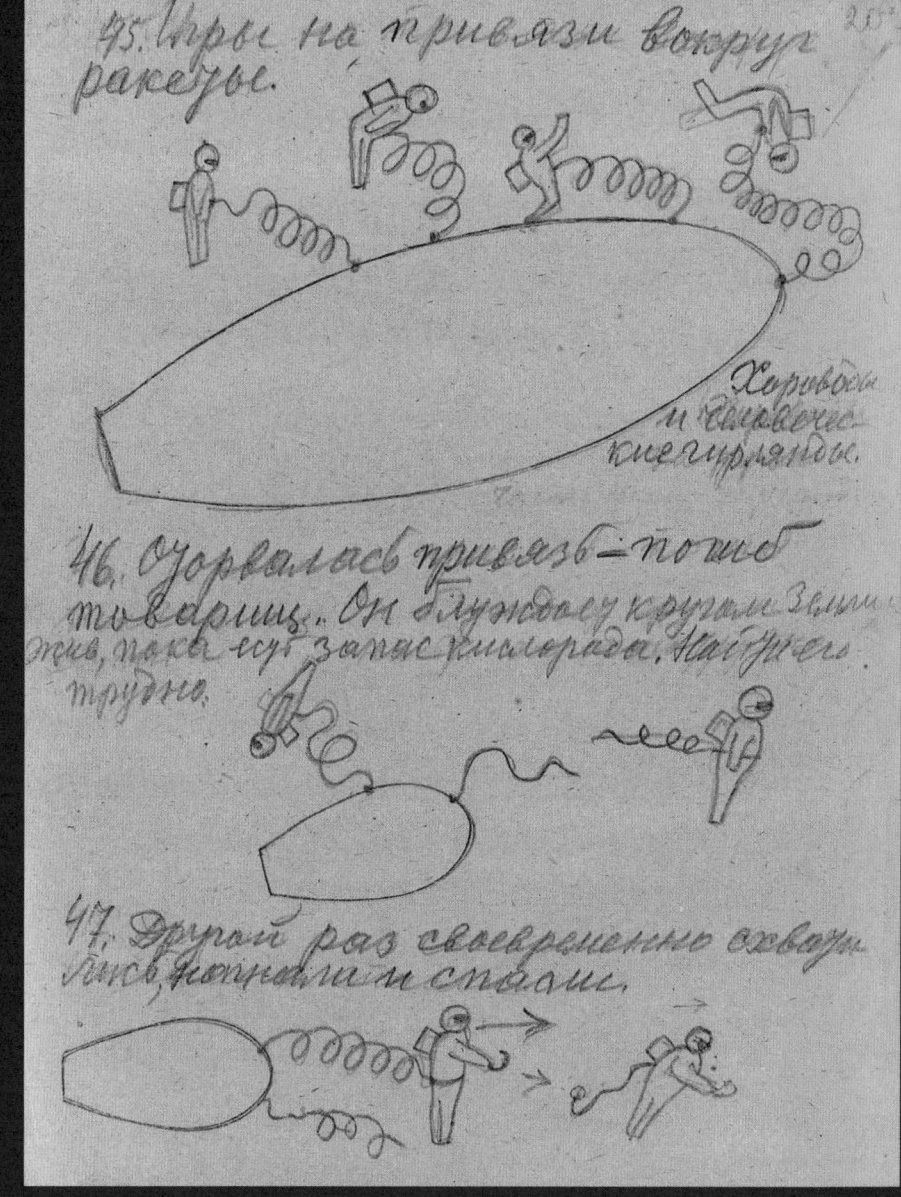

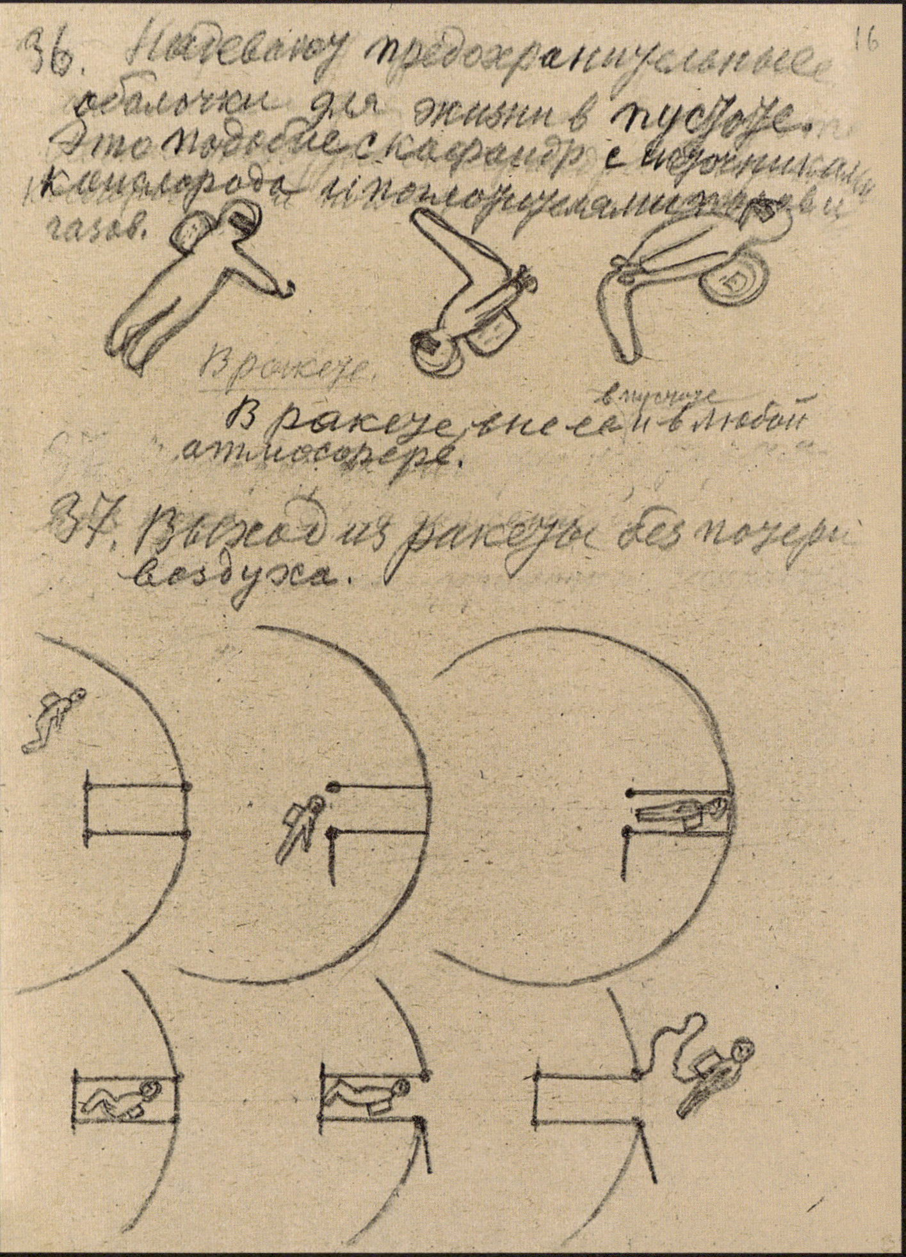

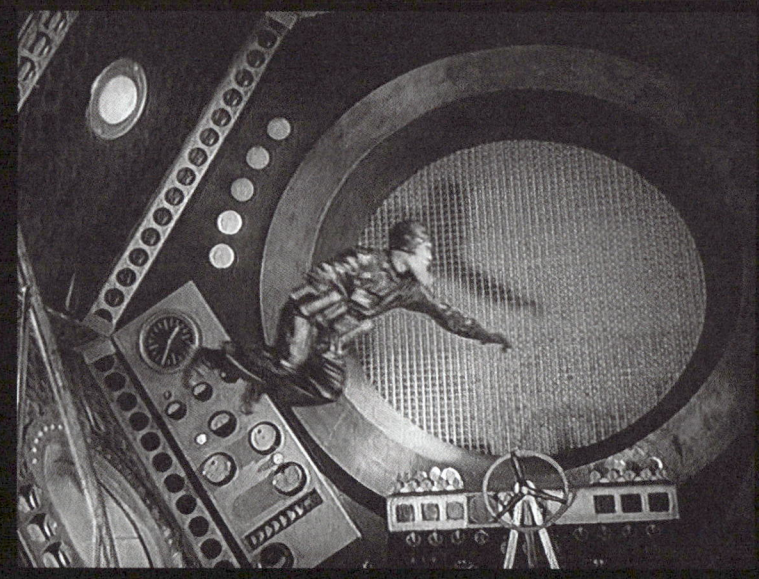

10.16 Drawing by Tsiolkovsky for the film *Cosmic Voyage* showing a cosmonaut exiting a rocket via an airlock, from his *Album of Cosmic Journeys* (1933).

10.17 Still from *Cosmic Voyage* (1936).

10.18 Monument to Konstantin Tsiolkovsky in Kaluga, 1983.

from the speculative. If you've had a wild idea about space travel, there's a good chance he thought of it a hundred years ago.

The obscure visionary Tsiolkovsky was to be embraced by the wider cultural milieu as political tectonic plates shifted violently after the First World War. The Soviet Russia of the next decade, like much of the world, was changing; radical ideas, science and technology were reshaping the globe. A new era of mass communication and mass transportation, aviation and exploration, was having a profound effect on both the real and the imagined. Not only could we now send messages across the Earth in an instant with the telegraph, but perhaps to and from other planets as well, as the opening scene of the 1927 Soviet film *Aelita: Queen of Mars* suggests. Cinema, radio and print were reflecting ideas of space travel, extraterrestrial life and techno-utopian/dystopian ideas to an excited public. The world was catching up with Tsiolkovsky's vision, and he was briefly hired as an early scientific advisor on the Soviet film *Cosmic Voyage* (1936), which was released shortly after his death. It's not surprising that this expansive outlook and interest in rocket propulsion extended far beyond Soviet Russia, notably with the scientist Hermann Oberth in Germany and the retiring American scientist Robert H. Goddard, who were both independently working on similar ideas at roughly the same time, and reaching similar conclusions to Tsiolkovsky, albeit in entirely different contexts.

Beyond the science, Tsiolkovsky's story fulfils an important, political function. That of the central protagonist for a 'founding father' narrative. Strong, simple narratives are vital for science and politics, and Tsiolkovsky fit the bill: his humble, rural existence away from the metropolitan centre; his crumpled Professor Calculus demeanour with his homemade tin ear trumpet, wire-framed spectacles, beard and bicycle, bearing a striking resemblance to the great sage Fyodorov himself. In the 1935 May Day parade, with Stalin in attendance, a pre-recorded speech by Tsiolkovsky was broadcast over loudspeakers across a muddy, rain-drenched Red Square. As important as the military strength on display was his voice – a statement of the new power of Soviet science and a vision of the future under Stalin.

In the Soviet Union we have many young pilots ... (and) I place my most daring hopes in them. They will help to actualize my discoveries and will prepare the gifted builders of the first space vehicle. Heroes and men of courage will inaugurate the first airways: Earth to Moon orbit, Earth to Mars orbit, and still farther; Moscow to the Moon, Kaluga to Mars!

Tsiolkovsky died in September of that year, long before any of his ideas were realized, but his status as a national symbol was assured. In 1957, after the launch of Sputnik, a statue was raised in Kaluga in his honour, and his modest wooden house is now preserved as a museum and place of pilgrimage for cosmonauts as part of their pre-flight rituals. Never underestimate the power that symbolism has on human history. It is through story rather than science (at least for now) that people become immortal.

  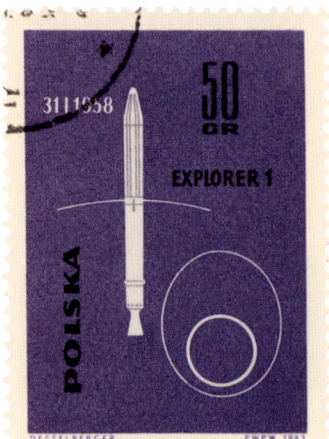  

10.19-10.28 Polish stamp series of 2018 dedicated to space exploration and the first satellites, with Tsiolkovsky's 1903 rocketship design leading the charge.

10.29, 10.30 Soviet model of a rocket based on the designs and notes of Tsiolkovsky. While the model grossly overestimates the living space available on board a rocket, it does convey an understanding of the physical constraints of space travel for that time.

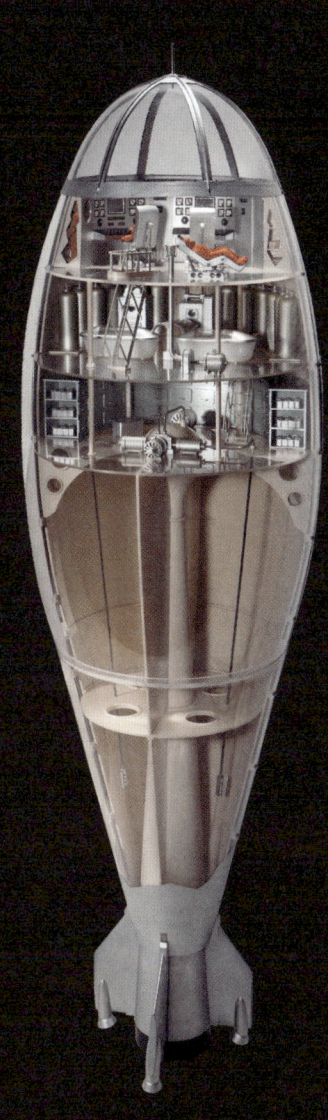

10.29

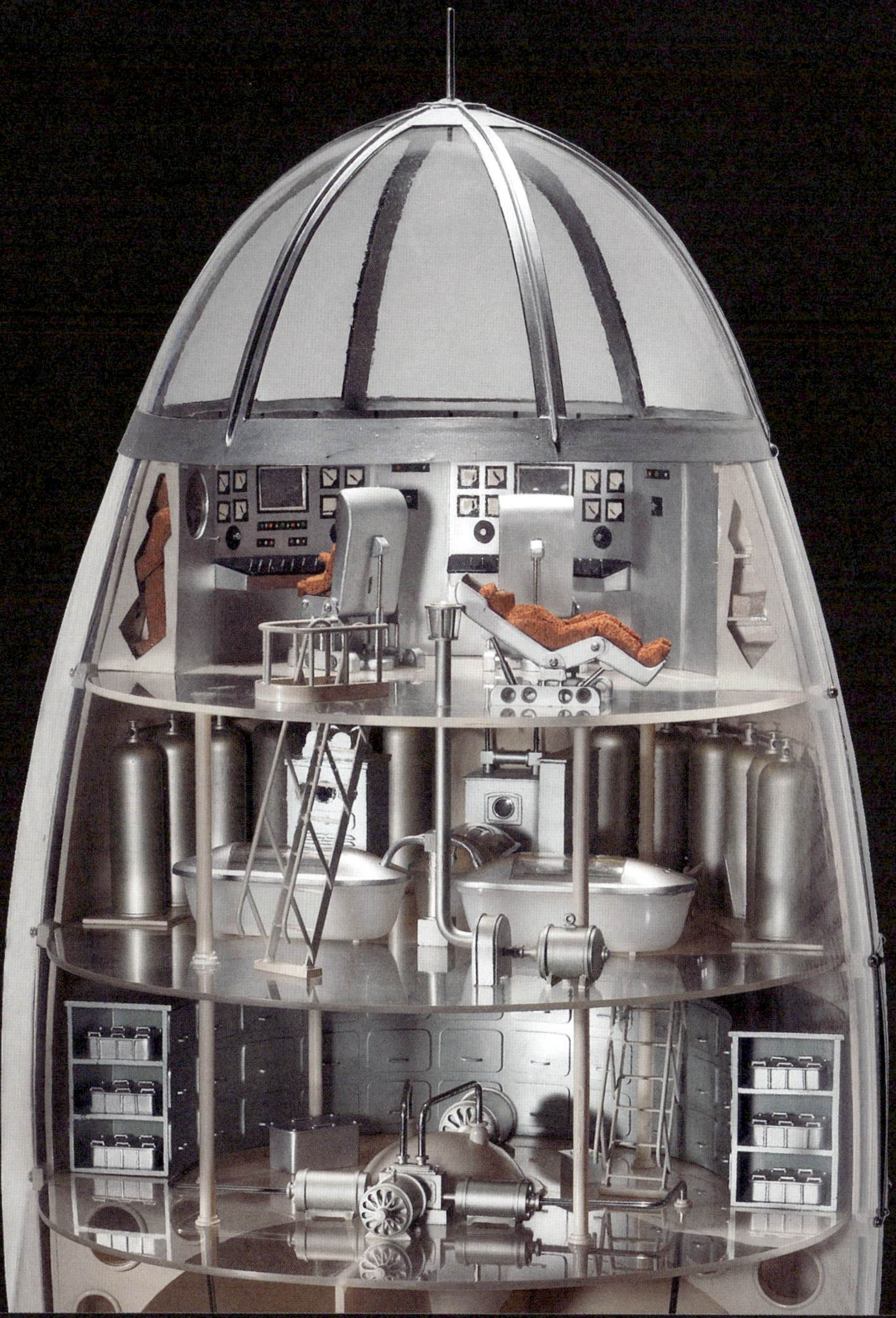

10.30

ROBERT H. GODDARD

On 19 October 1899, the seventeen-year-old Robert Hutchings Goddard propped his homemade ladder up against the big cherry tree in the garden of his extended family home in Worcester, Massachusetts, ready to cut away the dead branches. His mind was not on the job in hand but focused on more pressing matters, namely the *Boston Post*'s recent serialization of H. G. Wells's 'War of the Worlds in and Near Boston', a second-rate adaptation written to alarm its New England readers by relocating the famous Martian invasion. If Martians can make it all the way here, reasoned Goddard, how might we make a voyage there? What might such a machine look like? How might it work? From that moment, the problem of solving the practicalities of space travel consumed him for the rest of his life. His mind up until then had been immersed in other forms of inquiry and experimentation – in tadpole hatching, ornithology, sketching, photography, hydrogen balloon-making, rifle shooting, telescopes, electrical experiments and the wonders of his uncle George's tool shed. He read voraciously, favouring periodicals that offered the broadest window onto mankind such as *Scientific American*, *Wide World* and *Cosmopolitan*. His was a world that was witnessing the birth of the telegraph, the emergence of entirely new branches of physics and an astronomy enraptured by the possibilities of canals on Mars and vegetation on the Moon. A new century was here, and with it came an exciting future filled with wonder.

Goddard was no idle dreamer. If space travel was possible, he instinctively knew that whatever form it might take, physics and mathematics would be at the foundation, and Isaac Newton's third law of motion in particular: for every action (in our case propulsion in the vacuum of space – generally thought to be impossible at the time), there must be an equal and opposite reaction (something has to be thrown out the back in the opposite direction). He imagined a device working like the recoil of machine gun (reaction) as it fires towards the ground (action). Robert was an average student

11.1

by his own admission, whose education had been interrupted by illness, so he enrolled in maths and physics classes. His formal studies continued at Worcester Polytechnic Institute and then Clark University (also in Worcester), where he earned his PhD and spent most of his academic career. Among his many ideas, experiments, patents and papers were Futurist concepts ranging from Maglev-style transit systems (sealed cars powered by magnets, moving in a vacuum tube); gyroscopes in aircraft; the navigation of interplanetary space; mining hydrogen and oxygen from the Moon's surface; solar sails in space; ion propulsion systems and rocket planes suspended from high-altitude balloons, all the way to multi-generational migratory trips to the stars. His meticulously kept journals and diaries are a trove that ranges from the mundane to the practical to the wildly speculative.

In 1919 his most famous paper on rocketry was published, sponsored by a grant from the Smithsonian: *A Method of Reaching Extreme Altitudes*. In it, he included the calculations of the minimum mass needed to propel a weight of roughly half a kilogram (1 pound) to infinite height, and suggested the Moon as a marker. How would we know if a rocket landed on the Moon? With some flash powder, of course; the explosive material would ignite on impact and be seen by telescopes on Earth. This modest and sober addendum became front-page news, drawing unwanted attention to Goddard from both scientists like Hermann Oberth in Germany, who was working on similar ideas, and the popular press, hungry for sensational headlines. Goddard became the 'Moon Man', a moniker that he hated and which brought a flurry of correspondence from around the world from those eager to travel to space. Importantly, this obscure paper signalled for the first time the concept of space rockets leaving the pages of science fiction for those of respectable science journals.

Goddard knew that his mathematics needed to translate into an engineered form. But his groundbreaking experimental rocket Nell looked nothing like the spacecraft we know today.

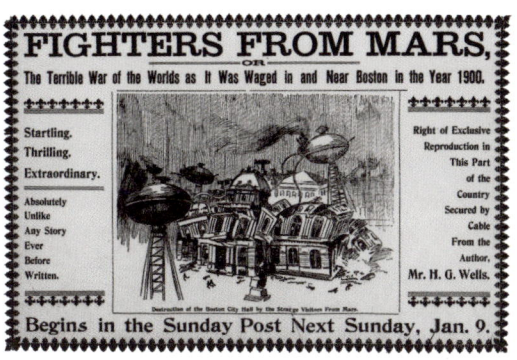

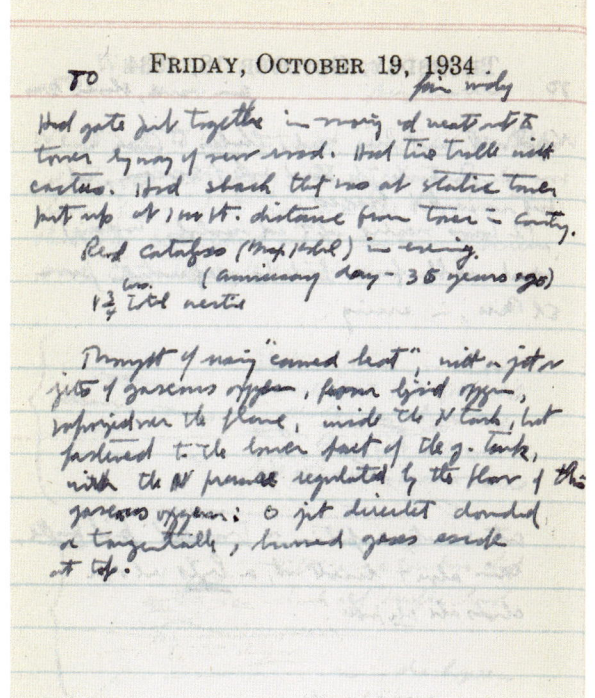

11.1 Dr Robert H. Goddard observing a rocket launch from his launch control shack in New Mexico, 1940, while standing by the firing control panel.

11.2 Advertisement for 'Fighters from Mars', or 'The War of the Worlds in and Near Boston', *Boston Post*, 3 January 1898.

11.3 Page from Goddard's diary, 19 October 1934. In every diary from 1899 onwards, Goddard would mark 19 October, the day he first thought of devoting himself to high-altitude work while trimming his cherry tree, as his 'anniversary day'. 'Had gate put together in am, and went out to tower by way of new road. Had tire trouble with cactus. Had shack that was at static tower put up at 1000 ft. distance from tower in country. Read catalogues (Max Kohl) in eve. (Anniversary day - 35 years ago) Thought of using "canned heat" with a jet or jets of gaseous oxygen, from liquid oxygen, vaporized over the flame, inside the N tank, but fastened to the lower part of the g. tank, with the N pressure regulated by the flow of this gaseous oxygen; O jet directed downward or tangentially, burned gases escape at top.'

Fig. 11.

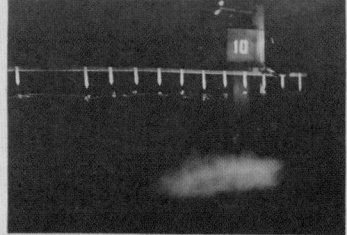
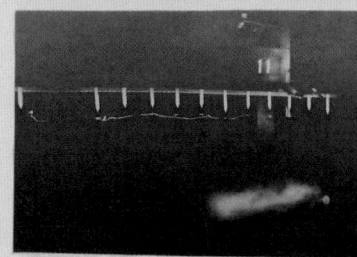
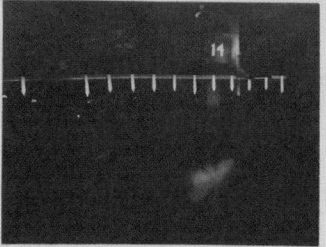

closing the primary switch at the left. The six-point switch at the right served to connect the tubes, in order, to the high-tension side of the coil.

The flashes were observed at a distance of 2.24 miles on a fairly clear night; and it was found that a mass of 0.0029 grams of Victor flash powder was visible, and that 0.015 gram was strikingly visible; all the observations being made with the unaided eye. The minimum mass of flash powder visible at this distance is thus surprisingly small.

From these experiments it is seen that if this flash powder were exploded on the surface of the moon, distant 220,000 miles, and a telescope of one foot aperture were used -- the exit pupil being not greater than the pupil of the eye (e.g., 2 millimeters) -- we should need a mass of flash powder of

2.67 lbs., to be just visible, and

13.82 lbs., <u>or less</u>, to be strikingly visible.

If we consider the final mass of the last "secondary" rocket plus the mass of the flash powder and its container, to be four times the mass of the flash powder alone, we should have, for the <u>final mass of the rocket</u>, four times the above masses. These final masses correspond to the "one pound final mass" which has been mentioned throughout the calculations.

The "total initial masses", or the masses necessary for the start at the earth, are at once obtained from the data given in Table VII. Thus if the start is made from sea-level, and the "effective velocity of ejection" is 7,000 ft/sec., we need 602 lbs. for every pound that is to be sent to "infinity".[6]

---

[6] A simple calculation will show that the total initial mass required to send one pound to the surface of the moon is but slightly less than that required to send the mass to "infinity."

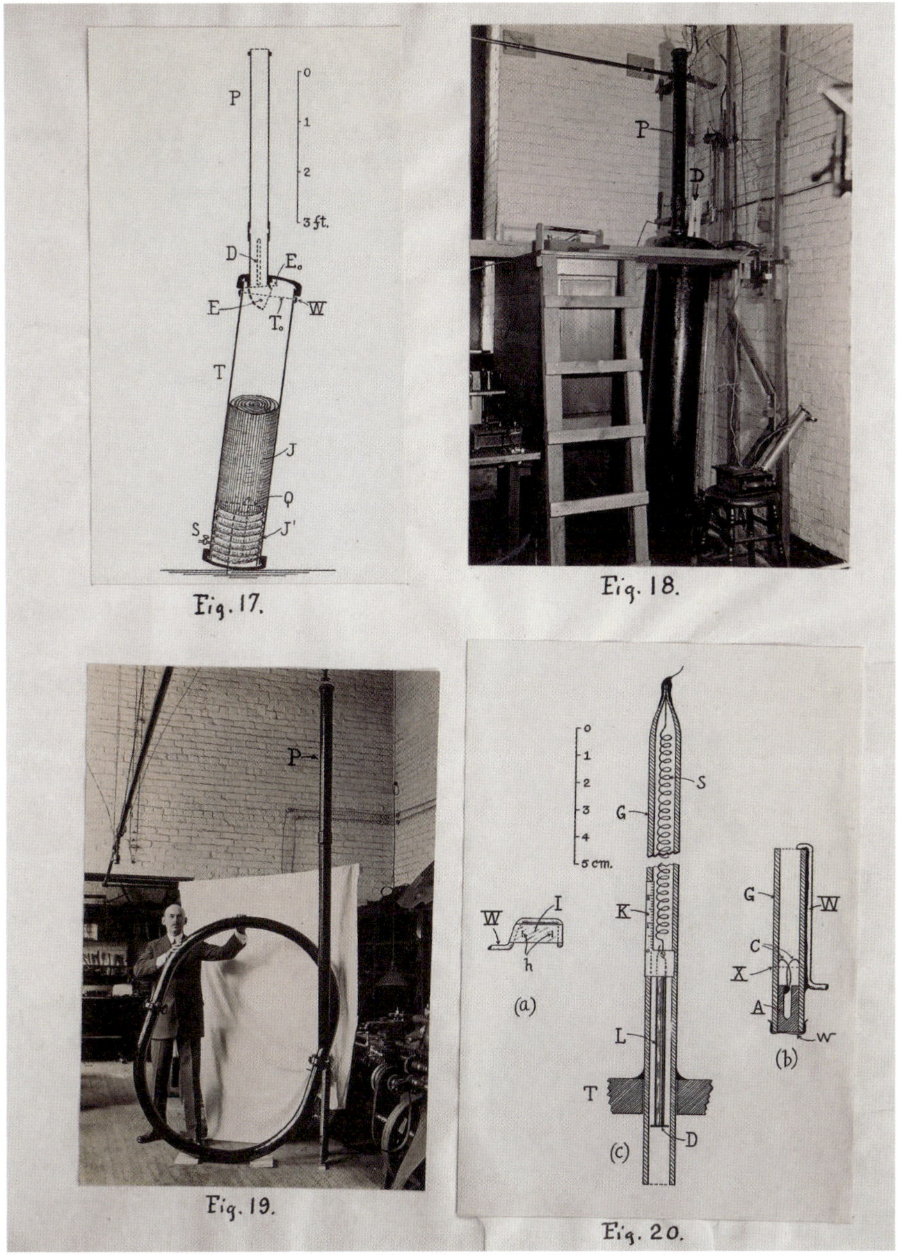

11.6

11.4-11.6 Pages from Goddard's *Method of Reaching Extreme Altitudes* [manuscript], c. 1919, detailing his experiments with flash powder. Note the tantalizing word 'infinity' at the bottom of the page. Despite his cautious tone, he proved that the sky was no longer the limit. Rockets were the key to unlock space travel.

11.7 Cutting from *The Virginian*, Norfolk, VA, 18 January 1920.

"Shooting at the Moon" Seems to Be Most Popular Pastime This Winter

11.7

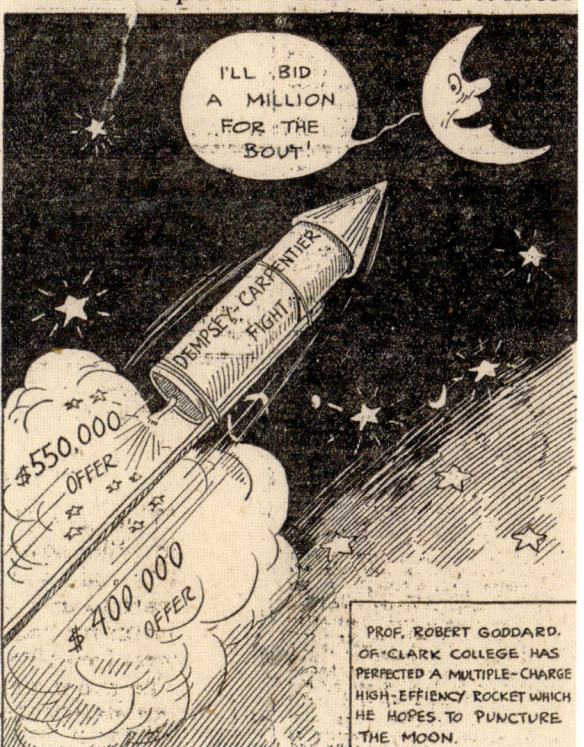

RA98 cone
1063 cone
tube
angle of blades

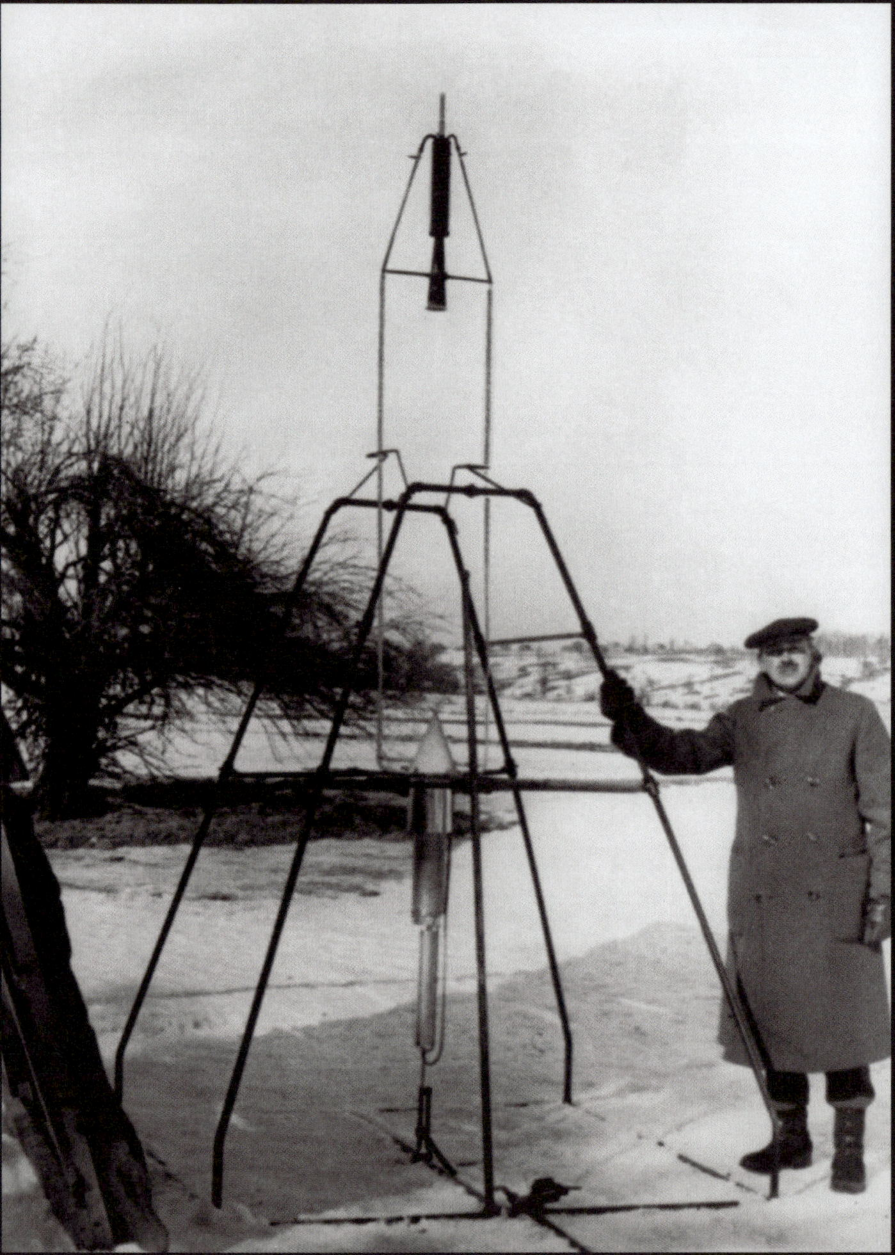

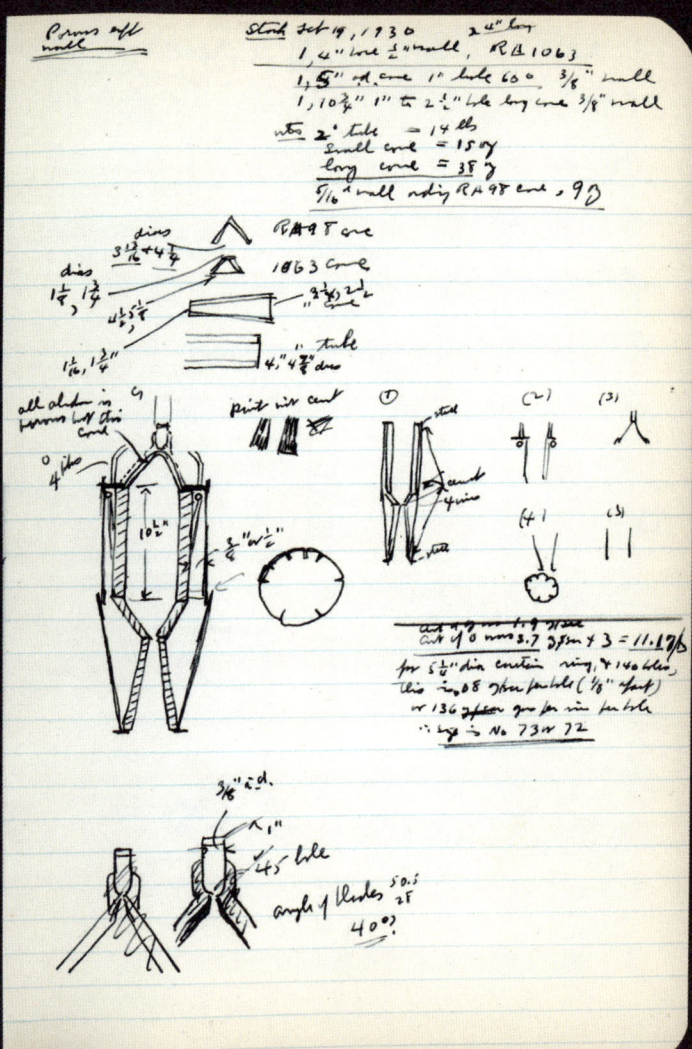

Named after *Salvation Nell*, a musical about a young woman lifted from the brink of ruin by the glory of the Salvation Army, it was an upside-down, spindly contraption. It was upside down because its slim motor was at the top of the structure and the two small gasoline and oxygen tanks were at the bottom, protected from the blast by an asbestos cone hat. These two sections were connected by two long fuel pipes, which also provided the rocket with its 4-metre (13-foot) skeleton structure. It's hard to see where the supporting frame ends and the rocket begins.

Date: 16 March 1926. Location: Aunt Effie's farm on the outskirts of Worcester. The weather was cold and the snowfall from the previous week was patchy and sugary underfoot. Destination: infinity (and beyond). Present: Robert Goddard in his long heavy military coat and cap; Henry Sachs and Percy Roope, Robert's machinist and assistant from Clark University; and Robert's wife of two years, Esther Kisk, who was tasked with filming the occasion. From the end of a long pole, the rocket was lit, and the world changed forever. A small flame appeared and burned with a steady roar, Nell settling itself before rising slowly into the sky and shooting off to the left like an express train, crashing into a patch of snow. Results: 12½ vertical metres (41 feet), covering a distance of 56 metres (184 feet) in 2.5 seconds. Esther said it looked like a 'fairy in flight; an aesthetic dancer'. Robert noted in his journal that the rocket seemed to have a mind of its own, as if it was saying, 'I've been here long enough; I think I'll be going somewhere else, if you don't mind.'

Nell was by no means the first rocket; crude rockets and fireworks had been around for centuries. But it was the first rocket that used liquid fuel – far more efficient and powerful than the explosive black powder that had been used up until this point. Goddard had launched the first prototype of a machine that could theoretically fly beyond the limits of aeroplanes and balloons, which needed an atmosphere to support them, and most importantly, when used with liquid oxygen, could fly in space.

In 1927, as his tests became more ambitious, he drew the attention of his increasingly disgruntled neighbours, but also of aviation hero Charles 'Lone Eagle' Lindbergh, fresh from his pioneering flight across the Atlantic and looking for his next adventure. Lindbergh visited the Goddards and helped secure funding from the philanthropist Daniel Guggenheim, which meant a move from New England to a full-time base in Roswell, New Mexico, a place with a dry, sunny climate – more suitable for humans and rockets alike – and far from prying eyes. The 1930s were 'golden years for the Goddards', as Esther noted. It was here in the American Southwest that the problem-solving of rocketry began in earnest – of steering and guidance using gyroscopes, of better fuel pumps and parachutes for recovery. The rockets were getting bigger, higher and faster: by the late 1930s, they were reaching altitudes of around 2,745 metres (9,000 feet). As the world drifted into war, Goddard tried to convince the military of the practical uses of his work, but received little interest. He did work for the US Navy, however, developing rocket-assisted devices that provided extra thrust for laden aircraft taking off.

11.8 Goddard and Nell, in the frame from which the rocket was fired on 16 March 1926, at Auburn, Massachusetts.

11.9 Page from Goddard's 'Rough Notes', 1927-28.

11.10 Goddard's diary entry from 17 March 1926, after the successful launch of Nell.

'We have seen in a Swedish newspaper that you intend to make a trip to the moon this summer, and since we are interested in the forthcoming trip, we hereby request that we may go along.'

TWO YOUNG MEN TO ROBERT H. GODDARD, STOCKHOLM, SWEDEN, 1926

11.11 Goddard and colleagues holding the rocket used in the flight of 19 April 1932.

11.12 Goddard rocket with four rocket motors. This rocket attained an altitude of 200 feet in a flight, November 1936, at Roswell, New Mexico. From 1930 to 1941, Goddard made substantial progress in the development of progressively larger rockets that attained greater altitudes and refined his equipment for guidance and control, his welding techniques, and his insulation, pumps and other associated equipment.

11.13 Goddard and colleagues at Roswell, New Mexico, 19 May 1937.

Dr. Goddard and colleagues holding the Rocket used in flight of April 19, 1932. They are, from l. to r., L. Mansur; A. Kisk; C. Mansur; Dr. R. H. Goddard; and N. L. Jungquist.

Photo Courtesy of Mrs. Robert H. Goddard

Goddard rocket with four rocket motors. This rocket attained an altitude of 200 feet in a flight, November 1936, at Roswell, New Mexico. (This picture is displayed through the courtesy of Mrs. Robert H. Goddard).

Dr. Robert H. Goddard and colleagues at Roswell, New Mexico. Successful test of May 19, 1937. Dr. Goddard is holding the cap and pilot parachute, parts of the successful operation.

Photo Courtesy of Mrs. Robert H. Goddard

Goddard's lifetime of poor health was catching up with him. He lived just long enough to see an American-captured German V-2 rocket up close and commented on the similarities of design, fuelling endless speculation on whether the Germans had spied on him or used his patents for their own benefit. Sadly, he did not live to see the ushering in of the space age with launch of Sputnik in 1957. It was only over a decade after his death from throat cancer in 1945 that the name Robert Hutchings Goddard was resurrected by a nation hungry for its own 'first' space story to rival the Soviets or the Germans. In 1959 he officially became 'the father of American rocketry' with the opening of the NASA Goddard Space Flight Center.

That Goddard was a brilliant scientist and engineer is under no doubt. Others were working on the same problem at roughly the same time, and achieving similar conclusions, albeit for very different purposes, but they were working on an industrial scale. It is extraordinary how much Goddard was able to achieve working with just his wife and a handful of assistants. However, his legacy goes far beyond rocketry. The records he left behind give us a priceless insight into the creative mind – that fine balance between the freedom to imagine without limits and the discipline of hard work and academic rigour. Robert H. Goddard knew it's not the dreaming that counts, it's what you do with the dream when you wake up.

'It's difficult to say what is impossible. For the dream of yesterday is the hope of today and the reality of tomorrow.'

ROBERT H. GODDARD, 1904

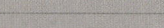

11.14-11.17 This is probably the rocket Goddard tried to launch on 23 September 1935 at Roswell, New Mexico, to demonstrate its capabilities to his supporters Charles Lindbergh and Harry Guggenheim. A technical problem prevented the flight, but earlier successes convinced them that Goddard was on the right track. Lindbergh persuaded Goddard to donate this A-series liquid-fuelled rocket to the Smithsonian collections.

HERMANN OBERTH

Hermann Oberth's destiny was written in the stars before he was born. In 1869 his grandfather Friedrich Krasser, a doctor and radical utopian socialist, made a prediction: in a hundred years his grandchildren would witness humans landing on the Moon. Exactly a hundred years later, in July of 1969, Krasser's grandson was invited to the Kennedy Space Center by his former protégé, Wernher von Braun, the architect of the American space programme. He was a guest of honour to witness Neil Armstrong set foot on the lunar surface – a moment in history which Oberth was instrumental in making happen. Of his rocketry 'founding father' contemporaries, namely the Russian Konstantin Tsiolkovsky and the American Robert H. Goddard, only Oberth lived to see the fruits of their visionary work.

Oberth was born in Transylvania (now Romania) on 25 June 1894. As a child he devoured the fictional space travel stories of Jules Verne. Like Tsiolkovsky, the inquisitive Oberth quickly realized that the unfortunate occupants of Verne's ballistic moon cannon (fig. 10.6) would be instantly killed by the force of the acceleration. This gave him pause. How else could we travel to the Moon? He concluded that the action/reaction principle of the rocket would be the only realistic means of getting there. Newton's mathematics showed that the rocket principle worked in the vacuum of space; it wasn't dependent on having air to 'push against', as was the popular belief. But he also knew that the explosive black powder that rockets used for fuel at the time wouldn't have the energy needed to propel an object into space: potent liquid fuels combined with liquid oxygen would be needed. He saw that multi-stage rockets would be the solution to reach the velocity needed for orbital space flight. The young Hermann Oberth had no idea, of course, that other minds had reached similar conclusions. As is often the case, when an idea is ready to be born, it's born simultaneously by multiple people.

By his mid-teens, Oberth had solved many of the fundamental tenants of rocket science. At least on paper. As well as consuming himself with the physics of rocketry, he was interested in the physiological aspects that a real space traveller might encounter. He tried to estimate the g-force tolerances an astronaut might experience by throwing himself backwards into a swimming pool from varying heights, and looked to simulate weightlessness underwater – all with various degrees of success!

He was persuaded by his father to study medicine rather than physics or engineering, but he still devoted his spare time to interrogating the problems of space travel. Towards the end of the First World War, he married Mathilde Hummel and left the world of medicine to concentrate on physics and mathematics.

'This is the goal: to make available for life every place where life is possible. To make inhabitable all worlds as yet uninhabitable, and all life purposeful.'
MAN INTO SPACE (1957)

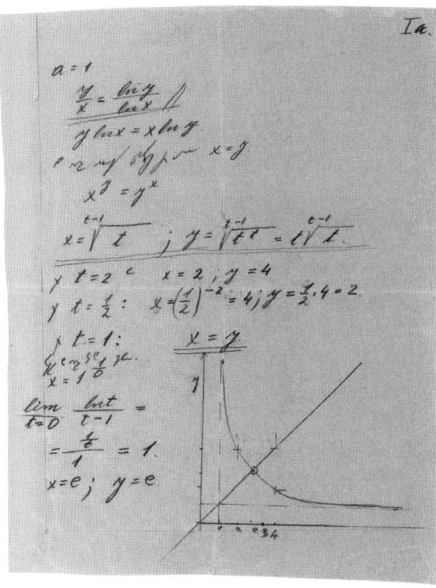

12.1 Hermann Oberth (front) with officials of the Army Ballistic Missile Agency at Huntsville, Alabama, in 1956.

12.2–12.7 Rare manuscripts from the German rocket pioneer Hermann Oberth, sent to 'Herrn [Mr] Joachim Ruff', 16 March 1971. In the covering letter, Oberth writes: 'I am enclosing three calculations I handwrote over the past few weeks. The first concerns a function that arises with the question how often a multi-stage rocket is to be separated when final load and velocity are given ... The second piece of paper concerns a process by which to transcribe a differential equation of the second degree into the inverse form ... The third paper addresses the following question: You have a curving road that you must cross at right angles only. Your destination is on the same side. What must the curve be to make it worth crossing twice? If you do not mind my messy notes, you may keep them.'

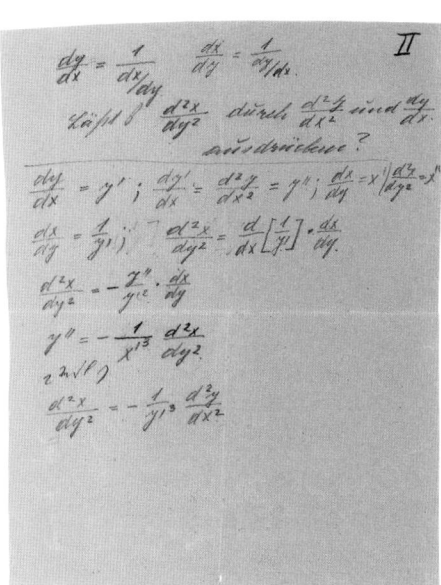
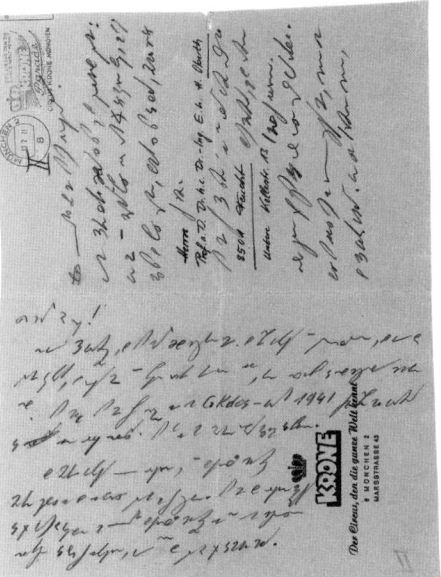
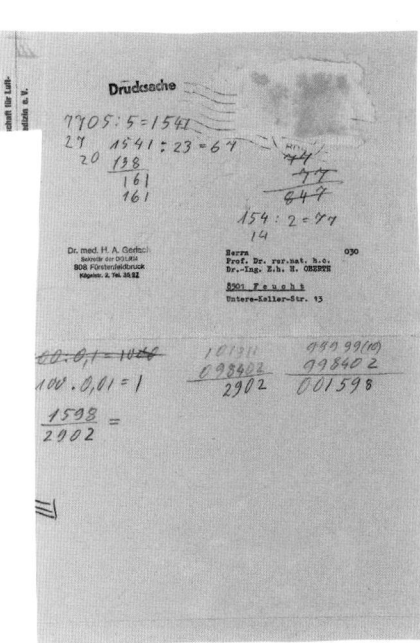
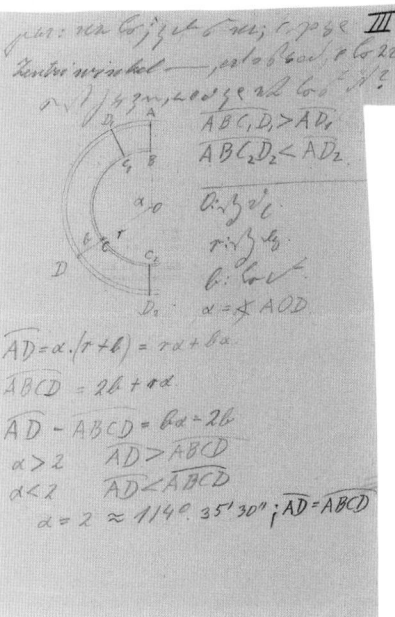

In the early 1920s he moved to study in Göttingen and then to Heidelberg, experimenting with building prototypes and setting out his theory of rocketry in earnest. His doctoral thesis was considered too avant-garde for academia, and so, with the encouragement of some of his professors, he decided to self-publish. In June 1923 the slim ninety-two-page volume, with its striking black cover and extraordinarily dramatic Fraktur typeface, was released: *Die Rakete zu den Planetenräumen* (The Rocket into Interplanetary Space). It would become the foundation of modern space flight. The book begins with four bold statements:

1. *In the current state of science and technology, it is possible to build machines able to ascend beyond the limits of Earth's atmosphere.*

2. *With additional refinement these machines will be able to attain such velocities that, left to themselves in space, they will not fall back to Earth's surface, and will even be able to leave the gravitational field of Earth.*

3. *These kinds of machines can be built to accommodate humans allowing them to travel into space (probably without physical problems).*

4. *Under certain economic conditions, the construction of such machines may even become profitable. Such conditions might arise within a few decades.*

Its pages laid out groundbreaking theoretical fundamentals of rocketry and space travel that backed up these statements. An expanded version called *Wege zur Raumschiffahrt* (Ways to Space Flight) was published in 1929, this time aimed at a general audience. The book contained many revolutionary ideas now taken for granted: the launching of rockets eastwards to gain velocity from the Earth's rotation; ways for spacecraft to return to Earth using the atmosphere to brake; jet rudders for steering; and gyroscopic stabilization. Oberth also foreshadowed spacesuits, radiation and meteorite concerns for astronauts, electrically propelled ion drives and beyond into speculative areas: commercial space stations and giant orbital space mirrors that could function as aids for navigation and even be used to influence the weather. *Wege zur Raumschiffahrt* became the intellectual impetus of the German rocket craze and crowned Oberth as the spiritual father of this new movement, particularly in the Verein für Raumschiffahrt. This German amateur rocket society included such luminaries as the French rocket scientist Robert Esnault-Pelterie, Austrian physicist Max Valier, who was experimenting with rocket-propelled cars, the German-American science writer Willy Ley and of course Wernher von Braun, who would be so instrumental in the Nazi V-2 and American space programmes. Unlike the work of quiet Robert Goddard, which went largely under the radar, Oberth's book started a European rocket revolution. What sets it apart is the breadth of its subjects – as well as the detail in which they are explored – and the inspiration it gave to others, particularly when it came to cementing ideas of space travel within wider culture.

In 1929 Oberth was hired as the scientific advisor for Fritz Lang's science fiction film *Frau im Mond* (Woman in the Moon), written by his wife Thea von Harbou. Lang's prophetic film about a group of unlikely moon voyagers was an extraordinary portrait of Weimar politics, framed within an aesthetic of industrial modernity and

12.8 Hermann Oberth *Die Rakete zu den Planetenräumen* (The Rocket into Interplanetary Space, 1923).

12.9 Oberth's model of a space rocket. Plate from *Die Rakete*.

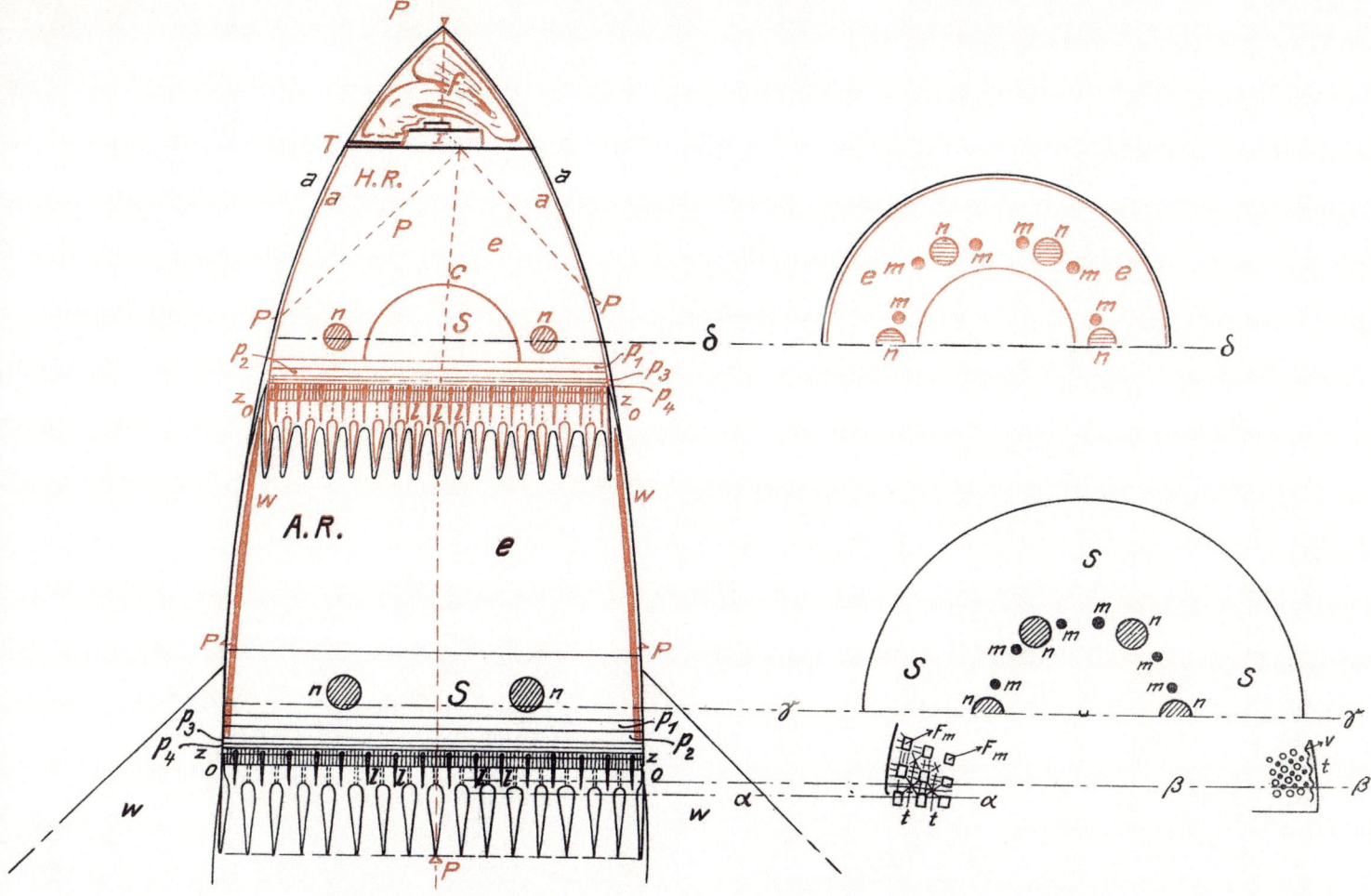

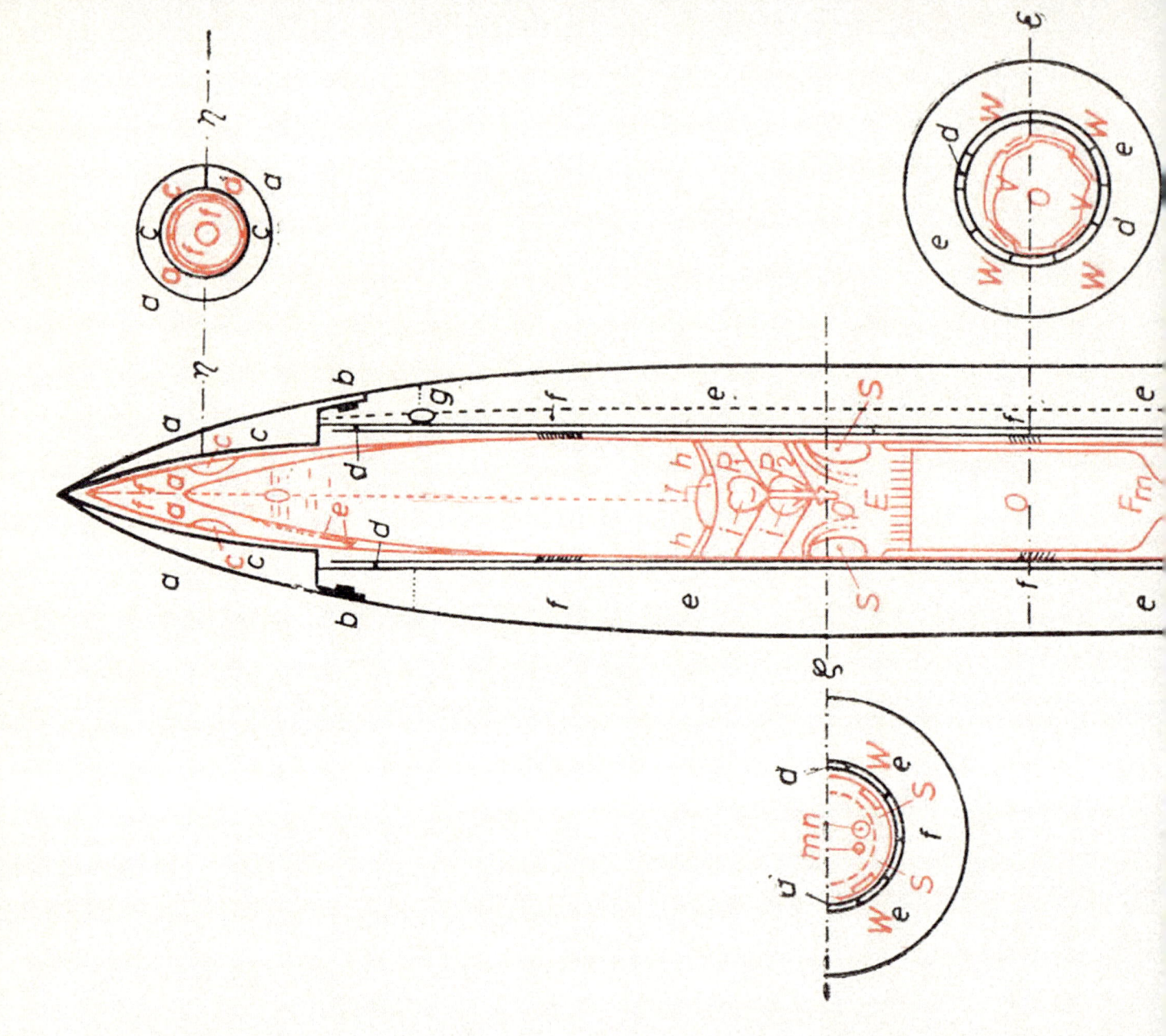

12.10 The Model B two-stage rocket, designed to escape gravity and go beyond the Earth's atmosphere. Plate from *Die Rakete*.

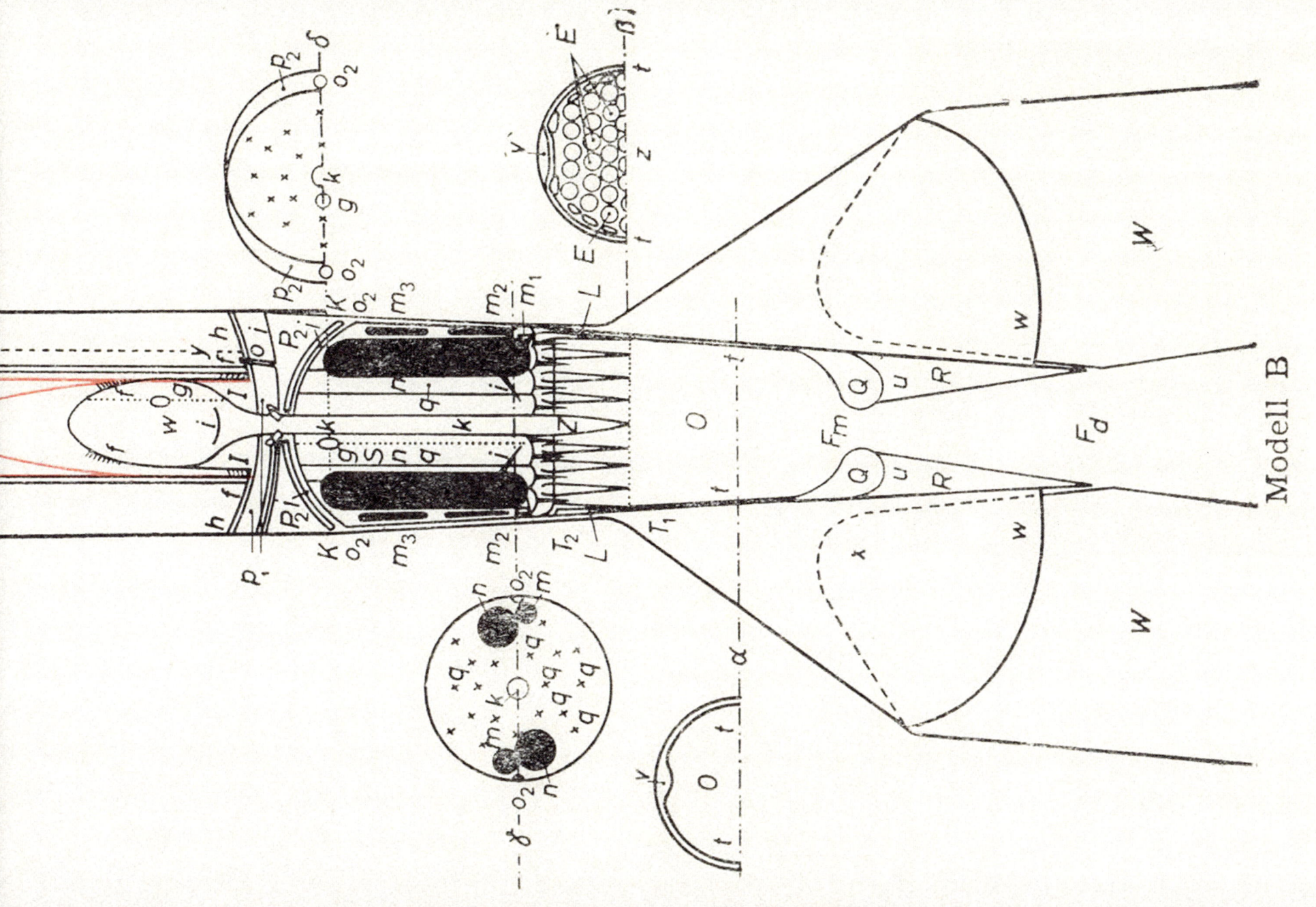

Oberth's revolutionary solutions to space travel. Lang wanted the film to be as accurate to the science as possible. Money was given to Oberth by the studio's publicity department to build an actual rocket that could be launched at the film's premiere. For Oberth and his assistants, this meant properly funded research and design, including experimenting with gasoline and liquid oxygen mixtures to power the rocket (which resulted in a burst eardrum for Oberth and damage to his left eye). The rocket wasn't finished on time, but it's interesting to note that it was the arts rather than academia that proved responsible for breakthroughs in the fiendishly complicated science of rocket combustion, as well as producing Europe's first liquid-fuelled rocket motor.

In *Frau im Mond*, many of Oberth's ideas are beautifully realized. We see graphics of the gravitational fields of the Earth and the Moon and detailed maps of the cratered lunar surface; the figure of eight orbital path to be taken to the Moon and back that was used by the Apollo missions. We see the first backwards rocket countdown, from 10 to 1, an invention not of science but of cinema. The astronauts in the film are launched lying face up on spring-loaded bunks to absorb the g-forces and wear spacesuits like diving suits, just as Oberth had envisaged. Most striking, however, is the view from the window of Spaceship Friede (Peace) as it sails to the Moon. Lang shows the audience the view of the Earth from space with the limb of the Moon scrolling beneath, identical to the famous Apollo 8 photographs of 1968, *Earthrise* (figs 20.17–20.18). The fictional crew pause for a moment to gaze back and contemplate its significance, just as humanity did when we saw the image for real. The first example of the overview effect, the sense of awe that astronauts experience looking back at the Earth from space. A cognitive Lagrange point where escape velocity meets escapism.

Cinema and space travel had come a long way since Georges Méliès's 1902 film *Voyage dans la Lune* (fig. 2.29), based on Jules Verne's famous story. Lang's film was the first grounded in a new reality. Science, fantasy and increasingly politics and economics were intertwining more explicitly than ever before. Oberth's utopian book showed a vision of the benefits of space travel. As storm clouds gathered and war approached once again, the journey to the stars that we were soon to embark on would tread a much darker path.

12.11 *Frau im Mond* German poster art, 1929.

12.12 Fritz Lang during the filming of *Frau im Mond*, 1929. The rocket capsule scenery behind him was built according to Oberth's designs.

12.13 Gerda Maurus in *Frau im Mond*, with plan of Spaceship Friede's orbital path from Earth to the Moon.

12.14 Still from *Frau im Mond* featuring the rocket capsule.

12.12

12.13 / 12.14

WERNHER VON BRAUN

On 8 September 1944 Wernher von Braun's handiwork came crashing into 5 Staveley Road, Chiswick, a leafy west London suburb. It left a 9-metre (30-foot) crater in its wake, destroying a dozen houses and killing pensioner Ada Harrison, a soldier called Bernard Browning, who was on leave, and three-year-old Rosemary Clarke. It was the first of the V-2 Wunderwaffe German rockets to hit London. As a 'weapon of wonder', it left a lot to be desired: it was cripplingly expensive to build, wildly inaccurate and possessed little strategic military value. Far from winning the war for Germany, it signalled the death throes of German victory. The vast 14-metre-tall (45-foot) V-2 rocket was much more successful as a weapon of the mind – for both the Allies on its receiving end and the Germans who spent over a decade building it. But to von Braun, it worked perfectly. Apart from landing on the wrong planet.

Baron Wernher Magnus Maximilian Freiherr von Braun (1912–1977) was the son of aristocratic Prussian parents. He was musically gifted, handsome and charming – attributes which would become as significant to his later celebrity status in America as his technical skills were in Germany. As a teenager he was swept along by the German rocket craze of the 1920s and was fascinated by Hermann Oberth's treatise of future space travel, *Die Rakete zu den Planetenräumen* (figs 12.8–12.10), In 1928 he joined the Verein für Raumschiffahrt (VfR), a rag-tag band of rocket engineers and early space-flight luminaries who were experimenting with new ideas in rocketry. Such was his academic brilliance, by the age of twenty-two he'd earned two engineering degrees and a doctorate in physics.

By 1932 the homespun activities of the VfR had attracted the attention of senior military figures, including Major General Walter Dornberger, a First World War artillery officer who was looking for the successor of the 'Paris Gun', the enormous siege gun that could throw an artillery shell 80 kilometres (50 miles). The Treaty of Versailles had prevented Germany rearming with ballistic weapons, but had said nothing about rockets. Rockets as a bomb delivery vehicle didn't exist, even as a concept. But with military interest came money, and for von Braun, whose talents were snapped up by Dornberger, money meant proper research and development.

By 1937 the twenty-five-year-old von Braun, now an official member of the Nazi party, found himself the technical director at Peenemünde,

'"Once the rockets are up who cares where they come down? That's not my department!", says Wernher von Braun.'

TOM LEHRER, 'WERNHER VON BRAUN' (1965)

13.1

a vast new military rocket range and research complex in a remote peninsular on Germany's forested Baltic coast. It was here under the leadership of Dornberger that the A3, a ballistic missile from von Braun's Aggregat series, was fired. These missiles were designed to launch a payload hundreds of kilometres beyond the scope of traditional ballistics. The A-4 became the most infamous, with a range of some 320 kilometres (200 miles) and the ability to carry a one-tonne warhead that could explode on impact. Its name was changed to the 'V-2' – the Vergeltungswaffen, or 'Vengeance weapon'. On 3 October 1942 the first successful test flight of the V-2 took place, the air filling with a rolling thunder as it left the test stand and grazed the edge of space, reaching an altitude of almost 90 kilometres (60 miles). This was the day the spaceship was born.

On the night of 17 August 1943, the RAF bombed Peenemünde after intelligence confirmed its true purpose, severely damaging its operations. The factory was moved to a less conspicuous underground facility in central Germany, near the town of Nordhausen, in what was an old gypsum mine. In 1944 the Mittelwerk factory was producing thousands of V-2s, with slave labour provided by the Mittelbau-Dora concentration camp complex. Here the further human cost of the V-2 programme was playing out in the tunnels where the rockets was constructed. Over 30,000 French, Polish and Russian prisoners of war died here, suffering in brutal and inhumane conditions.

13.1 German-born rocket scientist Wernher von Braun, director of NASA's Marshall Space Flight Center, at his office in Huntsville, Alabama, c. 1965. On a shelf behind him is a row of his rocket models.

13.2 Production of the V-1 rocket at the Mittelbau-Dora concentration camp, established in a tunnel inside Kohnstein mountain, Thuringia, 1944.

13.3 The aftermath of a V-2 explosion at Staveley Road, Chiswick, 9 September 1944.

By May 1945, Hitler was dead, and the V-2's short active service was over. Von Braun, Dornberger and other senior Peenemünde staff retreated to the Bavarian Alps, near the town

13.2

13.3

of Oberammergau. Below them lay the ruins of the Reich. Germany was being divided up, the Americans closing in from the West and the Soviets from the East. Both nations were looking for the crème de la crème of German scientists for their own post-war military ventures. Number one on that list was von Braun, who negotiated his surrendered to the Americans under the auspices of Operation Paperclip, a clandestine effort to procure the best of the German science intelligentsia to work on America's new rocket and space programmes. Their official Nazi histories were quietly overlooked. Communism was the new threat.

Along with the some 1,600 scientists, doctors and engineers claimed by the US came a valuable cache of V-2 rockets, and by 1946 the V-2 had gone from a weapon of war to the foundation of the American space programme. It was being tested and refined for scientific as well as military purposes. Clyde T. Holliday, an engineer from Johns Hopkins University, strapped a camera on a V-2 at the White Sands missile base in New Mexico and was able to take the very first image of the Earth from space (fig. 20.8). Within a decade both the Soviets and the Americans had built a new generation of rockets that could carry nuclear payloads across continents. The triple-use technology of the rocket had been established: an instrument of war, of science and of peaceful exploration. Von Braun was central to all three, notably as primary architect of the US Army's Redstone rockets, which launched both the first American satellite (Explorer 1) and the first American humans, through to NASA's Saturn V, which put twelve American men on the Moon. He was also able to recast himself to the American public. The 1950s saw a nation transfixed with stories and aesthetics of space travel. The glossy *Collier's* magazine published von Braun's exciting visions of a space utopia (fig. 13.9). On television he became the charismatic face of space thanks to Walt Disney, who produced a series of space-themed specials to promote his newly opened Disneyland, complete with its Germanic castle. The *Tomorrowland* segment of the show was the promise of 'things to come': *Man in Space*; *Man and the Moon*; and *Mars and Beyond*. With his good looks and thick German accent, he became an on-screen Disney character, the quintessential 'rocket scientist'. All of this was used to his advantage. Von Braun had the foresight to realize that to convince politicians to embark on wildly expensive plans like human space flight, the public would have to buy the ticket and come along for the ride.

13.4 Photograph of a German military group, including von Braun (to the right, in suit and tie) and Hitler, c. 1933. It is captioned on the verso in cursive German, giving the location as Kummersdorf, where von Braun began his rocket engineering career in 1932.

13.5 Press release issued by the Headquarters of the Seventh Army announcing the surrender of von Braun, 6 May 1945.

13.6 This V-2 was displayed at the war's end in Washington, DC, near 12th Street and Pennsylvania Avenue, NW. It symbolized not only the end of the war, but also the new shape of possible conflicts to come.

Wernher von Braun was undoubtedly brilliant. He was not just an engineer and a visionary, he had the rare relentless energy to make his dreams a reality. He was also a master opportunist who could manipulate a path for his own expedience, however difficult that path might be. Publicly he always defended his time as a high-ranking SS officer as being nothing more than a convenient means to a worthwhile end. But that asks us to ignore the fact that, with his full knowledge, thousands more people died making his V-2 in the slave labour concentration camp of Mittelbau-Dora than were ever killed by the missile itself. Von Braun simply chose to look the other way.

13.7 Wernher von Braun, 'Baby Satellite Sketch', 1953. This sketch formed the basis for artist Fred Freeman's illustrations in the 27 June 1953 issue of *Collier's* magazine. Von Braun's article on the 'Baby Space Station' appeared in that issue as part of the series 'Man will Conquer Space Soon!', and described a small satellite designed to stay in orbit for sixty days, carrying three rhesus monkeys, TV cameras, antennae, solar mirrors and Geiger counters. The present sketch provides a cross section of the satellite and three further views.

13.8 Barbara Diener, *Dr. von Braun's First US Driver's License Certificate*, 1946/2019.

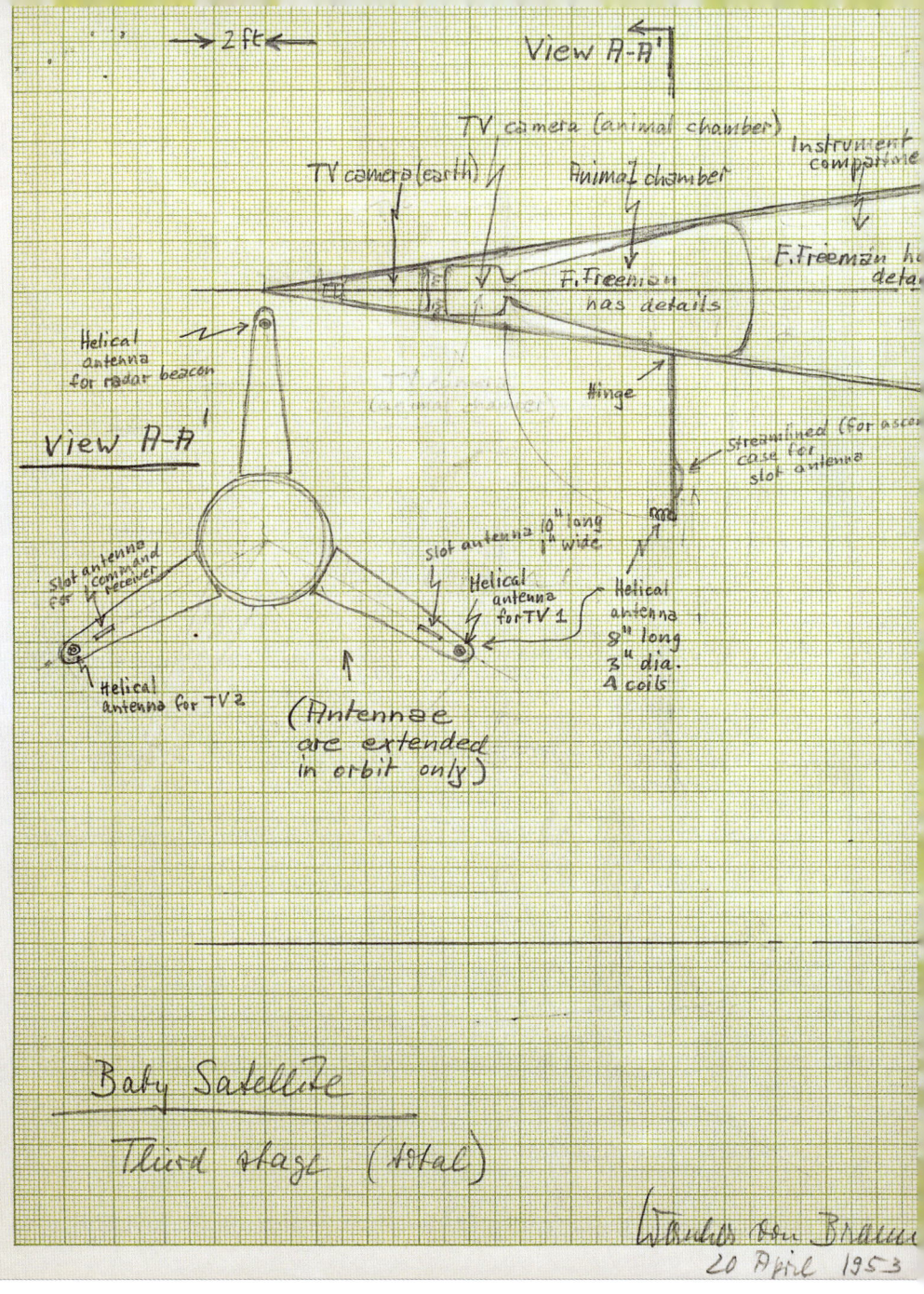

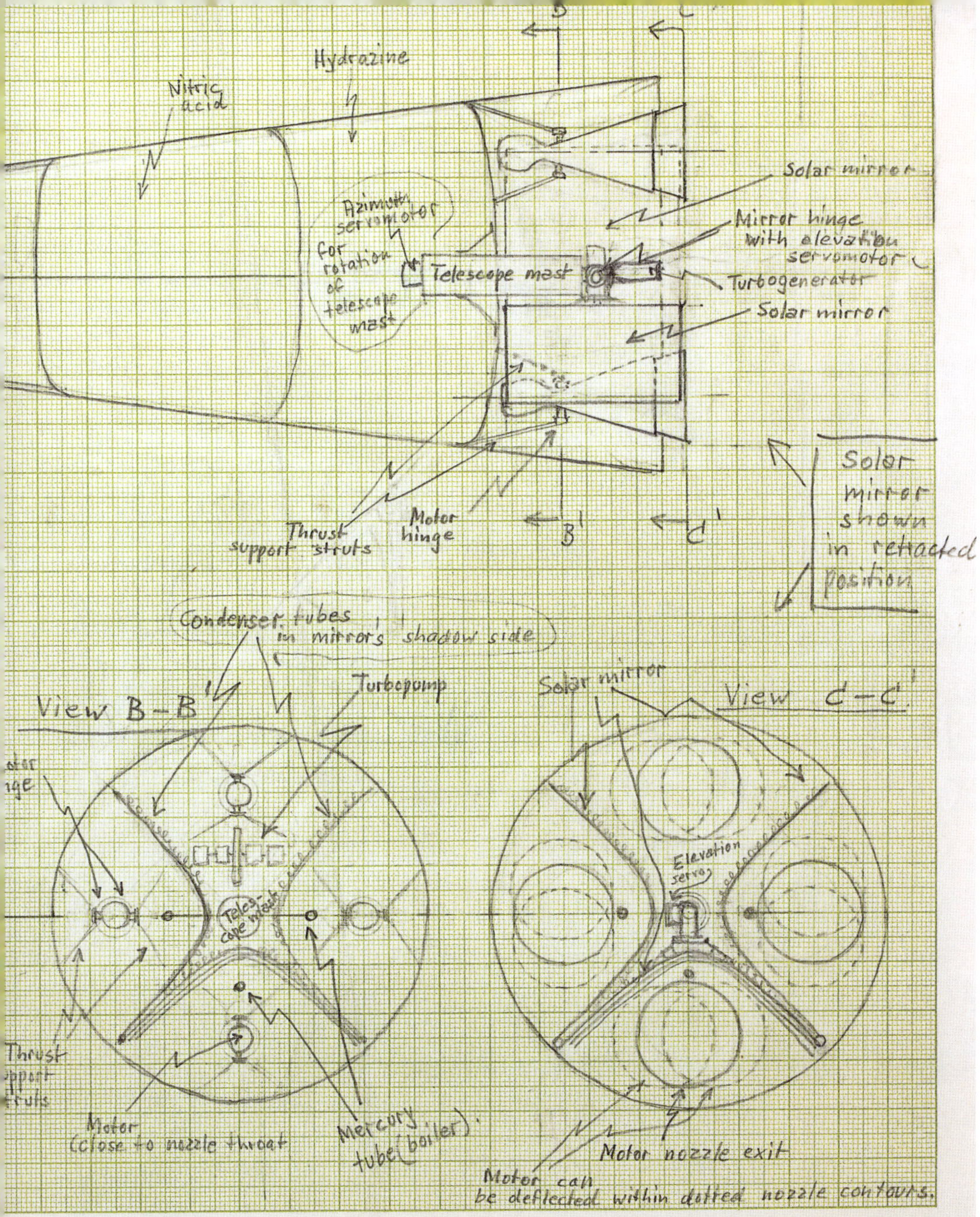

'For me Dora is a presence. English, French, Americans and Russians have shared the scientists and technicians who were our masters. And I could not watch the Apollo mission without remembering that that triumphant walk was made possible by our initiation to inconceivable horror.'

JEAN MICHEL, DORA: THE NAZI CONCENTRATION CAMP WHERE SPACE TECHNOLOGY WAS BORN AND 30,000 PRISONERS DIED (1975)

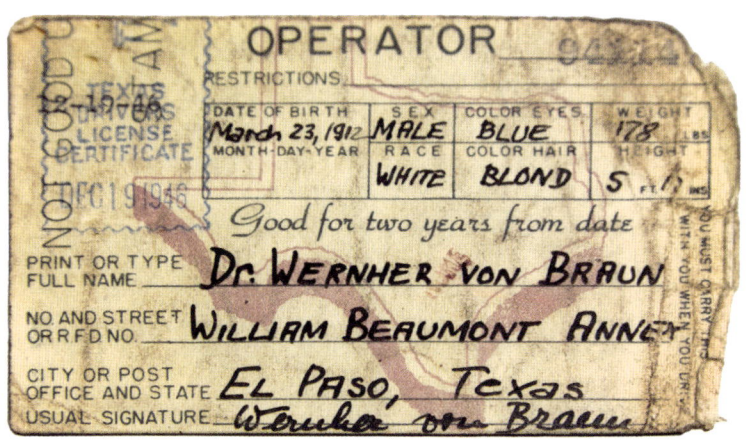

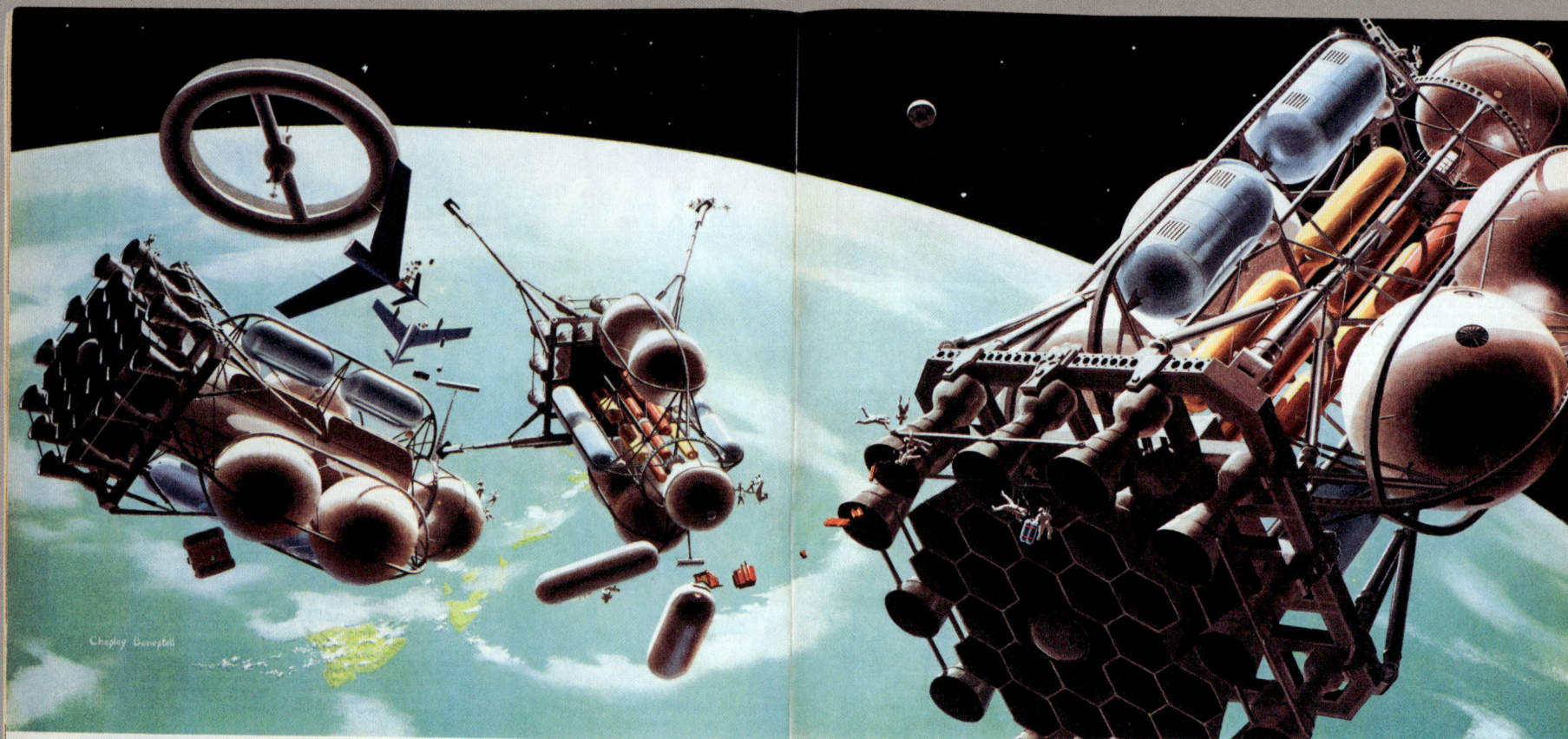

Weightless in orbit 1,075 miles above earth, workers in space suits assemble three moon ships. Hawaiian Islands lie below. Winged transports unload supplies near wheel-shaped space station top left. Engineers and equipment cluster around cargo ship lower left, passenger ships center and right

Man on the Moon
THE JOURNEY

By Dr. WERNHER von BRAUN
Technical Director, Army Ordnance Guided Missiles Development Group, Redstone Arsenal, Huntsville, Alabama

For five days, the expedition speeds through space on its historic voyage —50 men on three ungainly craft, bound for the great unknown

HERE is how we shall go to the moon. The pioneer expedition, 50 scientists and technicians, will take off from the space station's orbit in three clumsy-looking but highly efficient rocket ships. They won't be streamlined; all travel will be in space, where there is no air to impede motion. Two will be loaded with propellant for the five-day, 239,000-mile trip and the return journey. The third, which will not return, will carry only enough propellant for a one-way trip; the extra room will be filled with supplies and equipment for the scientists' six-week stay.

On the outward voyage, the rocket ships will hit a top speed of 19,500 miles per hour about 33 minutes after departure. Then the motors will be stopped, and the ships will fall the rest of the way to the moon.

Such a trip takes a great deal of planning. For a beginning, we must decide what flight path to follow, how to construct the ships and where to land. But the project could be completed within the next 25 years. There are no problems involved to which we don't have the answers—or the ability to find them—right now.

First, where shall we land? We may have a wide choice, once we have had a close look at the moon. We'll get that look on a preliminary survey flight. A small rocket ship taking off from the space station will take us to within 50 miles of the moon to get pictures of its meteor-pitted surface—including the "back" part, never visible from the earth.

We'll study the photographs for a suitable site. Several considerations limit our selection. Because the moon's surface has 14,600,000 square miles—about one thirteenth that of the earth—we won't be able to explore more than a small area in detail, perhaps part of a section 500 miles in diameter. Our scientists want to see as many kinds of lunar features as possible, so we'll pick a spot of particular interest to them. We want radio contact with the earth, too; that means we'll have to stick to the moon's "face," for radio waves won't reach across space to any point the eye won't reach.

We can't land at the moon's equator because its noonday temperatures reach an unbearable 220-degrees Fahrenheit, more than hot enough to boil water. We can't land where the surface is too rugged, because we need a flat place to set down. Yet the site can't be too flat, either—grain-sized meteors constantly bombard the moon at speeds of several miles a second; we'll have to set up camp in a crevice where we have protection from these bullets.

There's one section of the moon that meets all our requirements, and unless something better turns up on closer inspection, that's where we'll land. It's an area called *Sinus Roris*, or Dewy Bay, on the northern branch of a plain known as *Oceanus Procellarum*, or Stormy Ocean (so called by early astronomers who thought the moon's plains were great seas). Dr. Fred L. Whipple, chairman of Harvard University astronomy department, says *Sinus Roris* is ideal for our purpose —about 650 miles from the lunar north pole, where the daytime temperature averages a reasonably pleasant 40 degrees and the terrain is flat enough to land on, yet irregular enough to hide

PAINTING BY CHESLEY BONESTELL

'Today it is with the tribute of blood, sweat and tears paid to the Nazi scientists and technicians that our flight out into the Universe was possible. The escape into space had its beginnings in the burial of the living dead of Dora, who used to dream of impossible escapes.'

JEAN MICHEL, *DORA: THE NAZI CONCENTRATION CAMP WHERE SPACE TECHNOLOGY WAS BORN AND 30,000 PRISONERS DIED* (1975)

13.9 Wernher von Braun, 'Man on the Moon: The Journey', *Collier's* magazine, 18 October 1952, illustrated by Chesley Bonestell, *Assembly of the Moonships 1,075*

13.10 Cover of *Life* magazine featuring von Braun and a model of the moon rocket he designed, 18 November 1957.

GERHARD ZUCKER

14.1 Gerhard Zucker showing his 'rocket' at the International Air Post Exhibition, 1934.

14.2 Supplementary issue of *Die Rakete* in 1927 (Around the Earth in an Hour and a Half), the Verein für Raumschiffahrt's periodical.

14.3 Cover of *Popular Science*, July 1932.

14.4 Austrian engineer Max Valier in his rocket car, propelled with carbonic acid, in Burgwedel, near Hanover, Germany, 1930s. The car exploded after 50 metres (164 feet).

The German rocket craze of the 1920s and '30s found its way into some unusual places. None more so than the tiny Scottish island of Scarp, a stone's throw off the west coast of the Hebridean island of Lewis and Harris, now uninhabited,* but which used to be home to one of the most isolated communities in Britain. In 1934 there were ninety-five families living there, crofting, fishing and making Harris tweed. Conditions were harsh on this exposed granite outcrop. There was no running water or electricity, and crucially, no telephone. On the night of Sunday, 14 January, island resident Christina McLennan gave birth to a daughter. Health complications soon followed, and a fellow islander was quickly dispatched across the difficult half-mile stretch of water that separates the tiny island from Hushinish, a promontory on the southern end of Harris. But the Hushinish telephone was out of order. The son of the local postman drove 24 kilometres (15 miles) down a single track to Tarbert, the main town on Harris, to call a doctor. The doctor advised an immediate transfer to the nearest hospital, and so Christina was duly taken off Scarp across the rough water, strapped down to a makeshift stretcher fashioned out of an old door.

---

* A quick check on Rightmove tells me there's a house for sale on it. Tempting.

By Monday the cause of her condition was established: a second daughter was born. Two healthy twin girls who became known as Miss Lewis and Miss Harris, born in different places and in different weeks. The story reached the attention of young, charismatic German rocket enthusiast Gerhard Zucker, who immediately saw an opportunity – one that would turn a profit for himself, and one that would put Scarp on the map.

Herr Zucker was born in 1908 and grew up in the Harz mountains of northern Germany. His father owned a dairy business and helped fund his rocket experiments, inspired as he was by the exciting rocketry craze, Max Valier and the Verein für Raumschiffahrt.

The difference was that Zucker was neither a scientist nor an engineer. But what he lacked in technical rocket-building know-how he more than made up for in enthusiasm and creative business acumen. Germany was on its knees after the First World War, and the Wall Street Crash of 1929 reverberated through its economy. Bankruptcies were rife and levels of unemployment were catastrophic. Rockets were starting to generate interest not just in the context of space travel but as a military application to circumnavigate the Treaty of Versailles, which forbade German rearmament. There's a photograph taken on April 1933 of a twenty-four-year-old Gerhard Zucker immaculately dressed on Duhnen beach, near the town of Cuxhaven, standing next to his homemade 'air torpedo', a 4-metre (13-foot) rocket which he claimed was the largest rocket ever built. In reality it was not much more than a hollow metal tube powered by a handful of fireworks. A far cry from the serious business of German rocketry that would soon change the world. For better or for worse. The rocket launched 15 metres (50 feet) to the delight of the spectators before crash landing in the mudflats.

It wasn't bomb delivery or space flight that was foremost in his mind. Rather the use of rockets as a way of connecting hard-to-reach communities, and as an intriguing way to make some money. Rocket Mail as an extension of Airmail was an idea that had been touted for years, but for Zucker it was a way to make a profit in the area of 'philately' (stamp collecting). Anything that was connected to the rocketry craze (however tenuous the rocket), particularly if it was *flown*, was highly sought after, just as

it is today. Zucker would commission his own beautifully designed stamps and labels that could be attached to postal covers (big business for collectors and dealers), which he would fly on his publicized rocket demonstrations and then sell on.

In May 1934, after some test launches in Germany and around Europe, he was invited to London for the International Air Post Exhibition (APEX). He met a fellow German named C. H. Dombrowski, a stamp-collecting impresario who agreed to help finance his scheme and start

14.2

14.3

14.4

14.5 Stamp created for Zucker's Rocket Post, 1934.

14.6, 14.7 Rocket Mail labels by Gerhard Zucker for the 1934 APEX event in London.

14.8 The Rottingdean trial, *Daily Express* clippings, 7 June 1934.

14.9 Detail of the author's own Scarp/Harris envelope, featuring the commemorative frank.

an official UK-based rocket post company. On 6 June 1934, England's first Rocket Mail 'service' was launched on a hilltop on the Sussex Downs. There were six spectators in attendance – but most importantly a *Daily Express* journalist, to ensure maximum publicity. A thousand or so letters were adorned with Zucker's stamps, including six letters addressed to King George V seeking his seal of approval for the venture. The aluminium rocket was about 1 metre (3 feet) long with graceful stabilizing fins and a spring in the nose cone to cushion the blow on impact. But it was fuelled with the standard black firework powder, which he had had to source in the UK,† and launched along an inclined 4.5-metre (15-foot) metal 'rack' that was lubricated with butter. After a successful test launch, the letters were stuffed in and the rocket flew again. It roared into the sky and flew for a kilometre or so before falling to Earth. Like all of Zucker's rocket flights, performance was secondary to aesthetics. The letters were collected from the wreckage, taken to the Brighton Post Office and posted in the normal manner. No word followed from the king, but the story was duly publicized in the press.

And so to Scotland. In July that year, the team made their way to the Outer Hebrides for two demonstration postal service launches from Scarp to Harris. This was supported and attended by Sir Thomas Wilson Ramsay, the MP for the Western Isles whose job it was to look after the interest of its communities. Representatives of the press were also there, keen to witness the spectacle.

On 28 July the rocket was set up on the shore, with even more postal covers crammed into its body and nose cone than in the Sussex demonstration. After ignition, the rocket simply exploded, scattering the letters like confetti all over the beach. A few days later on 31 July, another attempt was tried, this time the other way round from Harris towards Scarp. But the result was the same. What seemed like abject failure for the scheme turned out to be highly profitable. The scorched, blackened and smoke-damaged letters with their commemorative stamps were branded with a dramatic extra: a 'damaged by explosion' frank on the back. To this day they are highly prized by stamp and space memorabilia collectors eager to part with their money.‡

After the infamous Scarp/Harris demonstrations, Zucker continued with his scheme on the Lymington golf course in Hampshire and then on to the Isle of Wight, before being deported back to Germany for reasons of national security. Over the years, much of Zucker's story has been further embellished with stories of spying, treason and fraud. He certainly spent a spell in prison in Germany in the 1960s for involuntary manslaughter following a lethal explosion during one of his demonstrations. But his passion for

---

† The export of such material from Germany was strictly *verboten*.

‡ I'm looking at my own one right now.

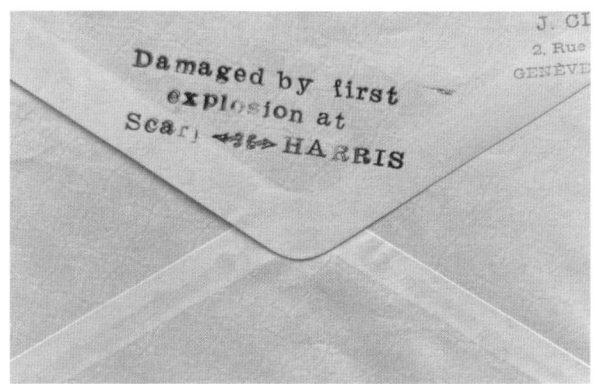

rocketry continued through his life. He served in the Luftwaffe in the Second World War before being invalided, and went on to work as a furniture dealer until his death in 1985.

As a rocket scientist, Zucker left a lot to be desired. As an entrepreneur who could spot an opportunity he had considerably more success. With his good looks and fine suits, he knew how to put on a show. A reminder that a diverting story and a healthy dollop of publicity (good or bad) can get you surprisingly far.

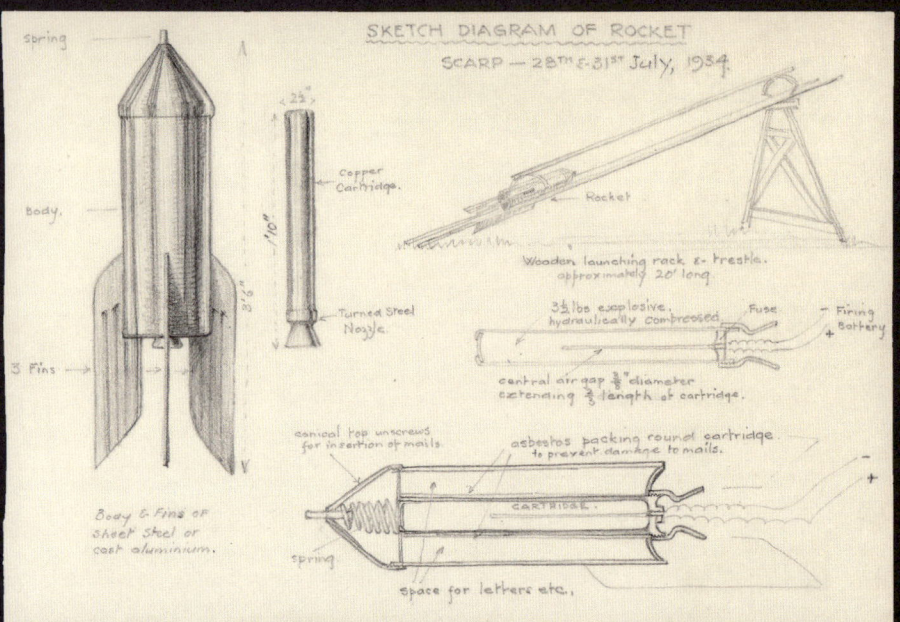

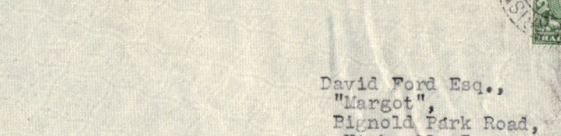

START OF UNSUCCESSFUL MAIL ROCKET TEST

The mail rocket invented by Herr G. Zucker (lying on ground and inset, left) being fired from Lymington golf course yesterday, carrying letters. It was hoped to reach the Isle of Wight, but the rocket fell on the mainland a mile away.

14.10 Gerhard Zucker, 'Sketch diagram of rocket', 1934.

14.11 Lymington trial, *Daily Telegraph* clipping, December 1934.

14.12-14.14 Burnt envelopes from failed trial, Island of Scarp to Harris, 1934.

14.15 Scarp to Harris rocket mail explosion newspaper clipping.

TENSE MOMENT.—A remarkable picture showing the explosion which occurred at the attempt to send off the first rocket at the Scarp to Harris rocket mail test. The effort miscarried, and thousands of letters were thrown into the air. Herr G. Zucker, the inventor, is seen holding the electric cable which set off the powder.—" Daily Record " photograph.

THE CHIEF DESIGNER

An Orthodox priest in full regalia blesses the Soyuz with holy water as it stands on the launch pad. As the rocket climbs, gaining speed and altitude, its four spent strap-on boosters succumb to gravity and fall away from its central core, forming a crucifix pattern in the sky for the observers on the ground. It's a phenomenon known as the 'Королевский крест', in honour of the man whose engineering legacy underpins human space flight. For years the identity of the architect of the grand Soviet space programme was hidden behind the Iron Curtain. He was known only as Главный Конструктор, 'The Chief Designer'. So protected was his identity that some in the West wondered if he was an individual at all. Perhaps the Chief Designer was a group of people? It was only after his death that the name of the figure responsible for so many Soviet space firsts was finally revealed.

Born in 1907, by his late teens the Chief Designer was studying aeronautical engineering in Moscow under the tutelage of legendary aircraft designer Andrei Tupolev. His interest in rocketry and the possibilities of space flight began to develop, and with a dedicated band of rocket enthusiasts known as the Group for the Study of Reactive Motion (GIRD) he began experimenting with liquid-fuelled rockets in forests near the capital. The group's successes caught the eye of the Red Army, and attentions were turned from ideas of space travel to weapons production. In 1938, during Joseph Stalin's Great Purge, he was arrested at his apartment by the secret police and severely beaten. He was denounced by some of his fellow engineers and imprisoned in the Lubyanka building on a string of trumped-up charges. There he was tortured before being sentenced to ten years in a Soviet Gulag, where he contracted scurvy and lost all his teeth, had his jaw broken, and is thought to have suffered a heart attack. The incarceration and deaths of many of the leading lights of Russian rocketry meant that research stagnated, falling behind the progress of the Germans. As war approached and engineers were increasingly needed, he was released early and moved to a prison for intellectuals and academics.

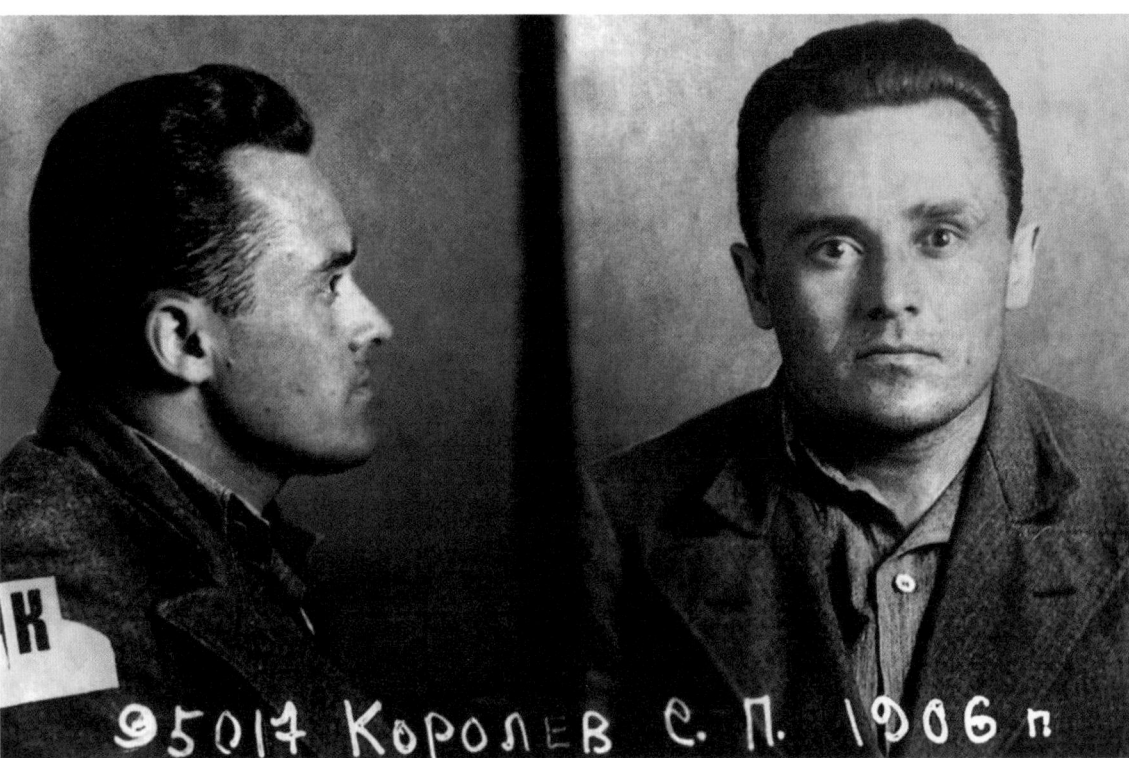

15.1

Immediately after the war, the last remaining German V-2 rocket scientists who hadn't been spirited away to America were relocated to the Soviet Union. Work began on reverse-engineering the recovered remnants of the V-2 missiles in a new Soviet missile programme. The aim was to build an Intercontinental Ballistic Missile (ICBM) that would have the power to lift a nuclear warhead across the North Pole onto American soil. But the Chief Designer saw the potential of such power far beyond designing weapons: he saw the scientific advantage and the national prestige of using a rocket to conquer space. In August 1957 the R-7 Semyorka, the world's first ICBM,

15.1 Mugshot of the Chief Designer after his arrest by the NKVD in 1938.

15.2 The first Soviet rocket with a liquid propellant engine, GIRD-X, was first launched in 1933. The rocket had been designed by a team of engineers under the direction of outstanding Soviet scientist Fridrikh Arturovich Tsander. Despite the death of Tsander on 28 March 1933, the project was completed by the team.

15.3 Colour lithographed poster after V. Victorov, Moscow, 1958: *Otchizna! ... Slava nauke, slava trudu! Slava sovetskomu stroju!* (Homeland! Glory to science, glory to work! Glory to the Soviet system!)

15.4 Soviet technician working on Sputnik 1, 1957.

15.5 A replica of the Luna 2 Soviet 'Easter egg', designed to explode on the Moon's surface spreading its individual pentagonal elements over a wide area.

15.4

was unveiled, standing nearly 35 metres (112 feet) tall with its central core stage flanked with four strap-on boosters. It was tested at the Baikonur Cosmodrome, a newly established rocket base deep in the remote steppe of Kazakhstan, far away from prying Western eyes. After a handful of failed attempts, the first artificial satellite, Sputnik 1, was successfully launched on 4 October 1957. The polished beachball-sized metal sphere with its four spidery antennae was released into orbit, falling around the Earth sixteen times a day at around 40,000 km/h (25,000 mph), broadcasting its presence with its famous *beep, beep, beep* radio signal. The space race had begun. In the deepening Cold War it was Russia who had claimed the technological high ground, deftly signalling that there was now nowhere on Earth beyond the reach of Russian influence. The Chief Designer was twice offered the Nobel Prize for his groundbreaking work, but was forced to turn them down in order to preserve his anonymity. The rocket engineer's work is forever tethered to the march of politics, culture and conflict.

Over the next decade the Chief Designer oversaw the golden age of Russian space history, a series of dramatic firsts that established their dominance in space and kept the Americans on the back foot: the first man and woman in space, the first dogs in space, the first space walk by an astronaut, the first two- and three-person crews. The Soviets, like the Americans, were also looking to the Moon as their next great challenge. In 1959 Luna 1 became the first object to leave the Earth–Moon system and Luna 2 became the first man-made object to touch the Moon itself. On board the spacecraft was a football-shaped 'pennant' made up of seventy-two titanium petals. Each one had the Cyrillic CCCP (USSR) stamped on it with the date 'September 1959'. The object was designed to explode on lunar impact, scattering the petals far and wide: the first flag on the Moon was a Soviet one.

With the Americans fully committed to President Kennedy's call to land a man on the Moon before the end of the 1960s, the Soviets embarked on their own secret plan. At its heart was the building of the N-1, a giant 30-metre (100-foot) rocket – the Soviet equivalent of the Saturn 5 – which relied on thirty engines all working simultaneously to get it off the ground. Each of its four test flights ended in failure, but it remained the most powerful object ever to fly until SpaceX's Starship fifty years later (fig. 0.4). The Soviet lunar-landing programme was eventually cancelled, a product of severe underfunding as well as the shifting political landscape, and it remained completely hidden until the collapse of the Soviet Union.

The invisible Chief Designer died while undergoing surgery on 14 January 1966, aged fifty-nine. On board the Russian segment of the International Space Station, among the icons and crucifixes, there are three photographs of the heroes of the Soviet space programme – the Trinity of Russian space history. The first is Konstantin Tsiolkovsky, looking every inch the prophet of the space age. The second is Yuri Gagarin, the first cosmonaut, smiling broadly and clutching a white dove. The third is the Chief Designer. The man who made it all possible: Sergei Pavlovich Korolev.

15.5

15.6 'Russia Launches First Earth Satellite 560 Miles Into Sky'. Front page of the *Los Angeles Times*, 5 October 1957.

15.7 *Daily News* (New York), front page depicting Russian cosmonaut Alexei Leonov, the first man to walk in space, 1965.

15.8 *Daily News* (New York), page 3: 'A 10-Minute Stroll in Outer Space'.

DAILY NEWS

FINAL

NEW YORK'S PICTURE NEWSPAPER®

10¢

Vol. 46, No. 229 New York, N.Y. 10017, Friday, March 19, 1965★ WEATHER: Mostly sunny and cool.

REDS SHOW OFF SPACE FLOATER

He Steps Outside for Cameras

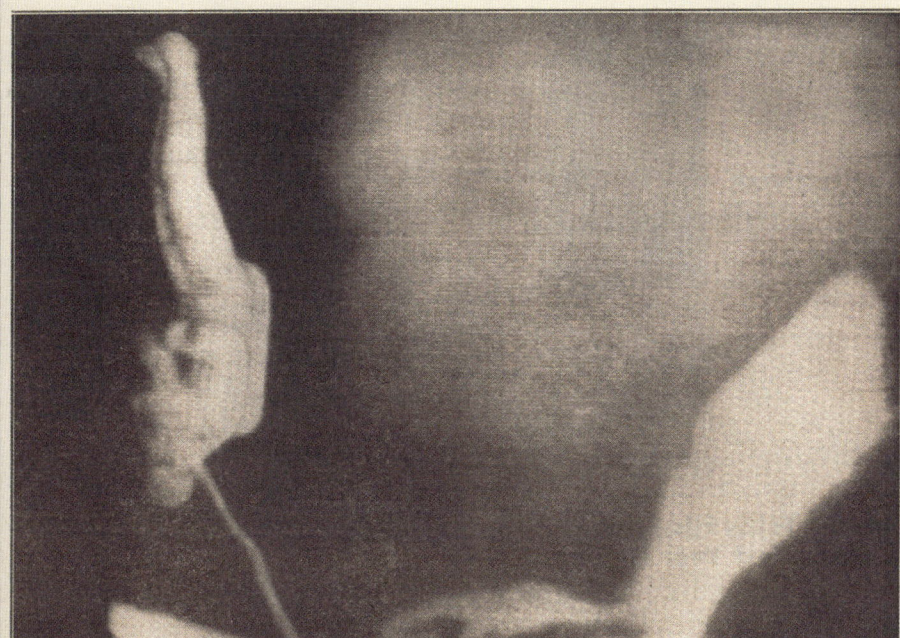

Doing the Swim in Space. Tethered to spacecraft, Soviet cosmonaut Lt. Col. Alexi Leonov floats hundreds of miles above the earth while traveling 17,500 m.p.h. Pilot, Col. Pavel Belyayev, remained at controls of orbiting capsule while Leonov took mankind's first "swim" in outer space. —Story p. 3; other pics. centerfold
(UPI Cablefoto from Moscow)

A 10-Minute Stroll in Outer Space

Col. Belyayev (l.) and Lt. Col. Leonov in their space helmets.
(Associated Press Cablefoto from Moscow)

Russian Takes Movies At 17,500 m.p.h.

Moscow, March 18 (UPI)—A Soviet astronaut stepped into space today and took a history-making flight at 17,500 miles an hour. His feat put the Soviets months ahead of the U.S. space program.

Television cameras on the Soviet spaceship Voskhod II flashed to earth pictures of Lt. Col. Alexei Leonov, 30, as he squeezed out of a hatch and appeared to float slowly away from the speeding capsule in his silver space suit.

Secured to the spaceship by a long lifeline, the spaceman stood on his head, did a near-somersault, took motion pictures and peered down at the earth hundreds of miles below him.

He stayed outside the capsule about 20 minutes, Tass reported, and spent 10 minutes floating free as far as 16 feet from the ship. A fellow astronaut, Col. Pavel Belyayev, 39, remained at the capsule's controls.

Moscow, March 18 (Reuters)—Pavel Belyayev, 39, commander of the new Voskhod spaceship, is the oldest cosmonaut yet launched into space.

Born June 26, 1925, he is nearly 10 years older than his space companion, Alexei Leonov, born May 30, 1934.

Both are members of the Communist Party, both are married and both are parachutists.

Belyayev was born in the Vologda region of northern Russia, and Leonov in the Kemerovo region.

After finishing school in 1942, Belyayev worked as a lathe operator in an arms factory, but joined the army a year later.

He was graduated from flying school in 1945 in time to be a fighter pilot in the war against Japan.

He and his wife, Tatiana, have two daughters, Irina, 15, and Ludmila, 10.

Leonov is a parachute instructor by training and is studying aviation engineering.

Moscow television said his father, Arkhip Leonov, 73, and his mother Yevdoria, 70, live on a pension in Kaliningrad with some of their nine children.

Leonov and his wife, Svetlana, have one daughter, Viktoria, 4.

U. S. Try in 1966

The United States does not plan to match Leonov's float in space until the fifth Gemini flight, sometime in 1966. The first U.S. two-man shot is scheduled for next Tuesday.

There was speculation Leonov's walk was aimed at preparing for a linkup of two orbiting craft on a future flight.

Every spaceman risks his life going into orbit, but Leonov, an engineer-designer with a wife and 4-year-old daughter, took the greatest risk to date.

If the lifeline had snapped, Leonov probably would have slipped away from the capsule and gone drifting to his death in space.

Outer space, with no atmosphere, gives him nothing to push or pull against to get back. The same lack of resistance made it possible for him to withstand the 17,500-mile-an-hour speed of the capsule.

New Altitude Record

Leonov's space float overshadowed other details of the flight. The capsule set a new altitude record. Its orbit took it to a peak distance of 307 miles, the highest humans have ever gone. The old record also was held by the Russians—255 miles in last October's triple flight.

The Voskhod (Sunrise) was launched from an undisclosed site at 10 A.M. Moscow time (2 A.M. New York time) for an expected one-day flight. It was orbiting the earth every 90.9 minutes.

At 9 P.M. (1 P.M. New York

Good Show Keeps Reds In the Lead

Washington, March 18 (UPI)—U. S. officials saluted Russia's new space spectacular and readily conceded today that it helps maintain the Soviet lead in manned flight.

The White House had no immediate comment. Press Secretary George E. Reedy referred all questions to the National Aeronautics and Space Administration.

Dr. Edward C. Welsh, executive director of the space council which advises President Johnson on U.S. space programs, gave this appraisal:

"This is a logical extension of the manned space program that they have. It helps them to maintain the lead they have over us in manned space flight. It shows that they have a space suit sufficiently well developed and designed to permit men to move outside their spacecraft."

U.S. officials were impressed with the Soviet achievement even though they had been expecting it for some time.

time), Tass announced, the spaceship had completed eight orbits. The agency said the Voskhod would not pass over Soviet territory between the eighth and 13th orbits, that at least 13 orbits were planned and that the spaceship would continue in orbit at least until tomorrow.

Belyayev, commander of the ship, is a former wartime fighter
(Continued on page 6, col. 4)

U.S. Readies Own Tandem

Cape Kennedy, Fla., March 18 (AP)—While Russian space twins soared through the skies, U.S. astronauts Virgil I. (Gus) Grissom and John W. Young staged a full-scale rehearsal today for their own tandem flight scheduled for Tuesday.

They donned their silver space suits and climbed into their Gemini 3 spacecraft to run through the entire three-orbit, 4-hour 52-minute mission. The bell-shaped spacecraft is perched atop a 90-foot-tall Titan 2 rocket.

Excessive Cloudiness Seen

While the astronauts made preparations, meteorologists predicted excessive cloudiness over much of Florida for Tuesday.

If the clouds are too thick during the three-hour launching period starting at 9 A.M., the shot would be postponed, because camera coverage could not be obtained.

The weathermen predicted favorable conditions for the Ranger 9 moon probe, slated for Sunday.

Grissom and Young had no comment on the Russian spectacular in which one of the two cosmonauts stepped out of their orbiting spacecraft.

Could Support a Man

The suits that Grissom and Young will wear on their flight are capable of supporting a man outside the capsule. But because of the cautious approach of the U.S. program, the first exposure of an astronaut to the space elements is not planned until the second manned Gemini flight scheduled in June.

On that flight, astronaut James A. McDivitt is to open his hatch and briefly poke his head into space. The first full emergence is not slated until the seventh flight, late next year. Young is to make a preliminary test next week when he disconnects his suit from the main cabin support lines, pressurizes it and uses self-contained oxygen and other systems.

"QUICK, FOLLOW THAT CAB!"
(UPI Telefoto)

Cartoon in yesterday's London Evening Standard pokes fun at U.S. situation in space race.

H. H. Fowler Is Named Treasury Head

Washington, March 18 (AP)—Henry H. Fowler, a Virginia attorney, was nominated by President Johnson today to be Secretary of the Treasury. He, a former under-secretary and has also served in other government posts.

Fowler will succeed Douglas Dillon, who has said he wishes to leave the Cabinet March 31 to return to his family's banking business.

Fowler, 56, was born in Roanoke, Va., and was graduated from Roanoke College. He obtained his law degree at Yale. During World War II he was assistant general counsel of the War Production Board and economic adviser for a U.S. mission to London.

Fowler was appointed under-secretary, the No. 2 Treasury job, by President Kennedy in 1961. He resigned last spring to return to private business in Virginia.

Russian scientists in Moscow monitor signals from Voshkod II.
(Tass foto via A.P. Cablefoto)

ACT 4.

ITER STELLARUM

'STARRY JOURNEY'
SPACE TRAVELLERS

SOVIET FIRSTS

12 April 1961, Vostok 1, Tyuratam, Kazakhstan. One of the hatch lights hadn't come on – the seal wasn't secure. At the top of the service gantry, engineer Oleg Ivanovsky and his assistants were frantically swapping out the bolts in an attempt to resolve the issue. Beneath them the modified R-7 Intercontinental Ballistic Missile (ICBM), designed for launching a thermonuclear warhead, was alive, creaking and groaning, cryogenic fuel steaming, poised on the launch pad like a dog straining at its leash. Inside the spherical Vostok capsule sat the composed twenty-seven-year-old First Lieutenant Yuri Gagarin (1934–1968), the handsome son of a carpenter and former apprentice foundry worker turned pilot. In his baggy orange spacesuit and oversized white helmet, he cheerfully whistled along to the music that he'd asked to be piped over the radio: 'Rodina slyshit', or 'The Motherland Hears', by Dmitri Shostakovich. His heartbeat on the electrocardiogram was reading a steady 64 beats per minute. A contrast to the lifeless mannequin who had completed the dress-rehearsal launch the month before. His instructions were simple. Observe. Don't touch anything. Remain alive. There was a lot that could go wrong technically and no certainties as to how the human mind would react to such an alien environment. Who knew what horrors might be in store for the first to cross that threshold? What effect such a profound journey might have? Seeing the Earth from space or being confronted with the infinite void of our galaxy could surely lead to some sort of cosmic insanity. At 9:07 a.m. Moscow time, with the hatch issue resolved, the rocket's first stage roared into life and began its eastward ascent, casting an eerie black shadow over what would soon be renamed the Baikonur Cosmodrome. 'Poyekhali!' was the first word spoken by Gagarin over the crackling radio: 'Let's go!' From this moment, human space flight was to leave the realm of fiction and fantasy and enter the world of politics and propaganda. That day he would break every record, flying higher, faster

16.1

16.1 Soviet cosmonauts during television show broadcast from the central TV station in Moscow, June 1963 (L-R): Vladimir Komarov, Yuri Gagarin, Valentina Tereshkova, Valery Bykovsky, Andriyan Nikolayev and Pavel Popovich(?).

16.2 First cosmonaut Yuri Gagarin at the Baikonur Cosmodrome, Kazakhstan, 1961.

and further than anyone ever had before. A human cannonball falling around the Earth like Isaac Newton's famous thought experiment (fig. 6.7).

The Soviets were leading the early space race for two important reasons: first, they had developed more powerful rocket boosters than the Americans. The boosters were needed to lift their heavier nuclear warheads, but they also made getting satellites, dogs and people into orbit much easier. Second, every aspect of the Soviet space programme was conducted in a thick fog of secrecy, with only state-controlled news of the great successes leaking to the outside world. They, on the other hand, could see America's space ambitions, successes and failures in full view. Working out how to steal NASA's thunder was easy – all they had to do was read the newspaper or tune in to the radio. After Gagarin's pioneering orbit the Soviet Union continued their pursuit of space firsts, looking to maintain their advantage. In 1963 Valentina Tereshkova (b. 1937) became the first woman in space, completing forty-eight orbits in her three-day mission in Vostok 6. It was so classified that her own mother had no idea her daughter was in space until her neighbour passed on the triumphant news after hearing it on the radio. The Vostok cosmonauts even had

16.3-16.9 Work being done on the Vostok 1 capsule in preparation for Gagarin's historic flight, 1961. Stills taken from a Soviet film about their space programme.

their own arboreal-themed code words in case of prying ears listening in to radio communications, such as palm tree ('I'm not feeling well'); silver fir ('Emergency descent request'); and raspberry ('My menstrual cycle has started'). Tereshkova's mission, which earned her the Hero of the Soviet Union medal, also suffered from a list of technical issues; all of them were hidden from the world for decades, but all of them were trivial compared to the 1965 Voskhod 2 mission undertaken by Alexei Leonov (1934–2019).

Leonov conducted the first space walk, leaving the two-person capsule to float above the Earth connected to the spacecraft by an umbilical cord. It almost ended in disaster. The ballooning of his pressurized spacesuit meant the suit stiffened. He lost all mobility and couldn't get back into the inflatable tunnel-shaped airlock. Trapped outside, he was forced to partially deflate the suit in the vacuum of space in order to regain movement. He almost didn't make it. But things were set to get worse as their re-entry sent them hundreds of miles from their designated landing area into a remote, densely forested region of Russia, known for its freezing temperatures and population of wolves and bears. Luckily, Leonov and his crewmate Pavel Belyayev were trained in wilderness survival (and equipped with a pistol).

Gagarin began his return to Earth as he passed over the west coast of Africa. In a single orbit (just under two hours), he'd taken in the oceans and continents crossing into night and back to day, Shostakovich lodged in his brain. But there was a problem: the two modules that made up Vostok 1 failed to separate properly, sending the spacecraft into a wild spin until the thick cabling that joined them eventually burned through in the fiery heat of re-entry. But the Vostok capsule had no capacity for a soft landing with its crew on board. Instead, as the module was falling through the lower atmosphere under its parachute, the hatch was jettisoned and the cosmonauts were ejected, falling to Earth under their own canopy. Gagarin landed in a field near Engels in Saratov Oblast – to the shock of a woman and her five-year-old granddaughter who were there planting potatoes – physically safe, and an instant hero.

The threads of history are delicate. Gagarin could have easily perished that day, his name forever lost in state secrecy. Instead he became a living icon, a symbolic monument to a world consumed by image and ideology. The flight launched him to quasi-religious status, the embodiment of Russian Cosmism. Man becomes Spaceman. But it could have been very different if the first US astronaut, Alan Shepard, had flown on 24 March 1961 as intended. American caution resulted in an extra uncrewed rocket test that day. Shepard finally flew his suborbital mission in the Project Mercury capsule Freedom 7 on 5 May 1961 (fig. 21.6), the wind taken out of his sails by the sensational news of Gagarin's orbit.* Soviet dominance in space had set the world alight.

---

* But news which at least helped motivate the American decision to aim for the Moon.

In the end, Yuri Gagarin's single flight in space didn't drive him mad, but global fame brought its familiar trappings. How could it not? Being the first is a currency that holds its value for an eternity, and becomes the price of immortality.

16.10 Gagarin and Tereshkova on a Soviet souvenir postcard. This postcard, which has the caption 'Visit the USSR', was issued in 1963 to promote tourism.

16.11 Colour poster issued by the Soviet State Publishing House of Decorative/Fine Arts in Moscow, Russia, to celebrate Tereshkova. The slogan says 'Glory to the first woman cosmonaut!'

16.12 Alexei Leonov, photographed during training in Moscow for his Voskhod 2 flight, 22 March 1965.

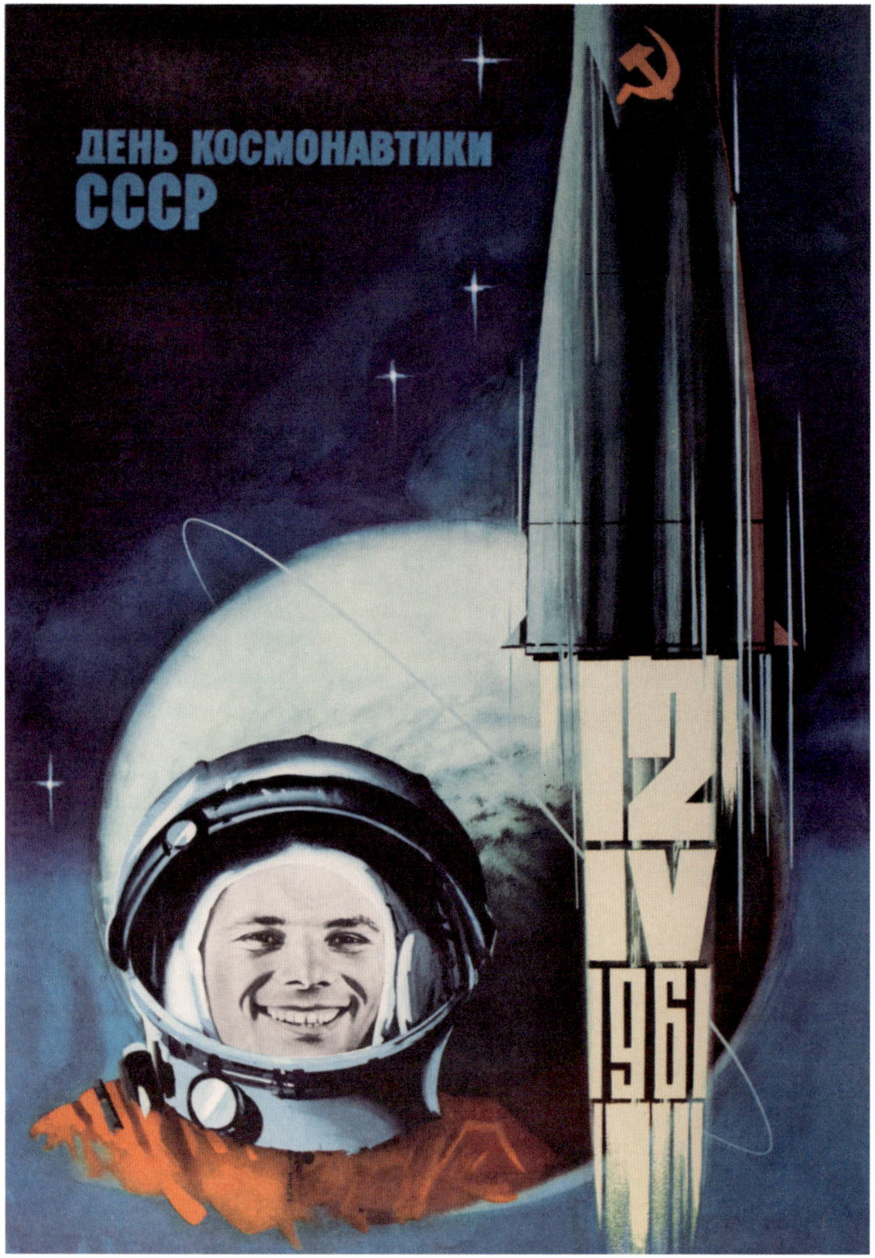

EXTRA

A WALK IN SPACE

And a Russian Takes It

New York Journal American
NEW YORK'S LARGEST EVENING NEWSPAPER
No. 27,930—DAILY THURSDAY, MARCH 18, 1965 10 CENTS
LATEST NEWS
(LATEST SCRATCHES)

Steps Off Spacecraft Carrying 2

The Russians Who Made History—Story and More Photos, Manned Flights Up to Now—Page 2. U.S. Space Timetable: Picture Story—Page 23

From The Journal-American's World-Wide News Services

MOSCOW, March 18—The Soviet Union claimed another great stride in the space race today, announcing that a cosmonaut for the first time had stepped out of a two-man space ship in orbit and then returned to it.

It was like taking a stroll in space.

Soviet television stations showed a film of the feat with the curve of the earth and the blackness of deep space visible in the background.

Tass, the Soviet news agency, announced that Lt. Col. Alexei Leonov, 30-year-old co-pilot of the spaceship Voskhod-2, went 5 meters (16.4 feet) from the capsule. He apparently was attached to it by a line or cable.

The official announcement said he made observations and then returned safely to the ship. He was wearing a special "autonomous life-support" suit, the announcement said.

LT. COL. ALEXEI LEONOV

Completed One Orbit

With him in Voskhod-2 was Col. Pavel Belyayev, 39, the flight commander. Tass said a powerful rocket sent them into orbit at 10 a.m. (2 a.m. N. Y. Time) and Leonov stepped from the capsule 90 minutes later, at the end of the ship's first orbit of the earth.

"He successfully carried out the range of prescribed studies and observations and safely returned to the ship. Outside the ship and after returning to the ship, Leonov feels well," Tass said.

"Leonov's emergency into outer space, his work outside the ship and return to the ship were transmitted to the earth by means of the ship's television

Turn to Page 2, SPACE

THE TWO ORBITING RUSSIANS LOOK ABOUT LIKE THIS. (ARTIST'S CONCEPTION OF OUR OWN TWO-MAN SPACECRAFT). CHART BELOW

NASA Photo from AP

Subway Eyewitness

By DONALD R. FLYNN and MIKE PEARL
Journal-American Staff Writers

Brooklyn police and the District Attorney's office know the identity of an eyewitness in the subway murder of 17-year-old Andrew Mormile, The Journal-American learned today.

The witness was one of the 10 to 12 persons in the first car of an IND 8th ave. subway train when young Mormile was killed, allegedly by towering Christopher Lynch, 17, who was charged with the murder yesterday.

Detectives hope to bring the witness in to give a statement today, said a source in the Brooklyn District Attorney's office.

This new witness' testimony would "clinch the case against Lynch," the source said.

Police already have one witness, 18-year-old Terry Toomer of 1221 Jefferson ave., Brooklyn, a pal of Lynch's who said he was there when the Mormile boy was killed.

This new witness "was not a companion of Lynch and Toomer," the source in the District

Turn to Page 6, SUBWAY

U. S. Space Rehearsal

CAPE KENNEDY, Fla., March 18 (AP).—A second full-dress rehearsal for America's first attempt to put two men into earth orbit faces Astronauts Virgil Grissom and John Young here today.

This time the astronauts will go through the motions of their scheduled March 23 launching with two Soviet cosmonauts overhead, zipping around the earth every 90 minutes in the spaceship Voskhod-2.

Apparently to steal the thunder of world publicity from the launching of America's two-man Gemini spacecraft, the Russians put Voskhod-2 into orbit at 2 a.m. (New York time) today.

The United States has successfully orbited two unmanned Gemini space capsules with the Titan-2 rocket from Cape Kennedy.

Major Grissom, an Air Force officer, and Comdr. Young, a Navy Lieutenant Commander, are scheduled to make three orbits in the capsule next Tuesday, the first of three manned Gemini flights slated this year.

Unlike any previous space feat by the United States or Russia, Maj. Grissom, the flight commander, is to maneuver the

Turn to Page 2, GEMINI

INDEX AND WEATHER ON PAGE 2 TODAY

J-A SPECIALS TODAY: Our Air-Odds For Cleaning It (Page 11); Answers to Taxpayers' Toughest Questions (Page 24); More of Pope John's Journal (Page 9)

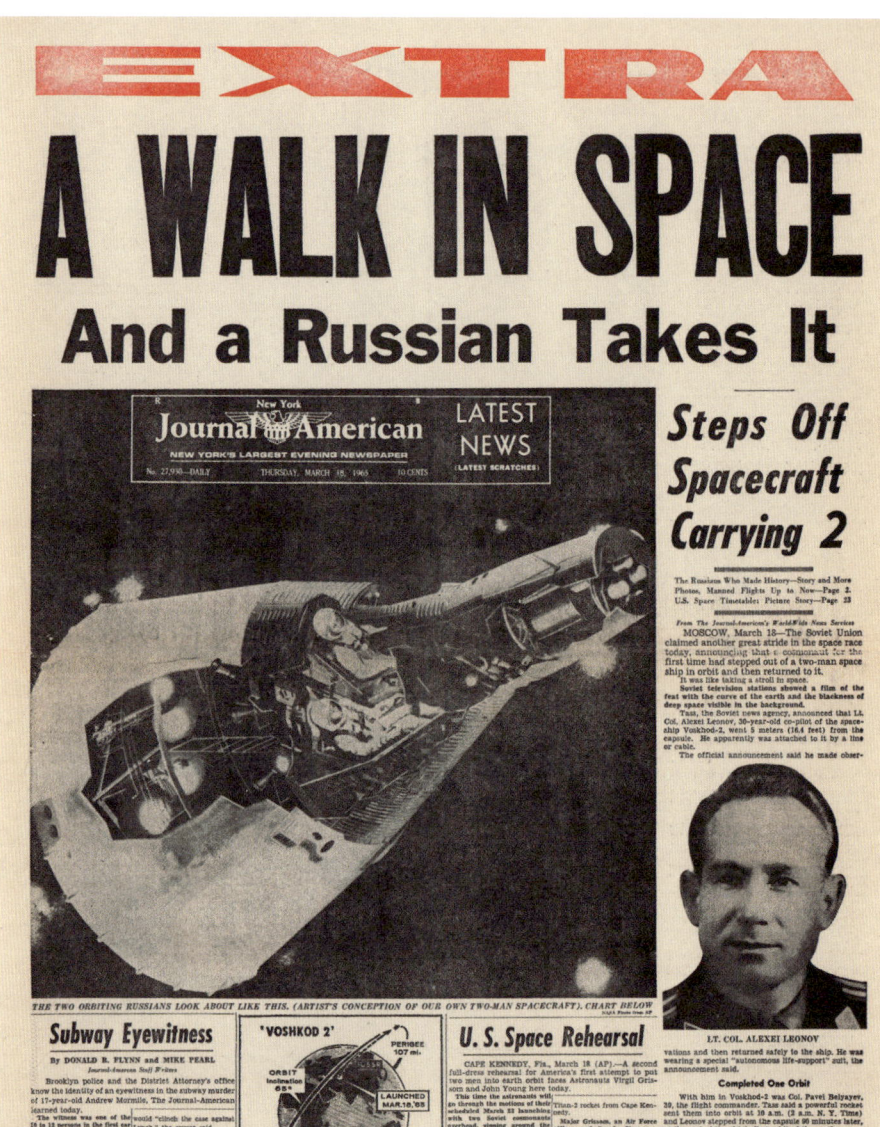

16.13

16.14

'The Motherland hears,
The Motherland knows,
Where her son flies
in the clouds …'

DMITRI SHOSTAKOVICH,
'4 SONGS, OP. 86: NO. 1,
RODINA SLYSHIT' (1951)

16.13 Poster celebrating the first crewed space mission, featuring Gagarin.

16.14 'A Walk in Space - And a Russian Takes It', *Journal American* headline following Alexei Leonov's historic first space walk in 1965.

16.15 Leonov outside of the Voskhod 2 spaceship during the world's first ever space walk.

16.15

16.16 The descent vehicle of Vostok 1 spacecraft, seen at the landing site 700 kilometres (435 miles) southeast of Moscow on 12 April 1961 after the first crewed space trip. The capsule landed empty as Gagarin had made a parachute jump at an altitude of 7,000 metres (23,000 feet).

16.17 Gagarin being greeted by the people of Moscow, 14 April 1961.

16.18 Gagarin poses on a beach with his wife, Valentina, and their daughter Yelena in June 1960, less than a year before he made history as the first person to travel into space.

16.19 Gagarin carrying a rocket cake with friends and family, 1964.

Yuri Gagarin's letter to his family, in case of death, 10 April 1961.

Hello, my sweet and much loved Valechka, Lenochka and Galochka,

Here I've decided to write you a few lines to share the joy and happiness I felt today. Today a governmental commission decided to send me first to space. You know, dear Valyusha, I'm so happy; I want you to be happy with me. An ordinary man has been trusted with such a big national task – to blaze the trail into space! Is there anything bigger to wish for? This is history, a new age!

The day after tomorrow is the launch. You'll be doing your usual things then. It's a very big task lying on my shoulders. I wish I had a chance to be with you for a little while before it, to talk to you. Alas, you are far away. But nevertheless I always feel you by my side.

I trust the technology completely. It will not fail. But it happens that a man falls on level ground and breaks his neck. Some accident may happen here too. I personally don't believe it will happen. But if it does, I ask all of you, but particularly you, Valyusha, do not waste yourself with grief. Life is life, and nobody is safe from being run over by a car.

Take care of our girls; love them like I do. Please, raise them not as some lazy mummy's girls, but real people who can handle anything life throws at them. Make them worthy of the new society – Communism. The state will help you do it.

As for your personal life, settle it the way your heart tells you, the way you feel right. I hold no obligation from you and I don't think I have a right for it.

This letter seems too gloomy. I don't feel like it. I hope you'll never see this letter, and I'll never have to be ashamed for this moment of weakness of mine. But if something goes wrong, you have to know it all. So far I lived an honest, rightful life; I served the people, even though this service was a little one.

In my childhood I read [legendary test pilot] Valery Chkalov's words: 'If you are to be, be the first.' Well, I'm trying, and I will try to the end. I want, Valechka, to dedicate my flight to the people of the new society, Communism, which we are about to become part of, to our great Motherland, to our science.

I hope in a few days we will be together again, and we will be happy. Valechka, please, don't forget my parents, and if you have an opportunity, give them a helping hand. Send them my warmest greetings and ask their forgiveness for my keeping them in the dark; they are not supposed to know anything.

This seems to be all. Goodbye, my dears. I embrace you all tight and kiss you.

Your daddy and Yura.

LAIKA

On the way we pass Gagarin's little cabin and the great Soyuz assembly building. Paint peels from the walls revealing a pleasing patina of yellows and blues like a photograph left too long by the window. Weeds escape from cracks in the concrete, defying gravity. Everywhere are small wiry dogs, pressed against walls or basking in sunny spots. They look happy, as if posing for a commemorative plate or cigarette packet. The first cosmonaut, 'Laika', was a stray dog from Moscow. Standing by the wire link fence we see the 'Gagarin's Start' launch pad. We can see the Soyuz rocket's breath in the cold. It will burst into life any moment, taking Tim [Peake], Tim [Kopra] and Yuri [Malenchenko] to the station. This is of no concern to our new friends, who sniff the frozen air. They've seen it all before. How beautifully appropriate that the strays, like the descendants of the canonized satellite dogs, are still here reporting for duty just in case. Or out of respect. Or perhaps they're ghosts. Obedient ghosts. Everything here is frozen in time and space and temperature. Dogs and distant camels and flies and people and train tracks and astronauts and concrete and instant coffee and murals and priests and rockets. The universe is filled with the lives of perfect creatures. Sergei and Konstantin aren't here to tell the dogs it's over, they can go home ...

AUTHOR'S JOURNAL FRAGMENT, DECEMBER 2016, BAIKONUR COSMODROME, KAZAKHSTAN

'Little dog lost to the rest of the World, up in your satellite basket curled ...' penned Denise Robins, the English romantic novelist – a notable contrast to the public outrage at the Soviets sending a dog into space. 'Why not use child murderers?' mused Lady Munnings, while various animal rights organizations implored callers who'd been jamming their switchboard to direct their fury to the Russian embassy in London instead. 'I don't understand. Dogs are supposed to be eaten not carted around through space!' was the apparent reaction of a Vietnamese farmer, as reported in an article in *Time* magazine. It was ironic that a sacrificial dog had been chosen as the first to demonstrate the survivability of orbital space travel. After the launch of Sputnik 1 in October 1957, Khrushchev was anxious for a follow up that would consolidate their early lead in the space race, and would coincide with the fortieth anniversary of the Bolshevik Revolution in November. What better replacement for the electronic metronome radio signal than a heartbeat?

'Laika', as she would be known, came from humble stock. Lyudmila Radkevich, then a young

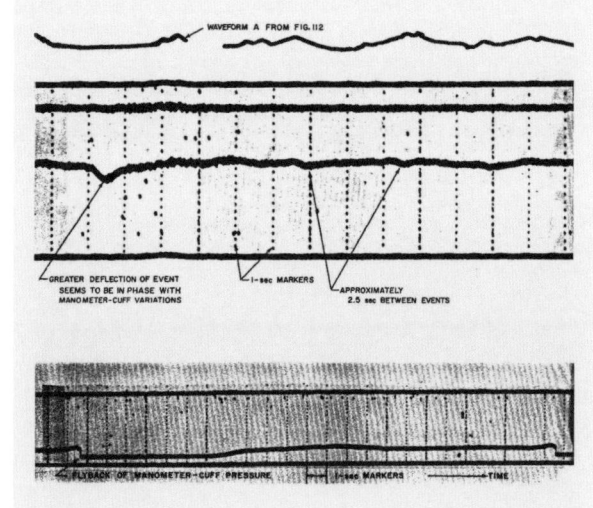

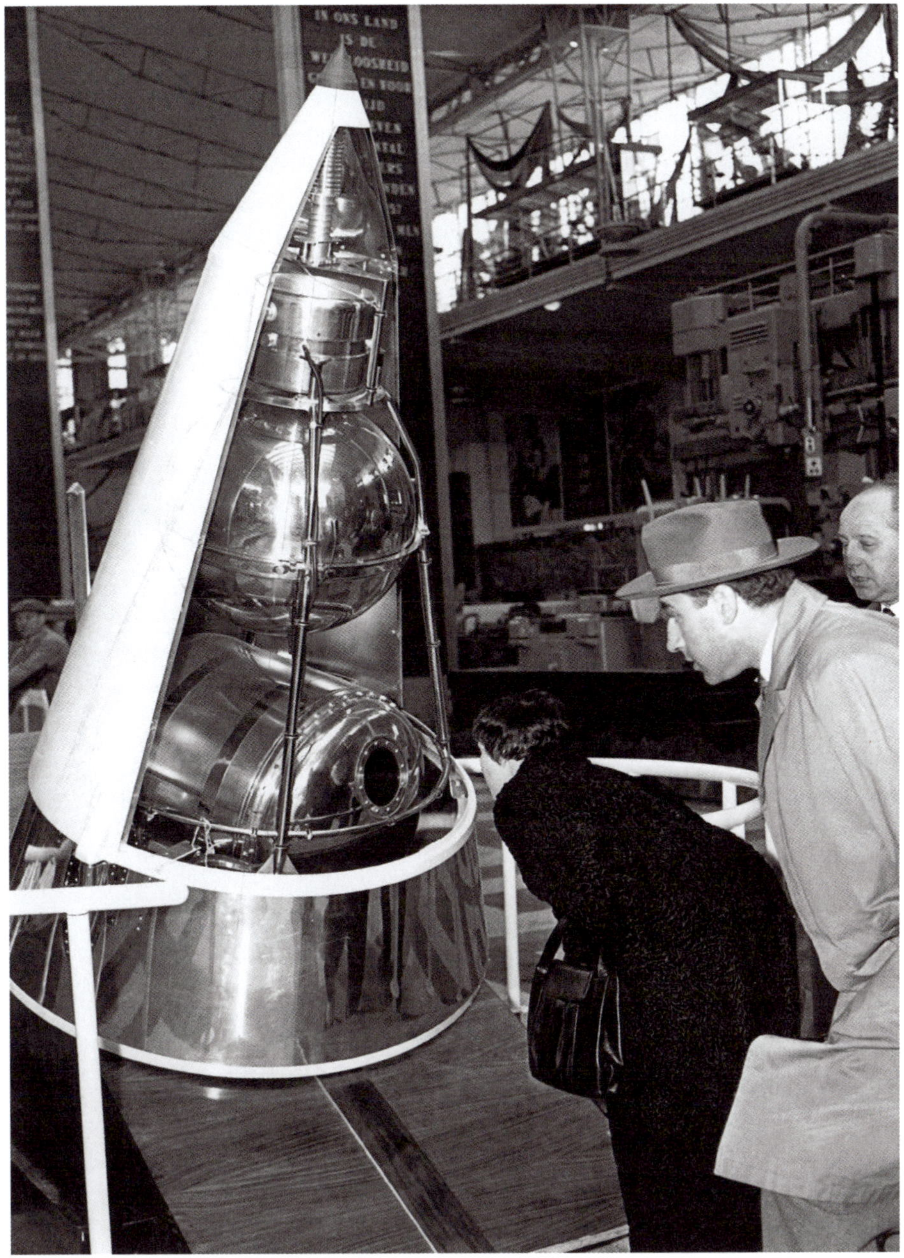

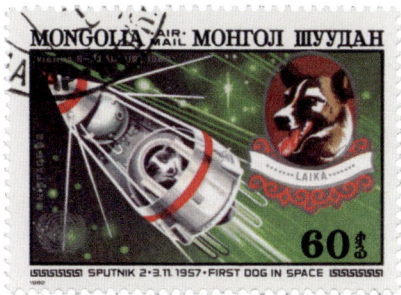

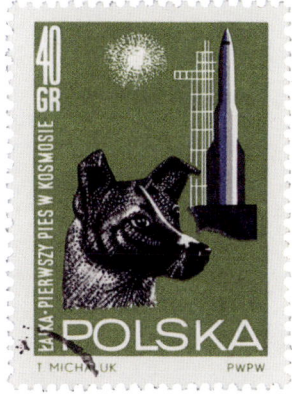

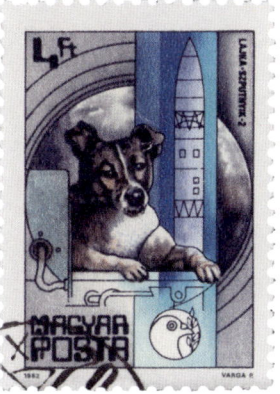

17.1 Laika, the first Soviet space dog.

17.2 The first Soviet biomedical data intercepted from space was from Laika. These are excerpts of the US tapes that recorded her heartbeat (top) and blood pressure (bottom), 3 November 1957.

17.3 A model of Sputnik 2 in the Soviet Pavilion at the Brussels Fair, 16 April 1958.

17.4-17.6 Commemorative stamps from Mongolia (c. 1982), Poland (n.d.) and Hungary (c. 1982).

17.3

17.4 / 17.5 / 17.6

researcher working at the Moscow Institute of Aviation Medicine, recalls driving around the side streets of Moscow with a ruler to measure up the stray dogs. They had to be no taller than 35 centimetres (14 inches) and lighter than 6 kilograms (13 pounds). Laika was chosen for her calm demeanour and her good looks – her small form fitted perfectly within the rocket, her dark mottled fur gave a useful contrast for the on-board photography, and her elegance and poise made her the ideal candidate to be immortalized in the Soviet agitprop that betrayed the gruesome reality of her death. She was the first of almost fifty satellite dogs to venture into the cosmos as our advanced guard. But like much of the Soviet space programme, decades would pass before details of her true fate came to light. There was never going to be a return journey, but the reports of a dignified and painless death that Soviet authorities were keen to promote were pure propaganda.

The complexities of designing a spacecraft that could sustain the life of a dog with any form of comfort were impossible to achieve in the time given to build it. No chance of curling up, it was standing room only. Laika was covered in electrodes to measure her vital signs, and a waste-collection bag was unceremoniously attached. After a three-day delay, during which she remained in the rocket at the Cosmodrome, at 5:30 a.m. on 3 November 1957 Laika began her ascension to the heavens, her heart rate spiking dramatically. The primitive cooling system didn't work properly, turning the hastily built space capsule first into an oven and then into a sarcophagus. She would have survived only a few orbits at most, returning to Earth five months later as a shooting star above the Caribbean.

The Soviet Union had their first canonized space explorer. A perfect creature that embodied the values of sacrifice and martyrdom. Enshrined in the realm of myth and legend.

17.7 Press conference with Russian 'Space Dogs' Malysjka, Linda and Kozyavka on 27 June 1957, four months before Laika's fateful journey in Sputnik 2.

17.8 'The way is open for a human being!': poster created by Konstantin Konstantinovich Ivanov of space dogs Veterok and Ugolyok, who participated in a twenty-two-day space flight, 22 February–16 March 1966.

17.9 A packet of Russian cigarettes depicting Laika and the Sputnik satellites, photographed 18 September 1958.

17.10 Laika stamps issued by Romania in 1957.

17.11

JOHN F. KENNEDY AND THEODORE C. SORENSEN

'We choose to go to the moon' has become an iconic line in political history for good reason. It's a powerful statement of intent and agency (an early NASA draft had written it in the past tense, 'chose'). It's reminiscent of the collective 'We the people' that opens the US Constitution. The Moon, which has always been a cultural symbol of the unobtainable, was now the prize for limitless American ambition. The line punches through the speech like the cymbal crash at the climax of a symphony. John F. Kennedy's virtuoso performance behind the lectern helps too. In his familiar Boston accent, Kennedy (1917–1963) pumped it out three times for emphasis over the applause of the Rice University audience, like the revving of a sports car engine. It's one of the great political speeches, rising and falling like an epic poem, filled with short simple clauses and rhetorical devices that give it a rhythm and momentum which sends shivers down the spine: *Of course we can go to the Moon!* It's built around universal themes and ideas – the allusion to outer space as a new ocean to be sailed. And comparisons are made with other great historic challenges: George Mallory climbing Everest 'because it's there'; Charles Lindbergh flying the Atlantic. And that most American of ideas, the opening of a new frontier. It's a speech about American exceptionalism in a time when the country was seen to be faltering. The shadow of Communism loomed, and America was looking to re-establish itself after Yuri Gagarin's historic first orbit and other Cold War flash points. 'We choose to go to the moon' was the perfect distillation of how a new president wanted the rest of the world to see America.

18.1 President-elect John F. Kennedy with Theodore Sorensen, right, discussing West Virginia's economic problems, December 1960.

18.2, **18.3** Excerpts from Ted Sorensen, 'We choose to go to the moon', with notes by JFK for the Rice University address, 1961.

- 8 -

strife, no prejudice, no hate in outer space. Its hazards are hostile to us all. Its conquest deserves the best of us all, and its opportunity for peaceful cooperation may never come again.

But why, some say, the moon? Why choose this as our goal? And they may as well ask: why climb the highest mountain? Why 35 years ago fly the Atlantic? *Why does Rice play Texas?* We choose to go to the moon in this decade, not because that will be easy, but because it will be hard -- because that goal will serve to organize and measure the best of our energies and skills -- because that challenge is one we are willing to accept, one we are unwilling to postpone, and one we intend to win.

It is for these reasons that I regard the decision last year to shift our efforts in space from low to high-gear as among the most important

- 9 -

decisions I expect to make in the Office of President.

In these last 24 hours *we* have seen the vast facilities now being created for the greatest and most complex exploration in man's history. *Plus* have felt the ground quake and the air shattered by the testing of a SATURN C-1 booster rocket, many times as powerful as the ATLAS which launched John Glenn, generating power equivalent to 10,000 automobiles with their gas pedals pushed to the floor. *we* have seen the site where five F-1 rocket engines -- each one as powerful as all eight engines of the SATURN combined -- will be clustered together to launch the Advanced SATURN vehicle, assembled in a new building to be built at Cape Canaveral as tall as a 48 story structure, as wide as a city block, and as long as two lengths of this field.

Kennedy's bold space ambitions were not simply about 'beating the Russians' as much as they were a political tool for extending the policy of Soviet containment. In fact, throughout his short presidency, he consistently hinted at a joint space programme with the Soviets as a strategy to soothe their vexed relationship and prevent the possible militarization of space – and to share the terrific cost. In his 1961 Inaugural Address he said, 'together let us explore the stars', and a year later, in an address at the University of California, Berkeley, he suggested that the 'sterile dogmas of the Cold War could be literally left a quarter of a million miles behind'. But it is in his address to the United Nations in September 1963 that he makes the point most emphatically:

Why, therefore, should man's first flight to the moon be a matter of national competition? Why should the United States and the Soviet Union, in preparing for such expeditions, become involved in immense duplications of research, construction, and expenditure?

Joint Soviet–American missions would happen, but Kennedy would not live to see them.

Kennedy's presidency will forever be defined by his extraordinary talents as an orator, and much of that success was down to skilful speech writing. All of his key speeches were written by his Special Counselor and closest of aides, Theodore 'Ted' Sorensen (1928–2010), whose literary style became the familiar voice of Kennedy. Sorensen had been at Kennedy's side

18.4 President John F. Kennedy speaks with Director of the George C. Marshall Space Flight Center (MSFC), Dr Wernher von Braun (left), during a tour of the MSFC at Redstone Arsenal, Huntsville, Alabama; Vice President Lyndon B. Johnson stands at right, and the Saturn C-1 rocket sits in background. President Kennedy visited the MSFC as part of a two-day inspection tour of National Aeronautics and Space Administration (NASA) field installations.

18.5 View of a static test-firing of a Saturn rocket booster at the MSFC; Kennedy and others watch from bunker in foreground.

18.6 Kennedy and astronaut John Glenn look inside the Mercury spacecraft Friendship 7, 23 February 1962, three days after Glenn's pioneering flight.

18.7 The Mercury Control Center in Florida played a key role in the United States' early space-flight programme. Located at Cape Canaveral Air Force Station, the original part of the building was constructed between 1956 and 1958, with additions in 1959 and 1963. The facility was officially transferred to NASA on 26 December 1963, and served as mission control during all the Project Mercury missions, as well as the first three flights of the Gemini Program, when it was renamed Mission Control Center.

18.8 Astronaut John Glenn entering Friendship 7 prior to the launch of MA-6 on 20 February 1962, when he became the first American to orbit the Earth. The MA-6 mission was the first crewed orbital flight boosted by the Mercury-Atlas vehicle, a modified Atlas ICBM (Intercontinental Ballistic Missile). It lasted for five hours and orbited the Earth three times.

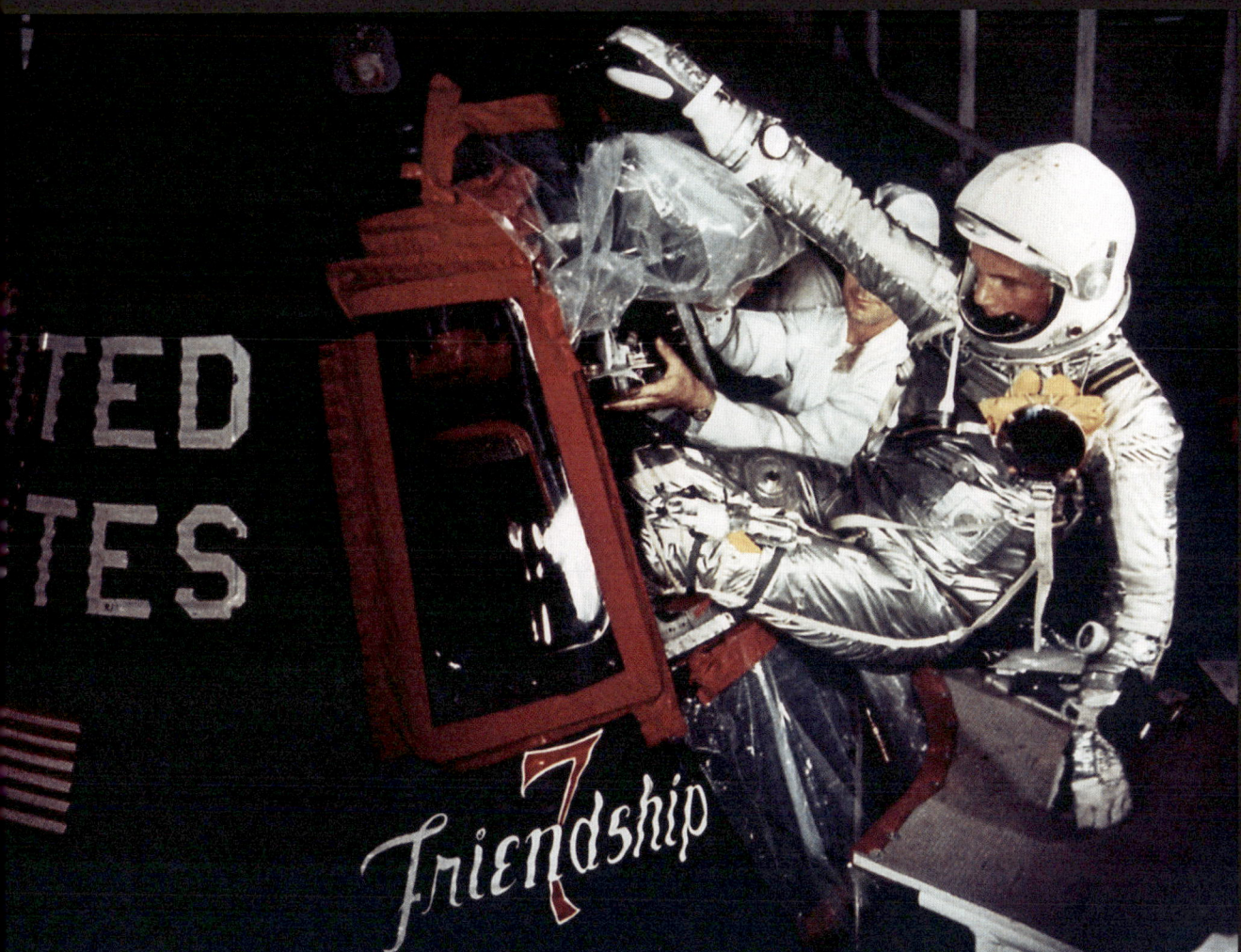

18.8

'Together let us explore the stars ...'
PRESIDENT JOHN F. KENNEDY, 1961

18.9 Robert Rauschenberg, *Retroactive II*, 1963.

18.10 President Kennedy speaks at Rice University, Texas, 12 September 1962.

since the mid-1950s, working quietly behind the scenes. As a young law graduate he had come to Washington to make a difference. Looking for his first job, he found political kinship in the Ivy League Kennedy, whose background was a million miles away from his own, and was impressed by his commitment to public service above all else. When Kennedy was a young senator, they toured every state in the Union together, and after working in such close proximity, Sorensen got to the point of being able to finish Kennedy's sentences. He was the perfect foil for Kennedy: he came from humble beginnings in rural Lincoln, Nebraska; he was a Unitarian to Kennedy's Catholicism; and he was the more progressive to the conservative-leaning Kennedy. Sorensen was also happy out of the limelight, while Kennedy played the flamboyant, charismatic political operator. Despite these differences, they were in political lockstep. Both shared a deep sense of duty and public service, as well as respect for the rule of law and the Constitution.

The secret to Kennedy's speeches was rooted in this close partnership. Sorensen would sit in on meetings to get a sense of the subject and what Kennedy would want to say. Back in his office, he'd write a draft. Kennedy would go through it and edit, add or subtract – we can see Kennedy's own pen on a draft of the 'We choose ...' speech, where he added his joke, 'Why does Rice play Texas?', a wink to the Rice University crowd and their football rivalry. The speech had originally been drafted by NASA and various other government agencies, but it's the simplicity of Sorensen's turns of phrase that have cemented its cultural legacy. The language is elevated in the tradition of great oratory – never alienating, always economical, to the point, and written for the voice to speak, not for the eye to read. In those few years in office together, Sorensen and Kennedy became a speech-writing double act, like the great music and libretto partnerships.

When Kennedy was assassinated, Ted Sorensen described it as one of the most traumatic days of his life. Despite Kennedy never seeing his space dreams achieved, it's his face that we associate with the success of Apollo rather than that of Lyndon B. Johnson, his successor (fig. 0.6), or Richard Nixon, who saw Apollo through to its conclusion. His speeches were a large part of that association, and over the decades they have helped fuel the Kennedy mythology. Sorensen never took public credit for writing them, such was his devotion to Kennedy's legacy. It was Sorensen who put the words in Kennedy's mouth, and Kennedy who breathed life into Sorensen's words.

What makes the Rice University speech so enduring? Ultimately, an effective speech comes from the ideas and conviction it presents. Kennedy and Sorensen were the great political duo who promised the Moon to the world. The power of the spoken word, perhaps more than anything else, assured that promise would be kept.

NASA

The National Aeronautics and Space Administration (NASA) opened its doors on 1 October 1958, almost exactly a year after the launch of Sputnik, the first Soviet satellite.

The launch took place during the International Geophysical Year (IGY), an eighteen-month science initiative which gathered many of the world's leading scientists to accelerate research on the Earth, the atmosphere and its relationship with the Sun. New advances in rocketry were playing a role in better understanding the physics of the mysterious upper atmosphere. Both America and the Soviet Union had announced plans to launch small satellites as part of the project. Capitalizing on the political advantage that presented itself, the Soviets pulled the rug from under the Americans by secretly launching first, which left a red-faced President Eisenhower playing catch up. America quickly responded in December with the launch of the US Navy's Vanguard TV-3. It reached an altitude of roughly 120 centimetres (4 feet) before violently exploding on the launchpad, sending the tiny grapefruit-sized satellite into a nearby Florida swamp. It was an inauspicious start for the Americans in what became the space race, but they finally got off the starting blocks with the successful deployment of the Explorer 1 satellite on 1 February 1958.

Sputnik had no scientific purpose other than alarming the West with its famous beeping radio signal. Its legacy is that of a provocative gesture; a message that nowhere on Earth was out of range for Soviet rockets. It was this perceived 'missile gap', as well as a growing understanding of the potential for scientific research, that motivated Eisenhower to create a broader civilian administration to oversee

'To provide for research into problems of flight within and outside the Earth's atmosphere, and for other purposes.'

'NATIONAL AERONAUTICS AND SPACE ACT OF 1958', PUBLIC LAW #85-568, 72 STAT. 426. SIGNED BY THE PRESIDENT ON 29 JULY 1958

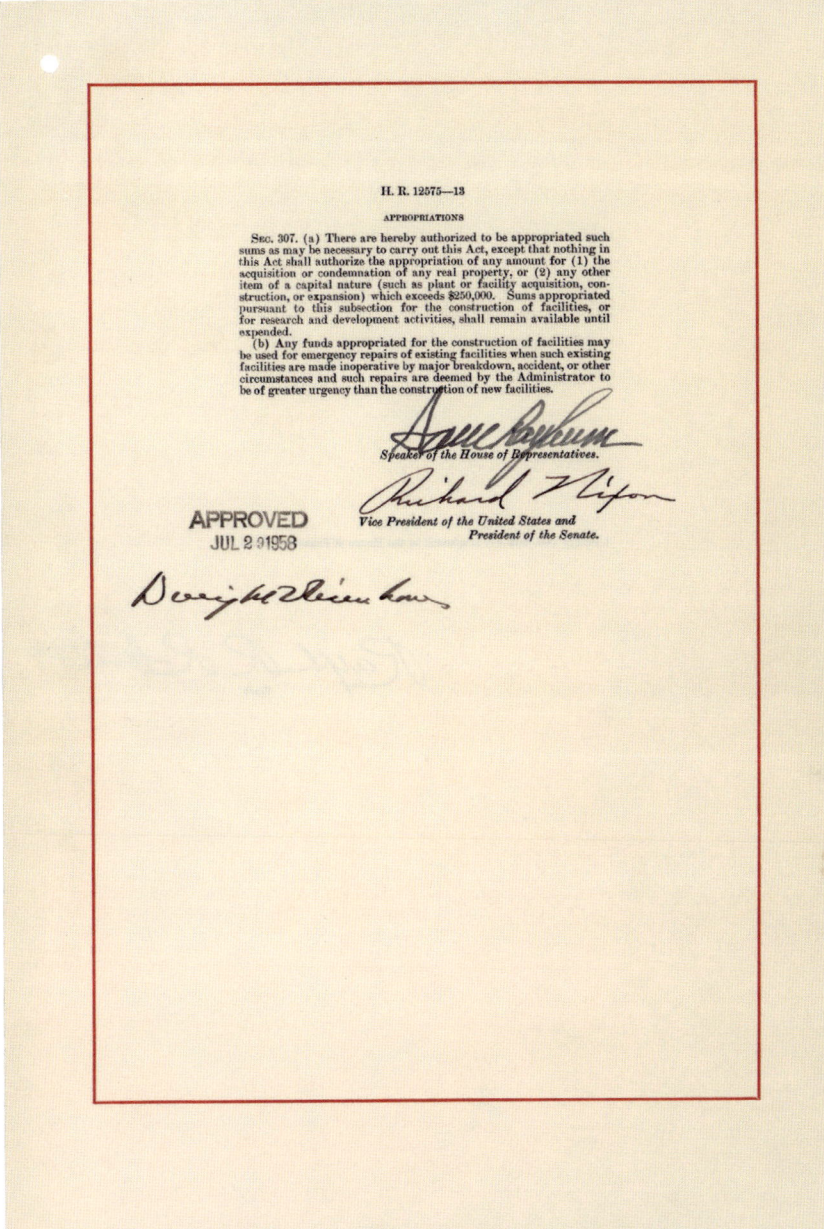

19.1 December 1957: Toronto citizens read of the failure of US Vanguard missile that was launched at Cape Canaveral, Florida.

19.2 The aftermath of a failed launch of the Vanguard rocket, which was intended to put America's first satellite into orbit.

19.3 Act of 29 July 1958 (National Aeronautics and Space Act of 1958), Public Law 85-568, 72 STAT 426, which provided for research into the problems of flight within and outside the Earth's atmosphere.

America's long-term space interests. This new space administration would absorb the current National Advisory Committee for Aeronautics (NACA), which oversaw all things aviation, and wrestle the nation's space activities away from the three branches of the military. T. Keith Glennan became its first administrator and Hugh Dryden, the former director of NACA, his deputy.

A brand-new agency with lofty goals needed a strong visual identity, so in September 1958 an open call went out to employees for suggestions for a new official seal. James 'Jim' Modarelli was an artist and designer who at the time was the head of the Research Reports Division at what would become the NASA Glenn Research Center. When visiting NASA Ames, he had been inspired by the futuristic aesthetics of a wing for a supersonic aircraft concept that had been put on display. It was a cambered arrow shape with a cobra-like upturned nose, designed by the aeronautical engineer Elliott Katzen. Meanwhile, another NASA designer, Harry DeVoto, had begun sketching some ideas: the name of the agency around a circle (as all seals had to be round), a blue star field background and a globe with an orbiting object. Modarelli adapted this idea, adding Katzen's wing shape as a bright red swoosh known as the 'vector' – a nod to the agency's continuing commitment to aeronautics as well as space.

Several designs were submitted, but Glennan personally chose Modarelli's as the winner. From there it went to the Army Institute of Heraldry for some minor tweaks to meet the standards

'On a disc of the blue sky strewn with white stars, to dexter a large yellow sphere bearing a red flight symbol apex in upper sinister and wings enveloping and casting a gray-blue shadow upon the sphere, all partially encircled with a horizontal white orbit, in sinister a small light-blue sphere; circumscribing the disk a white band edged gold inscribed "National Aeronautics and Space Administration U.S.A." in red letters.'

EXECUTIVE ORDER 10849, 27 NOVEMBER 1959

needed for presidential approval, followed by an assessment by the Commission of Fine Arts, a board of art experts who advise the government on matters of aesthetics. The design was passed – albeit reluctantly, with some criticizing it as resembling 'current commercial advertising' – and the official seal was rubber stamped by the president on 27 November 1959.

The seal was to be used for official usage, such as on flags and medals and on the front of lecterns, but Modarelli was also asked to come up with a simplified, informal insignia that could be used more widely. He kept the main elements of the seal – the blue starry background, the elliptical orbit motif and his cherished swooshing red wing – and across the middle, 'NASA' was written in a custom wide-lettered serif font. It was this design that affectionately became known as the 'meatball', one of the most globally recognized examples of graphic design in history. Throughout the 1960s, as NASA's human spaceflight programmes transfixed the world, the logo became synonymous with an exciting Buck Rogers vision of America's new confidence in space; a symbol of the space age that seeped beyond the walls of NASA into wider popular culture. Notably, the legendary April 1965 edition of the magazine *Harper's Bazaar*, which became a quintessential symbol of the Swinging Sixties, taking the new NASA aesthetic and blending it with fashion, art, music and literature. The English model Jean Shrimpton graced the front cover with a pink florescent space helmet headdress. Inside the magazine, she dons the iconic silver Project Mercury spacesuit, insignia patch in plain view, lent by NASA to art director Ruth Ansel. To the public, it was Jean Shrimpton who became the first woman in space. The meatball had escaped from its role as a government insignia to become part of the language of a progressive cultural revolution.

The insignia remained in service until the mid-1970s, when it was retired courtesy of the Federal Design Improvement Program, whose purpose was to refresh the graphics of various government departments and agencies creating a more uniform, modern and efficient message. The New York design firm Danne & Blackburn set to work on revitalizing NASA's visual identity with a now-infamous logotype, known to its detractors (and then to the whole world) as the 'worm'. The concept was straightforward: 'NASA' in flowing, simplified red type, with the crossbar of the 'A's removed to create a thrusting nose-cone shape. It had a purpose: modernity, precision and uniformity. It was innovative and futuristic.

19.4 NASA Lewis Research Center, 'We're now NASA', 30 September 1958.

19.5 Supersonic Wing Wind Tunnel Model, the inspiration for the NASA insignia.

19.6 Final version of the NASA seal.

19.7 The NASA insignia designed by James Modarelli.

19.8 Richard Danne and Bruce Blackburn's infamous NASA logotype.

Pages from the *NASA Graphics Standards Manual* (1976):

19.9 'Press Kits/Directories'
19.10 'NASA Uniform Patches'
19.11 'Motor Vehicle Identification Demonstrations'
19.12 'NASA Aircraft Markings'
19.13 'NASA Spacecraft Markings'

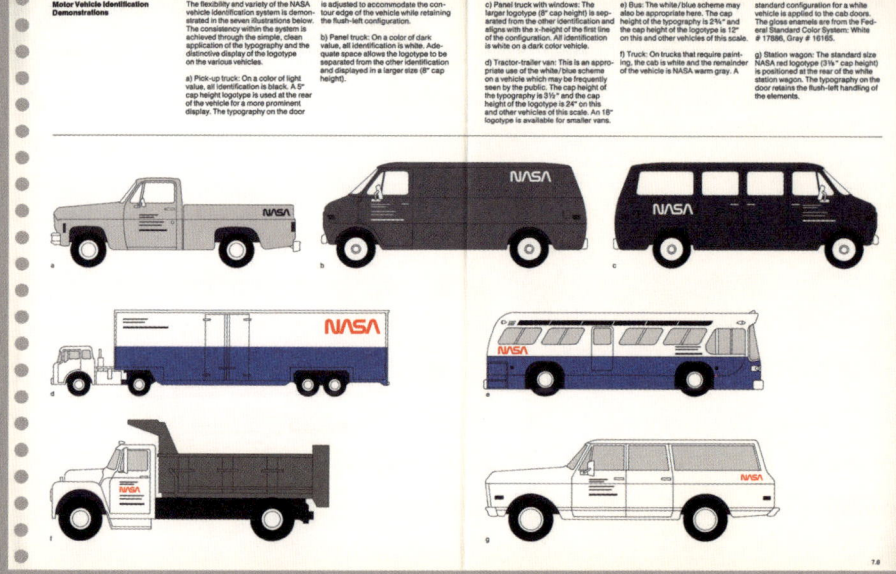

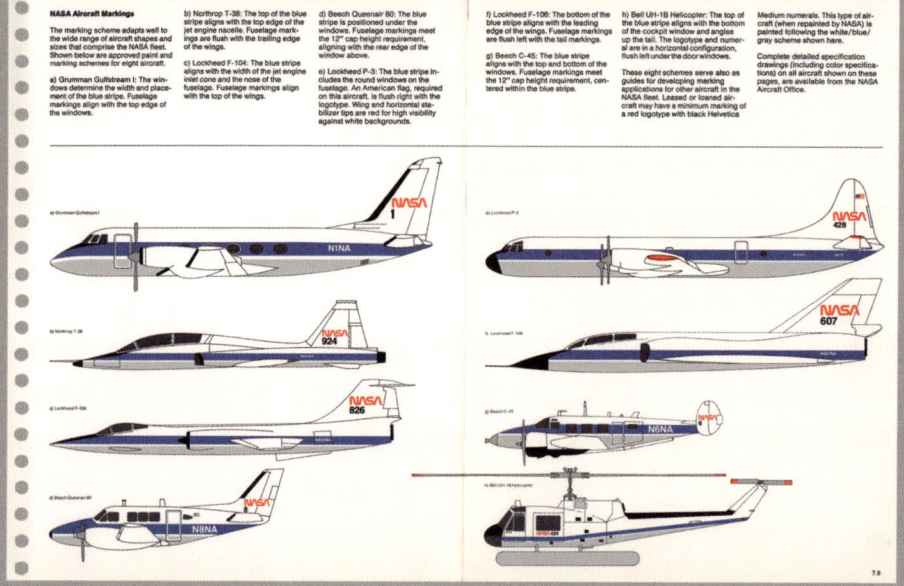

NASA Spacecraft Markings

The marking of NASA spacecraft vehicles is essential, critical, and difficult. It is quite important that any identification or markings which appear on spacecraft be consistent with the overall goals of the NASA Unified Visual Communications System. These vehicles represent tangible evidence of many of NASA's most interesting programs. As such, they are the focus of considerable public and media attention and should be marked in simple but effective ways.

Another important consideration is that the vehicle be marked so that it can be identified from different angles, whether in a launch mode or in outer space.

Of course, the overriding consideration is that the markings not interfere or impede the scientific mission of the craft. This principle applies to maintenance as well as the operational qualities of the craft when performing in space. This objective is very achievable as demonstrated on the Space Shuttle shown below.

Only a few isolated areas were designated for graphics by flight engineers and scientists. Working within these serious constraints, the Shuttle Orbiter is fully marked with all of the basic identifiers: The NASA Logotype, the American flag, United States, USA, plus the name of the particular craft. Helvetica Medium is the typeface used on the spacecraft.

Note that the NASA Logotype appears in NASA Gray so as not to conflict with the red of the American flag. The flag is equal to the height of the capital letters on the side, top, and bottom of the craft. The placement of these identifying elements is responsive to technical requirements as well as being harmonious with the basic shape and form of the Shuttle.

On the following gatefold you will see examples of other spacecraft which employ one or several of the available markings. Though they vary in size, shape, and configuration, they nevertheless maintain a strong overall relationship within the NASA Unified Visual Communications System.

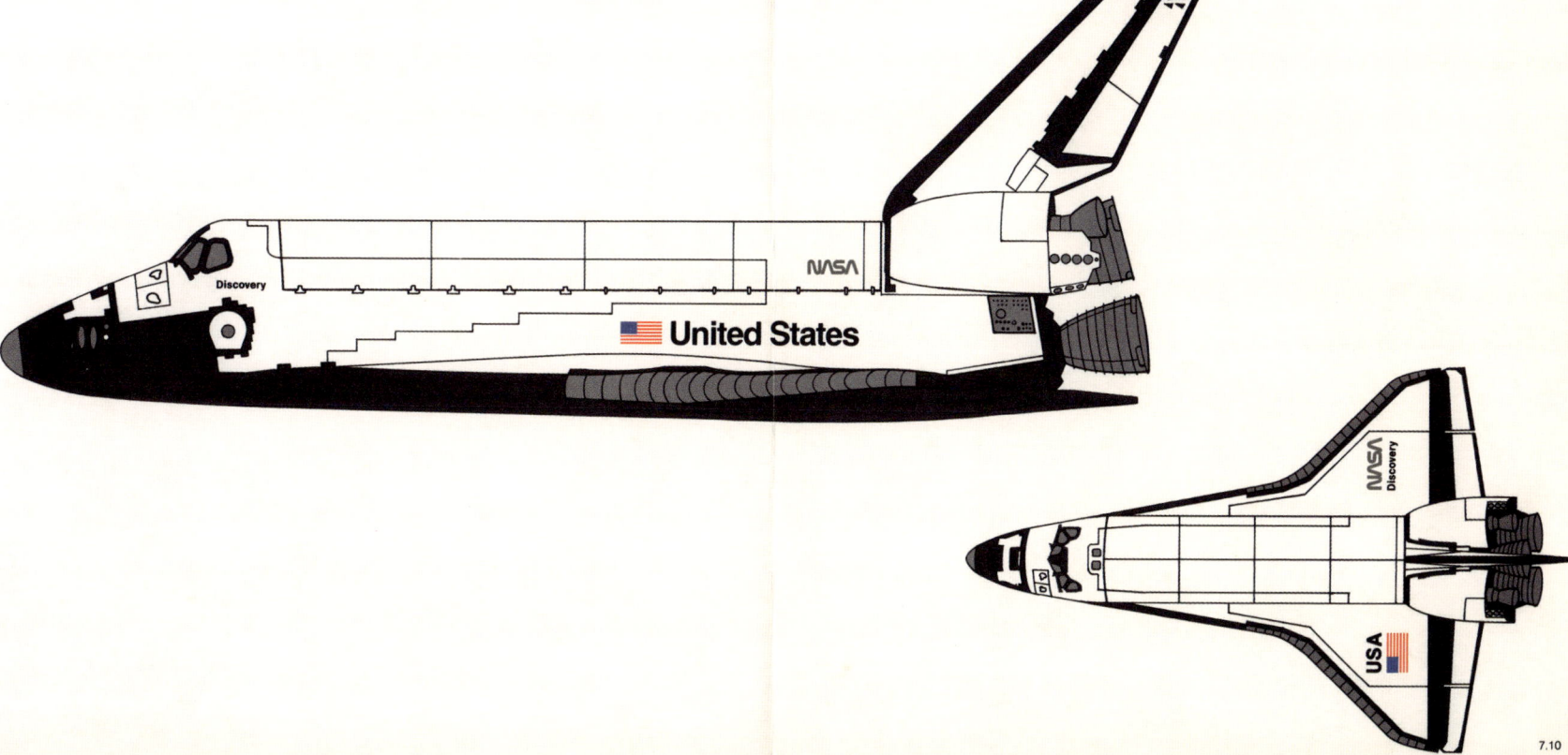

It was easy to standardize across the agency's assets, and much easier to reproduce in a world before digital printing. The change also marked the beginning of the impending ubiquity of Helvetica, the corporate typeface that has dominated the commercial visual landscape for the past half-century. Loved and hated, the logotype became NASA's new insignia. For the wider public, it was most visible on the Space Shuttle orbiters and on the flight suits of the astronauts, but as such it became increasingly associated with the Challenger disaster of 1986. In 1992 the new NASA administrator Dan Goldin demanded that the old, cherished meatball logo, a symbol of NASA's values and heritage, be reinstated:

The affectionate Meatball will replace the slick NASA logo, and slowly it will die into the horizon and never be seen again.

It was a decision designed to raise the morale of an agency that had seen better times. But the worm didn't die. In 2020 there it was, gracing the slim white SpaceX Falcon 9 rocket that launched astronauts Doug Hurley and Bob Behnken to the International Space Station after the Shuttle American launch hiatus. It was a dramatic cameo, heralding a new chapter in American space flight. The Instagram age has elevated the logotype insignia into the very watchword of beautiful design, helped along by the successful remastered edition of the 1976 *NASA Graphics Standards Manual*, published in 2017, which graces the bookshelf of many a design-conscious millennial.

The NASA logos haven't changed, but we have. Both the meatball and the worm now coexist, speaking in different languages to different generations yet saying the same thing: NASA is much more than a space agency. The acronym transcends national and cultural boundaries. Few people, other than those (like you) with a broader interest in space, could say with certainty what the letters even stand for. But everyone on Earth knows what NASA means. It's a visual language, and its history has become synonymous with something far deeper in human culture. NASA is simply a shorthand for space exploration. It represents the spirit of adventure.

19.14 The SpaceX Falcon 9 rocket with the infamous worm logotype at NASA's Kennedy Space Center in Florida.

19.15 Five astronauts and two payload specialists were aboard the Space Shuttle Challenger on 28 January 1986: (back row L-R) Ellison S. Onizuka, S. Christa McAuliffe, Gregory B. Jarvis and Judith A. Resnik; and (front row) Michael J. Smith, Francis R. 'Dick' Scobee and Ronald E. McNair. McAuliffe and Jarvis were the payload specialists, representing the Teacher in Space Project and the Hughes Company, respectively.

NEIL ARMSTRONG AND THE LEGACY OF APOLLO

20.

'He thinks, he acts, 'tis done.'

NEIL ARMSTRONG'S HIGH SCHOOL YEARBOOK

Between 1969 and 1972, an extraordinary thing happened: in one giant leap of intellectual creativity, humans from planet Earth successfully bridged more than 385,000 kilometres (240,000 miles) of empty space and set foot on another world. For a brief moment we became multi-planetary, fulfilling the dreams of countless earlier generations of poets and philosophers. In the intervening years, no other human endeavour has captured our imagination so absolutely. Its minutest details pored over in forensic detail. Project Apollo was the result of the combined effort of some 400,000 people. A moment in a troubled century when something monumental gripped the attention of the entire world.

How should we think about Apollo as it transitions from living memory into legend? What did it mean then? What does it mean now with over fifty years of hindsight? Was it a defining moment in our species' evolution? Has intelligent life on some other far-away planet had the will or capability to do such things? It should certainly be remembered for the perspective it gave us about our place in the cosmos. In 1948 Clyde T. Holliday, an engineer from Johns Hopkins University, strapped a camera to a V-2 rocket in White Sands, New Mexico, and captured the first grainy image of the Earth from space. A view that became more vivid and startling during the Apollo flights. It's often been said that the *Earthrise* photographs (figs 20.17–20.18), taken during the Apollo 8 flight in 1968, helped kick-start the environmental

---

20.1 Apollo 11 commander Neil Armstrong inside the Lunar Module while it rested on the lunar surface, 20 July 1969.

20.2 Cover of T. A. Bergstralh, *Photography from the V-2 rocket at altitudes ranging up to 160 kilometers* (1947). The report contains the first published photographs of the Earth taken from space, as shown in the following pages.

20.3 The V-2 rocket No. 21 on its launching platform, showing one of the cameras mounted in the midsection.

20.4 (OVERLEAF) A composite picture of four separate photographs taken at an altitude of 162 kilometres (101 miles), covering approximately 1.3 million square kilometres (500,000 square miles) of southwestern United States and northern Mexico.

20.5, 20.6 (OVERLEAF) 'Photograph 11', taken 227 seconds after take off at the same altitude of 162 km, includes an overlay (**20.6**) showing landmarks in New Mexico, Arizona and the Gulf of California.

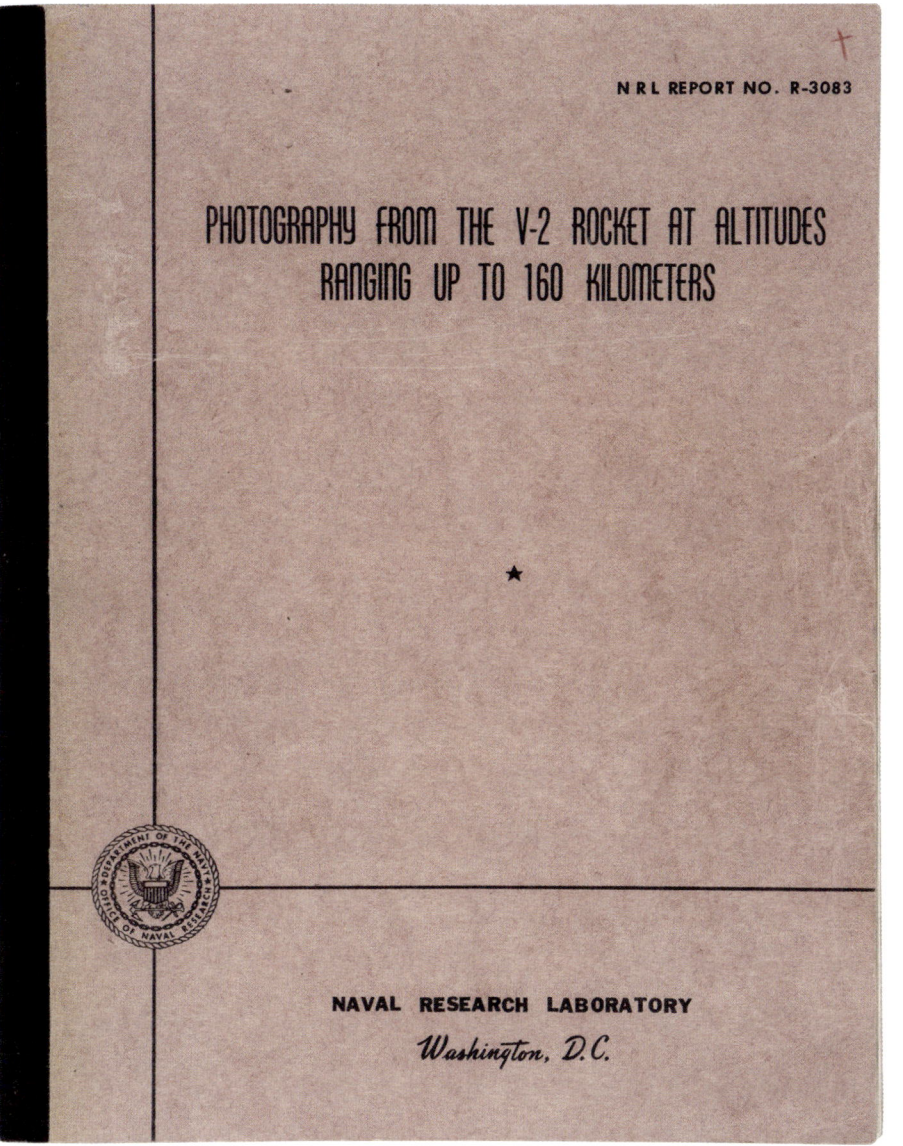

20.2

20.3

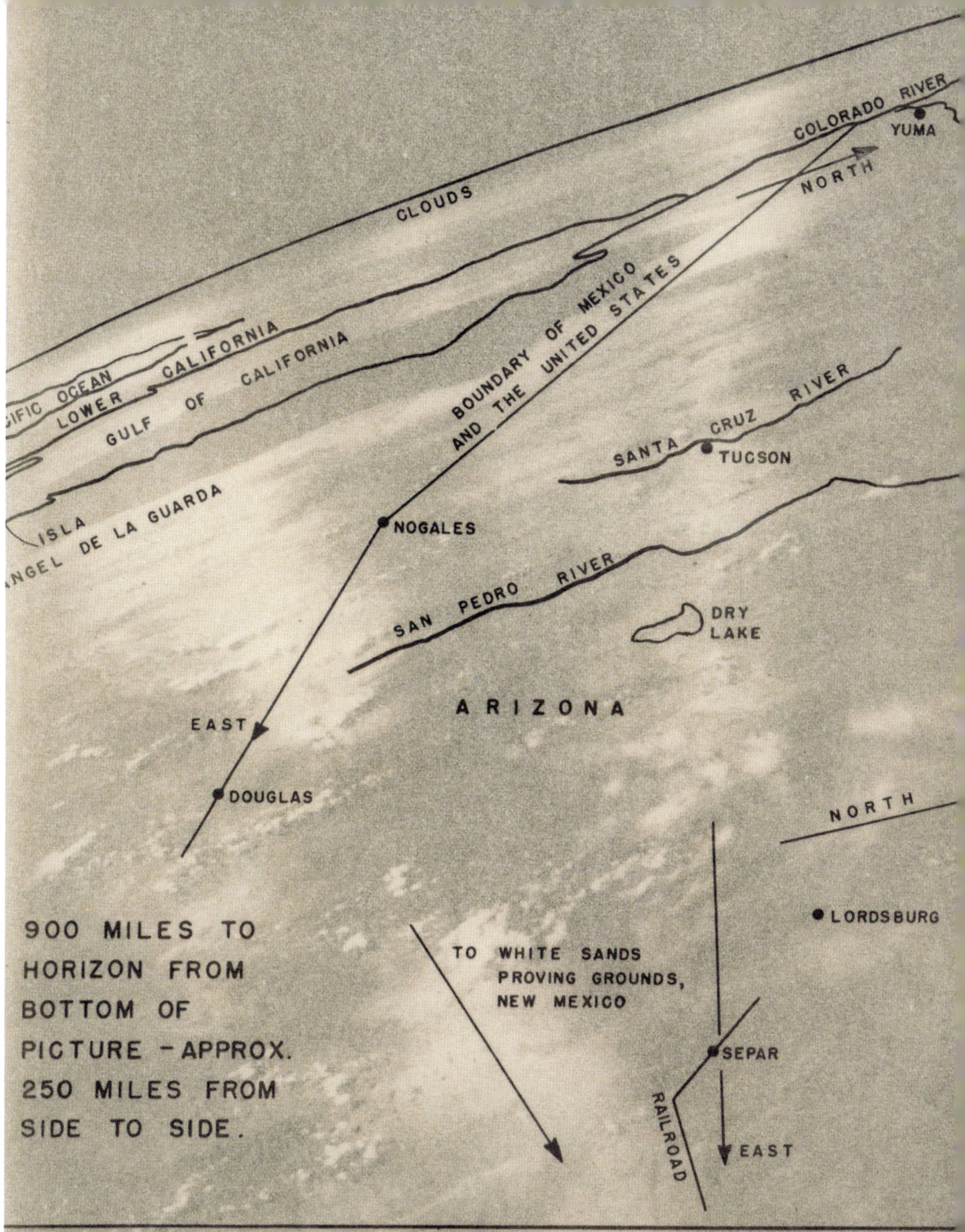

Figure 11. A photograph taken 227 seconds after takeoff at an altitude of 162 kilometers (101 miles). The rocket was then within a quarter of a kilometer of the peak of its trajectory. The camera was pointed southwest.

20.7 Press photo of Holliday, with the following message typed on the reverse: 'Editors; hold for release Sunday, Oct. 24th ... Mr Clyde T. Holliday, of the Applied Physics Laboratory, Johns Hopkins University, located at Silver Springs, Maryland, is shown inspecting two of the cameras used in the filming sequence.'

20.8 'Rocket Camera photographs Earth's Curvature', one of the first of Holliday's images circulated to the Associated Press, 3 October 1948, with the following message typed on the reverse: 'At an altitude of about 57 miles, a sequence camera in Aerobee rocket fired at White Sands Proving Grounds made this picture plainly showing the curvature of the Earth and the ground haze around it.'

20.9 On 26 July 1948, Holliday assembled over 200 photographs taken by an automatic camera on a V-2 some 100 kilometres (60 miles) up. A separate set of photographs was taken from a Navy Aerobee rocket around 110 kilometres up just over an hour later. After three months' work of matching and stitching, the two dramatic panoramas on the image were released on 19 October.

20.10 Motion picture frame from footage taken during V-2 flight No. 56, launched from White Sands Missile Range, 18 November 1949. The photograph was taken from an altitude of about 65 kilometres (40 miles), less than two minutes after launch. The curvature of the Earth is not yet apparent.

movement, highlighting our planet's isolated fragility, a sentiment that continued with the *Blue Marble* photograph taken by the Apollo 17 crew, which showed the whole illuminated face of Earth – one of the most important photographs ever taken, and one which became a fitting full stop to the Apollo Program.

We can see Apollo in the context of a world in flux after the Second World War. Countries and ideologies jostling for position, establishing their dominance over this new frontier. As the frost was hardening between America and the Soviet Union, fears of space becoming a militarized domain were very real. At the same time, space exploration was seen as a potential political unifier – a way of keeping the enemy closer in a 'peaceful coexistence'. Beyond the geopolitics, there was the messy social politics. Was it worth the huge expenditure? There were no projects (other than wars) that were quite as expensive as Apollo. Why weren't the billions of dollars being spent on more important or equitable things? At the time America was bogged down in the Vietnam War and was facing civil rights unrest across a number of fronts. The assassination of Martin Luther King in 1968, just a few months before Apollo 8 flew to the Moon and back, served as a casus belli for an intensification of already strained racial tensions. Apollo generated its fair share of resentment and public protest, as encapsulated by Gil Scott-Heron's famous protest poem 'Whitey on the Moon' (1970).

We should also certainly remember Apollo for its science; each mission was designed to extract as much knowledge as possible. On the surface of the Moon, the astronauts became researchers as well as ambassadors. Retroreflectors were installed that enabled scientists on Earth to measure the Moon's precise distance using laser ranging, alongside a variety of equipment designed to study the lunar environment. The collection of 382 kilograms (842 pounds) of lunar material gathered over the six missions also gave geologists a deeper understanding about the formation of the solar system. Much of the material remains in storage, untouched for the benefit of the scientists of the future. But as valuable as that data was, it wasn't valuable enough to justify sustaining the enormous cost. Apollo was foremost an engineering project, conceived from a long list of practical unknowns: would a huge, single 'direct ascent' rocket take us there and back? Or would a 'Lunar Orbit Rendezvous' with a smaller lunar lander be a better solution? How would all the different systems required work together? And how might we build a computer capable of managing all these systems in a world where microcomputers don't exist? The punishing deadlines forced us over a variety of technical hurdles that we might not have cleared otherwise. How many decades did Apollo leap-frog us into the future?

Above all, Apollo's legacy is its power; it's one of the great sagas of our age. Its real-life dramatic arc feeds our story-addicted minds, a great quest unfolding elegantly in scenes and acts. In it we see the skilled writer's familiar techniques: the battle against the improbable odds; the race against the clock to reach the Moon before the decade was out. There are good guys and bad guys, depending on which side of the Iron Curtain you're on. There are dragons to be slain and tragedies to be grieved over, most notably the devastating Apollo 1 fire that killed astronauts Ed White, Gus Grissom and Roger Chaffee. There are moments of tension and suspense as we root for the shipwrecked crew of Apollo 13 to make it safely to shore. And of course, the great climactic scene in July 1969 when that first small step was taken on the Moon's surface. It's no wonder the project continues to provide endless material for Hollywood and historians alike.

20.11 Close-up view of Apollo 17 lunar sample number 72415,0, a brecciated dunite clast weighing a little over 32 grams (about 1.14 ounces).

20.12 Project Apollo Lunar Landing Flight Techniques, 14 January 1960.

20.13 Neil Armstrong in the Lunar Module simulator situated in the Flight Crew Training Building at the Kennedy Space Center, 16 June 1969.

20.14 Armstrong floats safely to the ground as LLRV no. 1 crashes at Ellington Air Force Base, 6 May 1968.

20.15 (OVERLEAF) Armstrong is seen here packing lunar samples – one of the few photographs that show him on the Moon.

20.16 (OVERLEAF) Lunar Module pilot Edwin 'Buzz' Aldrin walks on the surface of the Moon. Armstrong took this photograph with a 70mm lunar surface camera – we can see him reflected in Aldrin's visor.

20.17, 20.18 (PP. 184-85) These two images from the *Earthrise* sequence show the Earth peeking above the Moon's horizon. The lunar terrain pictured is in the area of Smyth's Sea on the nearside, but for the true view of the astronauts, you need to rotate this book 90° to the left.

Apollo's drama is carried on the shoulders of its protagonists. Of the twelve men who walked on the Moon, one in particular has carried much of its weight – the consequence of being the first. Neil Armstrong (1930–2012) was a clean-cut engineering student from a small town in Ohio.

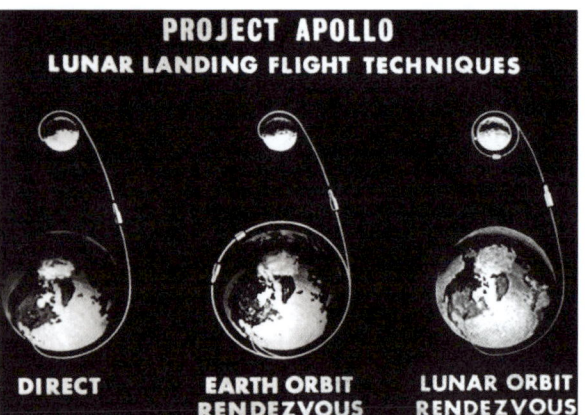

He served as a naval aviator in Korea before flying the experimental hypersonic X-15 rocket plane and joining the astronaut corps. To the public, he represented the American values of hard work, courage, humility and duty. His own character was woven into Apollo's wider plot as he became famed for the clear-headed way he dealt with risk and danger. As commander of Gemini VIII's rendezvous and docking mission, he wrestled the spacecraft from the wild spin it encountered as it was docked with the Agena target vehicle. While training to fly the Apollo Lunar Module (LM), he ejected calmly from the malfunctioning Lunar Landing Research Vehicle, nicknamed the 'flying bedstead', and parachuted back to Earth among the flaming wreckage. He was back in his office that afternoon doing paperwork.

But Armstrong's finest moment came with the eyes of the world watching: landing the Lunar Module Eagle on the surface of the Moon. With warning alarms sounding and seconds away from running out of fuel, he manually navigated the spacecraft over a field of boulders and managed to safely place it down in the Sea of Tranquillity. As CAPCOM Charlie Duke said over the radio from mission control in Houston, 'Roger Tranquillity* ... you've got a bunch of guys about to turn blue. We're breathing again.' The world breathed too. On 20 July 1969 Armstrong took that small step off the LM and placed his foot into the powdery lunar surface, for the benefit of all mankind. He thought, he acted. It was done. Armstrong had faced personal challenges getting to that moment: the loss of his young daughter to cancer, as well as the deaths of close friends in the astronaut corps, particularly Ed White. But life after Apollo was no plain sailing. Armstrong's quiet, considered demeanour was often interpreted as being stand-offish or reclusive by a public constantly clamouring for his attention. He was certainly reserved: he spoke carefully and was uncomfortable with his notoriety. At the same time, he accepted the historic role he'd been cast in, understanding that Apollo was something paid for and owned by the public. Armstrong wanted to be remembered as an engineer and a pilot, not as the astronaut hero. The further Apollo drifted into mythology, the firmer Armstrong kept his feet on the ground.

Apollo's legacy is a Rorschach test. In it we see what we want to see: adventure, waste, triumph, conspiracy or heroism. It stands out as an anomaly in history, a moment when 400,000 ordinary men and women came together to achieve the impossible. The astronauts who embarked on that extraordinary journey were ordinary men. And it was ordinary men and women who made it happen. Neil Armstrong was the perfect representative: the hero with a thousand faces, however reticent a hero he might have been.

---

* 'T(w)angquillity'.

RYAN NAGATA

Making things from scratch forces you to think harder and to look closer. There are few who have thought harder and looked closer than the American artist, prop-maker and cosplay legend Ryan Nagata (b. 1981). His name has become synonymous with his extraordinary spacesuit replicas, which have graced private collections, museums and films such as the Neil Armstrong biopic *First Man* (2018) and the romantic comedy *Fly Me to the Moon* (2024). Nagata's obsessive spacesuit-making quest began at the age of fourteen while watching *Apollo 13*, during which he became fascinated by the form and aesthetics of the Apollo spacesuits. Leaving the cinema, he had an epiphany: it would be so cool to try and make one of those. How hard could it be? He's spent a lifetime finding out.

A spacesuit isn't really a 'suit' at all. It's been described as a wearable spacecraft of the smallest possible dimensions. Like a spacecraft, it has to support the narrow parameters of human life in the most extreme environment of all. It has to provide oxygen to breathe; get rid of carbon dioxide; regulate extreme temperature fluctuations; and protect against the micrometeoroid flux density in the vacuum

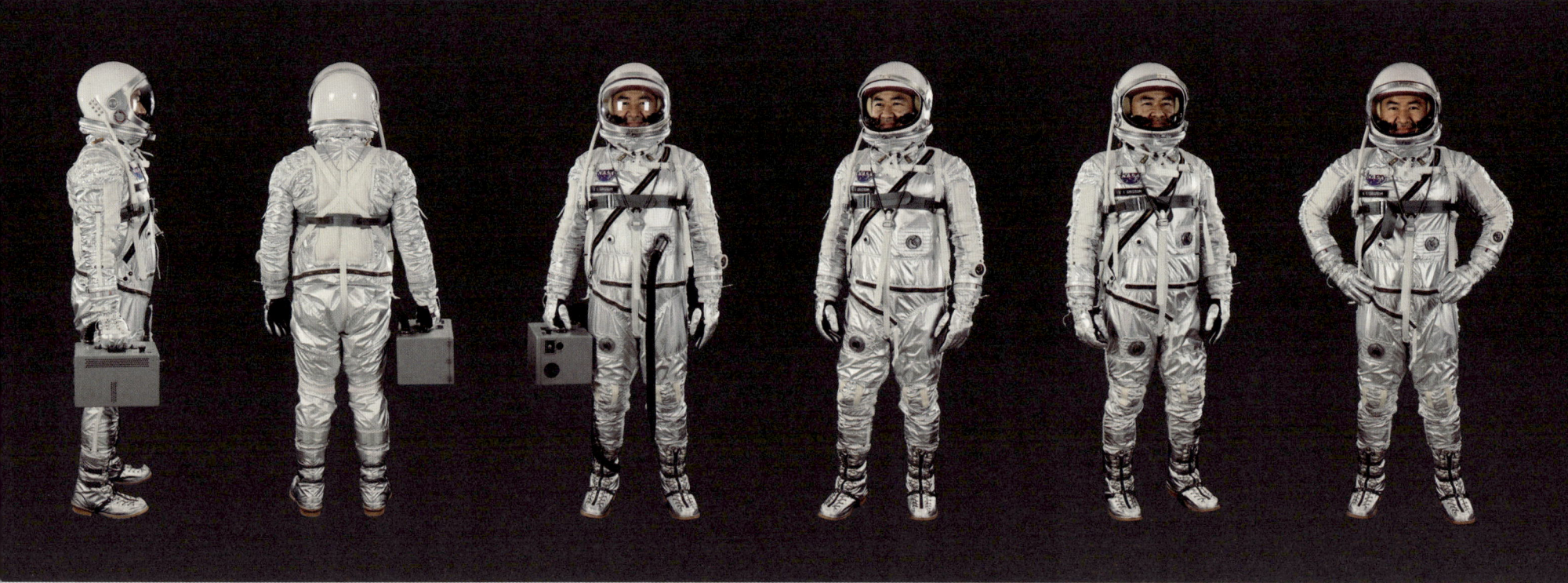

21.1 Ryan Nagata modelling his high-fidelity replica of a Navy MK IV full pressure suit (the forerunner to the famous Project Mercury silver spacesuit), made with leftover aluminized nylon fabric from *First Man* (2018).

21.2 Nagata modelling his replica of Wiley Post's 1934 pressure suit, made for the Stafford Air and Space Museum.

21.3 American aviator Wiley Post, the first pilot to fly solo around the world, was also known for his work in high altitude flying. He's pictured here wearing the third pressure suit he helped develop, c. 1934.

---

of space or on the lunar surface. It has to maintain an airtight bubble of atmospheric pressure flexible enough for an astronaut to be mobile, while keeping a constant volume – no easy task for spacesuit designers dealing with the unpredictability of soft materials that need to be sewn, glued or taped rather than precision-welded. The Apollo spacesuit stitching, sewn by a highly skilled team of seamstresses at the International Latex Corporation (see Chapter 22), was as important in getting Americans to the Moon as the rocket engines or guidance computer. In short, the spacesuit is an engineering marvel.

Spacesuit history begins with the birth of aviation, and in particular with the 1930s record-breaking aviator Wiley Post, whose bizarre-looking homemade pressure suit is credited as the grandfather of the spacesuit. As we took our first steps into the stratosphere and beyond, the high-altitude pressure suit began to evolve. The drab sage green US Navy Mark IV was treated with an aluminium coating to lend it that silver Buck Rogers look for the X-15 rocket plane pilots and Project Mercury astronauts. As America's attention turned towards the Moon, a whole new wave of spacesuit ideas, designs and concepts were trialled and abandoned, tested and re-tested. At NASA's Ames Research Center in the late 1970s, designer Hubert 'Vic' Vykukal was developing extraordinary prototypes: hard suit concepts that looked like futuristic suits of armour, with stove-pipe joints and hand-painted graphics. These eventually gave way to the famous A7L pressure suits worn on the Moon by the Apollo astronauts, which were made possible by the post-war explosion of exotic man-made materials like nylon, Dacron, Teflon, Mylar and neoprene (fig. 21.14). The spacesuit has long been an important signifier in broader popular culture, and a key ingredient of science fiction films since

21.4 The first group of astronauts announced by NASA, selected in April 1959 for the Mercury Program. Front row (L-R) are Walter M. Schirra Jr, Donald K. Slayton, John H. Glenn Jr and M. Scott Carpenter. Back row (L-R) are Alan B. Shepard Jr, Virgil I. Grissom and L. Gordon Cooper Jr. Slayton and Glenn's spacesuit boots weren't ready for this *Life* magazine shoot. Two pairs of ordinary work boots were bought from a local store and were swiftly spray-painted silver. Nobody will ever know!

21.5 This is the spacesuit worn by astronaut Alan Shepard during the first crewed space flight launched by the United States on 5 May 1961.

The Mercury spacesuit was a close-fitting, two-layer, full pressure suit developed by the BF Goodrich Company from their Mark IV pressure suit, as used by the US Navy. It was selected by NASA in 1959 for use in Project Mercury, and during the course of the Mercury Program underwent minor modifications, primarily in the shoulders.

21.6 Astronaut Alan B. Shepard Jr being helped into the lower half of his pressure suit.

21.4

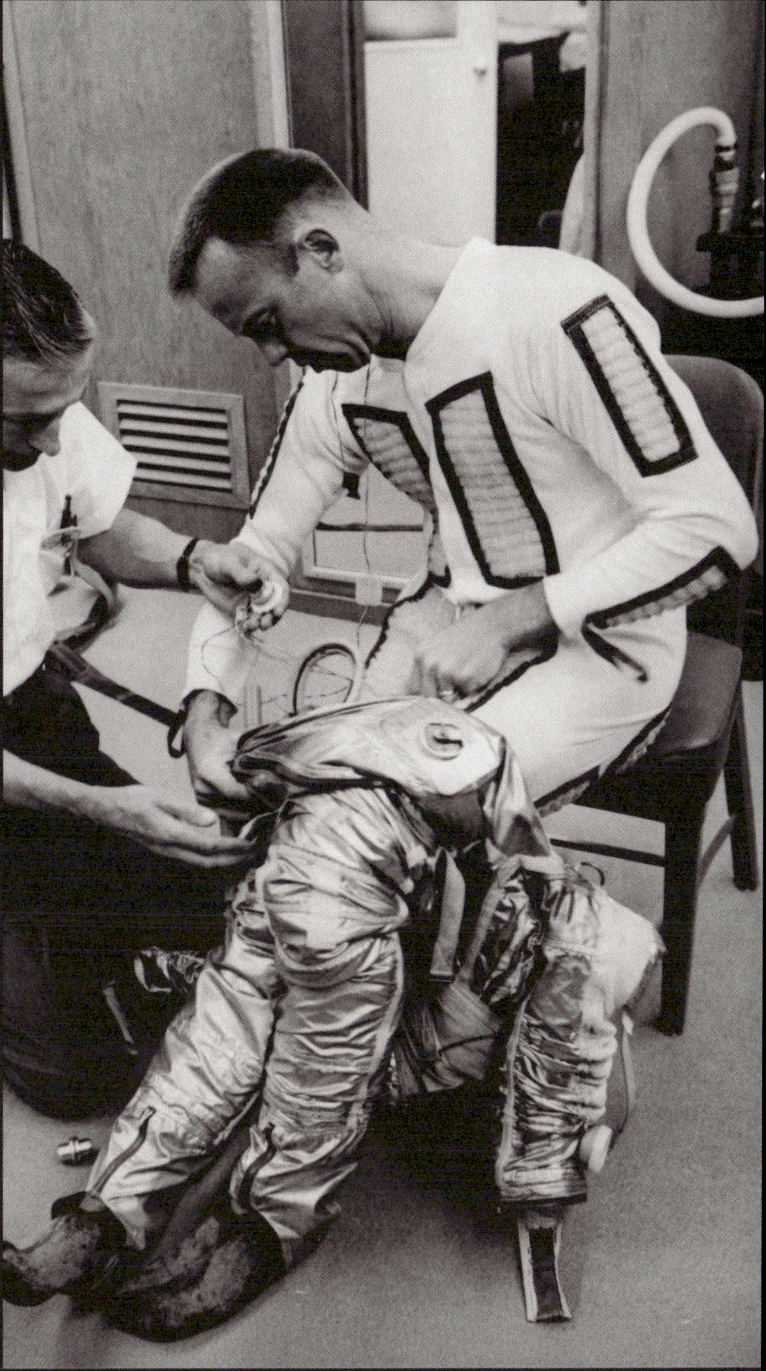

21.5

21.6

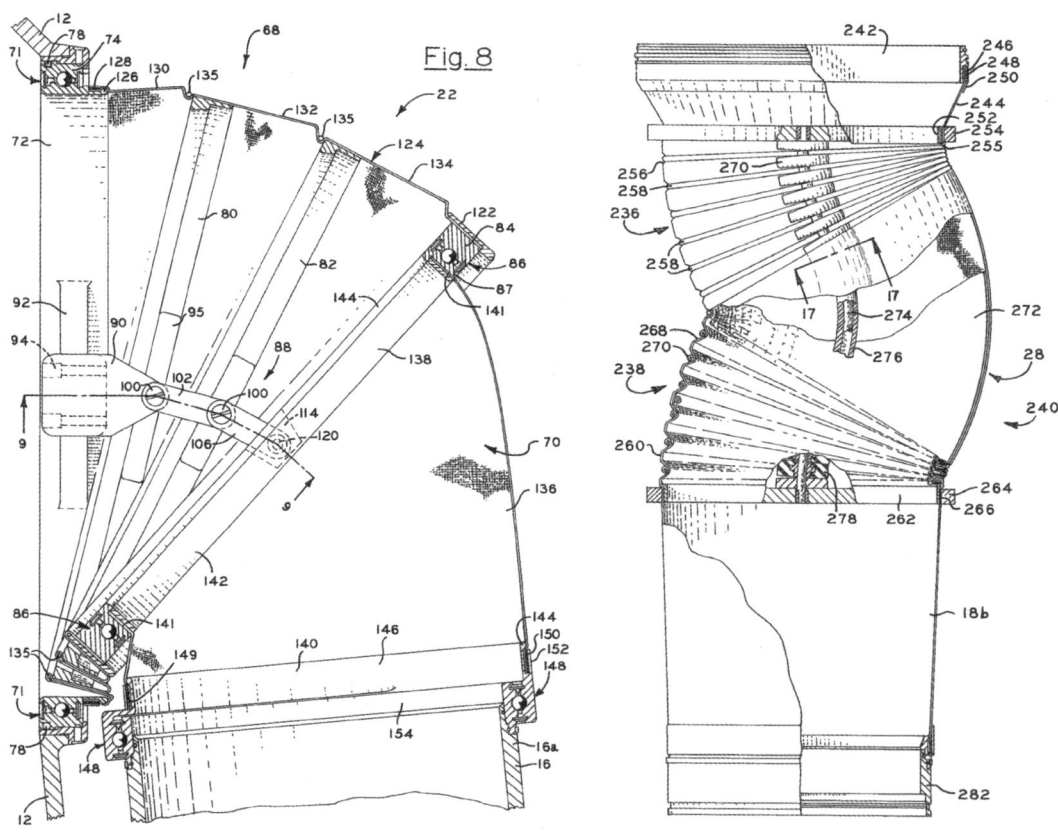

21.7 An X-ray of the EX1-A Apollo Applications Project suit from 1968 shows its 'toroidal joint system'. The system permitted the maximum amount of movement for the astronaut.

21.8, 21.9 Hubert 'Vic' Vykukal's spacesuit mobility joints, U.S. Patent #4,091,464, issued 30 May 1978.

21.10-21.13 Vykukal models the AX-3 spacesuit at the Ames Research Center on 22 July 1977.

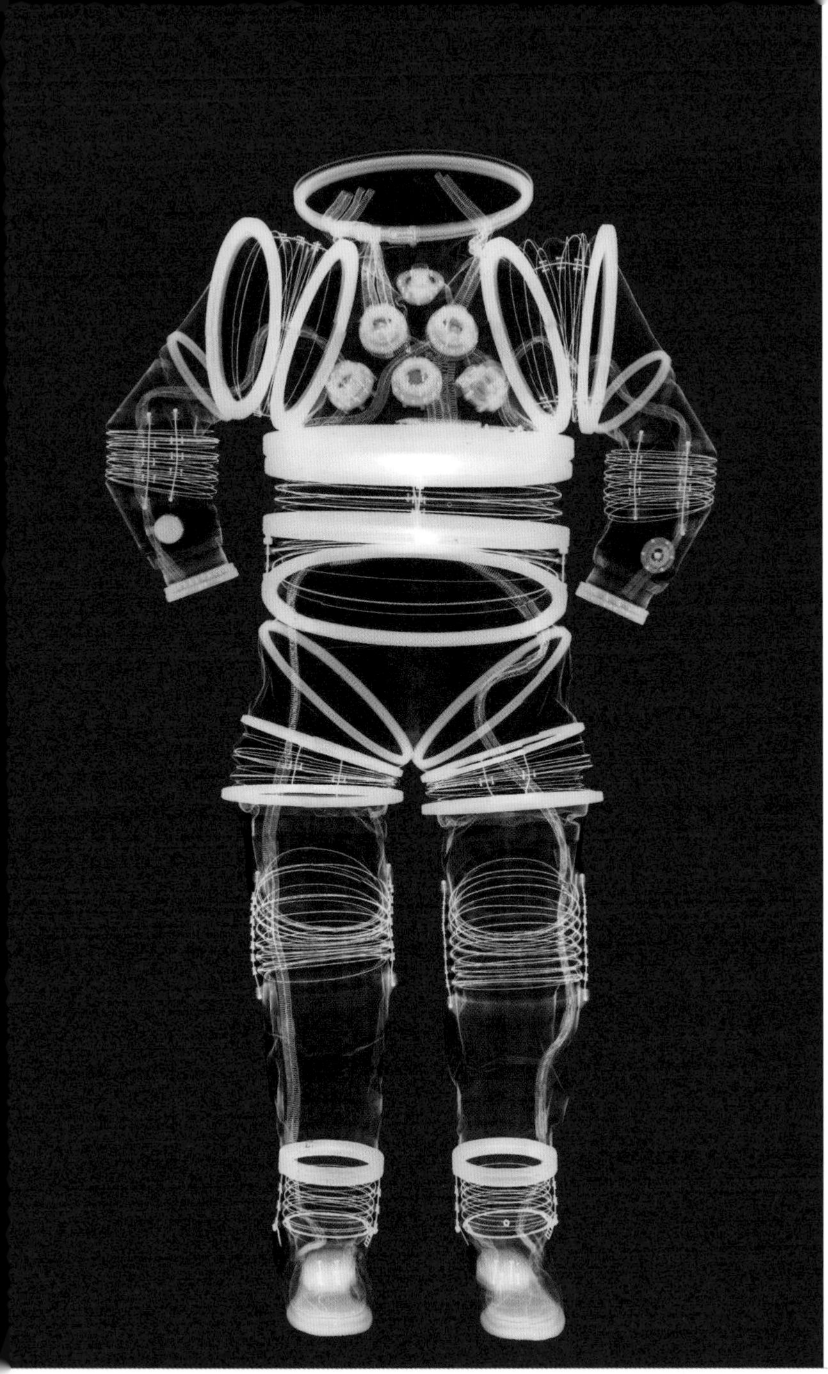

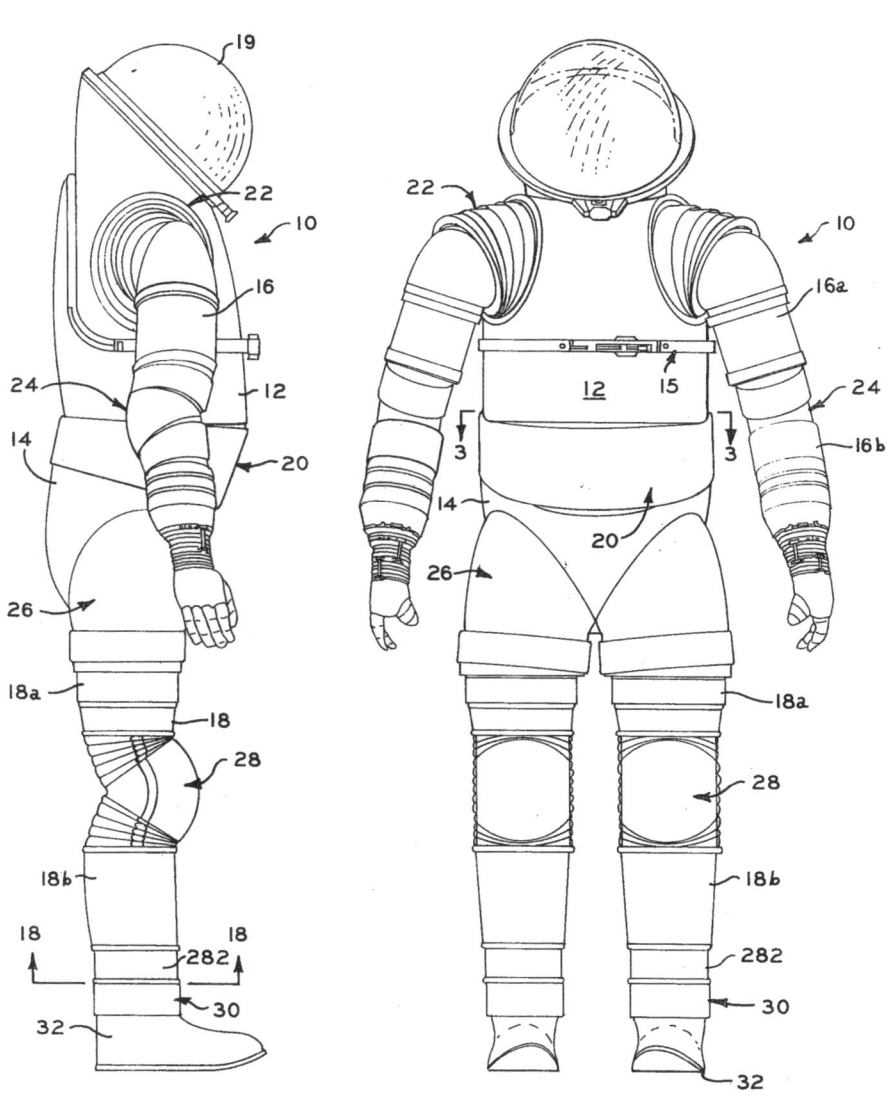

21.9

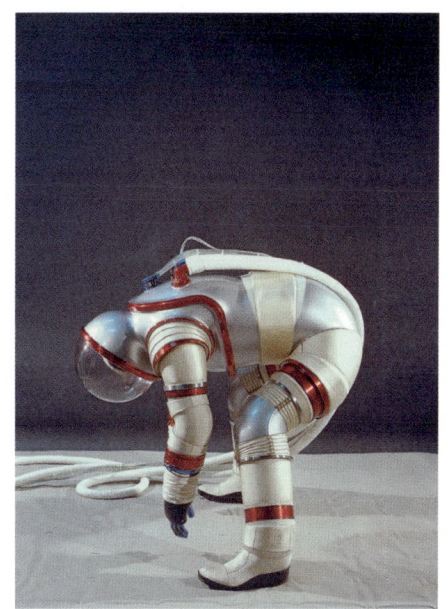

21.10–21.13

the advent of cinema. It's not by chance that the new, sleek SpaceX suits were designed in part by the Hollywood costume designer Jose Fernandez (fig. 32.8). The next generation of moon suits have been engineered by the aerospace company Axiom, who have teamed up with fashion house Prada in the design process, continuing the tradition of form, function and aesthetics.

Nagata's high-fidelity replicas have gone through their own evolution, better replicating the original suits' heft through his closer matching of the fabrics and mastery of the various complex and often-forgotten technical processes needed to make them. They are a masterclass in attention to detail. Thanks to Instagram, his work has found a huge audience, which has highlighted the spacesuit as an important cultural object that bridges engineering, craft, material science, exploration and history. But his replicas have also been embraced by the aerospace companies who made the original suits, as well by the conservationists Cathleen Lewis and Lisa Young at the Smithsonian in Washington, DC, who have done much to preserve and document the legacy of the spacesuit.

In Nagata's workshop at the back of his suburban Los Angeles house stands his crowning achievement to date: a hyper-realistic replica of the multilayered and fiendishly complex A7L Apollo spacesuit. This time it's not just skin deep. It's as close to the real thing as possible. The famous white Beta cloth outer layer has been removed like the flayed skin of a Leonardo da Vinci anatomy sketch, revealing the spacesuit's hidden inner workings – its muscle, blood, guts and sinew. The pressure bladder, the cabling, the bellowed shaped convolute joints, the machined, etched and anodized metal disconnects and collar ring. All accurate and faithful. It's a testament to what's possible if you really put your mind to something. A next-level exercise in engineering archaeology.

This spacesuit might just keep you alive on the Moon. You'll certainly look good finding out.

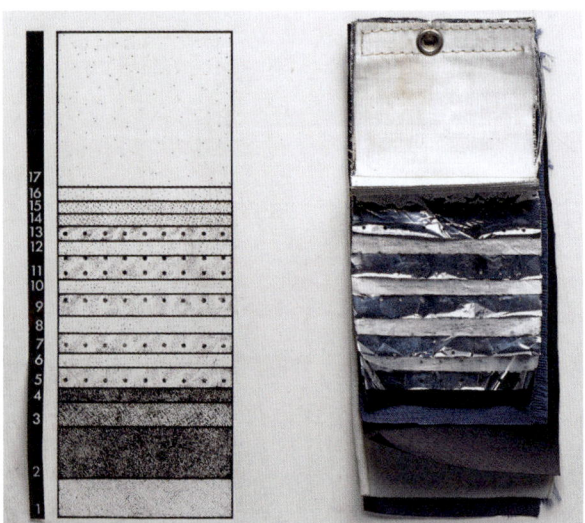

21.14

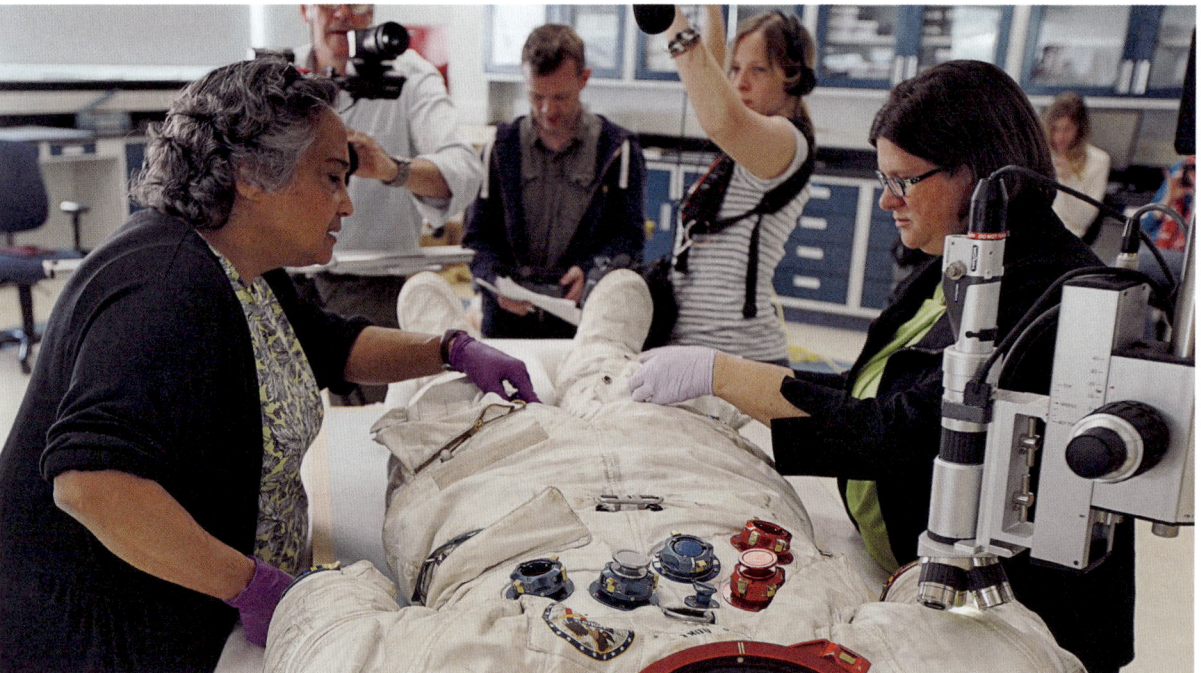

21.15

21.14 Section of the Apollo Extravehicular Activity (EVA) spacesuit used on NASA's Gemini missions of 1965 and 1966. The three inner layers form the pressure retention garment, including a lightweight, heat-resistant Nomex material followed by a gas-tight bladder of neoprene-coated nylon and a nylon restraint layer to stop the pressurized garment from ballooning outwards. The next twelve layers make up the Thermal Micrometeoroid Garment, and the final two outer layers, made of Beta cloth and Teflon fibre, give protection against abrasion.

21.15 Cathleen Lewis and Lisa Young examining the lunar dust still trapped in the fibres of Charlie Duke's Apollo 16 spacesuit at the Smithsonian, 2014.

21.16 An X-ray of Alan Shepard's Apollo 14 spacesuit allows curators and conservators to 'see' inside space clothing - a task that had previously been done by peering through the neck or the wrist with a flashlight.

21.17 A labour of love. Ryan's homemade, hyper accurate Apollo 11 A7L spacesuit replica with the outer white Beta cloth layer removed. Seven years of planning and countless hours of work went into this extraordinary project.

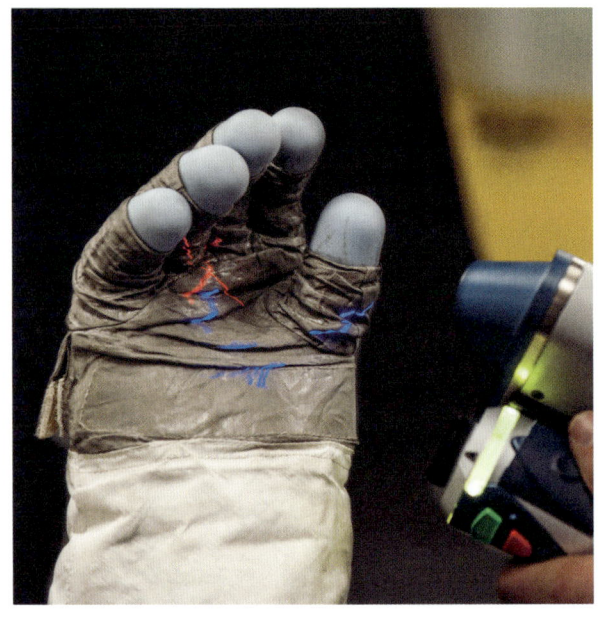
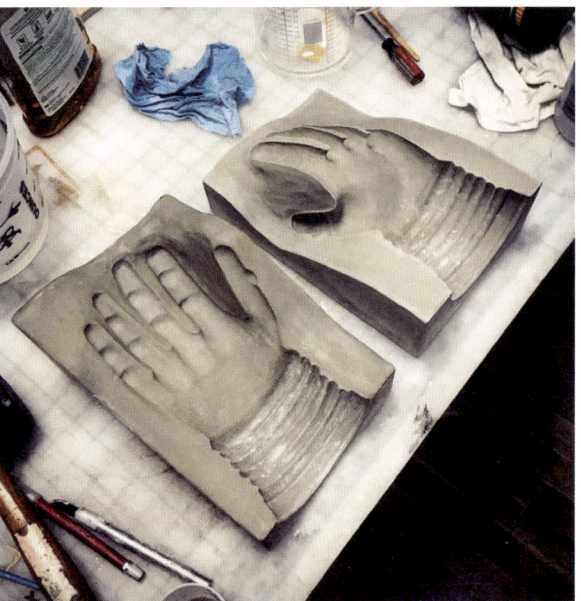
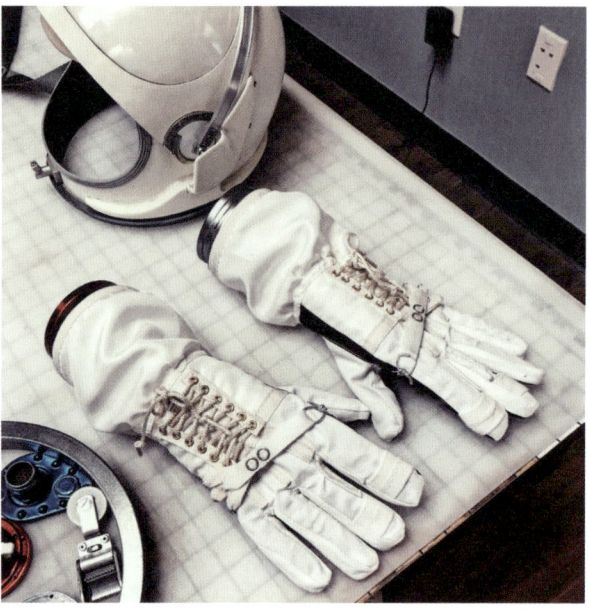
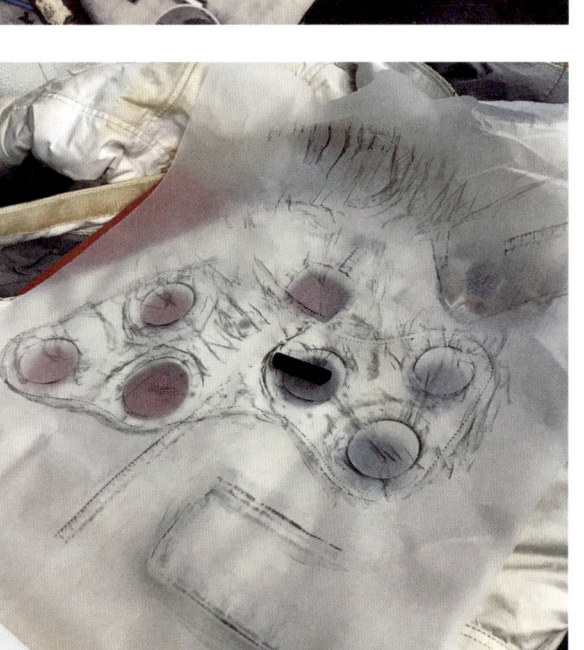

21.18 A Faro laser arm scanner captures 3D data of Neil Armstrong's Apollo 11 EVA glove at the National Air and Space Museum, Smithsonian, 2016.

21.19 Plaster mould made by Nagata for slip casting Apollo intravehicular gloves in neoprene.

21.20 Nagata's replicas of the Gemini gloves.

21.21 Pattern pieces by Nagata for his Mercury gloves.

21.22 A rubbing that Nagata made of a real A7L-B spacesuit at ILC Dover in 2019.

21.23 Nagata wearing his replica Gemini G4C spacesuit.

21.24 Nagata modelling his high-fidelity replica of an Apollo A7L-B spacesuit in his workshop.

21.24

'LITTLE OLD LADIES'

The old Chinese saying 'men plough, women weave' was as true among the 400,000 people who worked on Apollo as it was anywhere else in America in the 1960s. Gender and class divisions of labour in the new space industry were a representation of wider society. On the whole, hi-tech jobs were performed by educated men, repetitive work that required dexterity and concentration, or work associated with craft and domesticity, was done by women. There are of course notable exceptions in the Apollo photographic archive: Poppy Northcutt, the mission planner who looks incongruous among the exclusively male faces in the Houston mission control; or Margaret Hamilton, the MIT software engineer whose photograph standing next to a tower of computer code that she developed for the Apollo Program has become an important cultural marker in conversations around gender stereotypes.

The work done by the seamstresses for Apollo was as vital as that of the astronaut or the rocket scientist. The International Latex Corporation in Delaware was a prime contractor for NASA, responsible for making the spacesuits. Women who had learned to sew at high school or from their mothers or grandmothers had been drafted into the space programme from Playtex, where they had been working on sewing production lines making girdles and baby clothes. Others had come from local companies, fabricating luggage and boxing gloves. A new set of skills was required to make the multilayered, custom-made spacesuits, gluing, taping and sewing exotic space-age materials: Beta cloth, Nomex, Kapton, Dacron, aluminized Mylar film and Chromel-R, materials that were fragile, difficult and expensive to work with, and which had to be treated with a heightened level of care and respect. It was a steep learning curve for the seamstresses. They were tested not just in their sewing prowess, but in their temperament, their ability to concentrate and to work under extreme pressure: mission-critical cutting and sewing to within 2 millimetres (1/16 inch) accuracy; working at a very slow, steady pace on the industrial Singer sewing machines to avoid damage to the material and to ensure their work was flawless.

22.1

22.2

'We would use rockets not yet designed and alloys not yet conceived, navigation and docking systems not yet devised in order to send a man to an unknown world ...'

CARL SAGAN, 'THE GIFT OF APOLLO' (1994)

```
TAPE WEDB RUN  6 VERIFICATIØN C4.18A TEST  RUN DATE 10/23/68 EDIT DATE 23ØCT68  SDS TIME  1640 AGC TIME  1639.56 ID 777  PAGE  1

PRØGRAM        VERB 21     NØUN 15          R1  37777   R2          R3  00025   FLASH 0    DSPTAB +11 00000
PRØGRAM        VERB 21     NØUN 15          R1  37777   R2          R3  00025   FLASH 0    DSPTAB +11 00000

      SDS  REFERENCE  MATRIX                     AGC  REFERENCE  MATRIX                     S/C  ATTITUDE  MATRIX
 -.41222304   .79159081  -.45106113         -.41222304   .79159081  -.45106113         -.41226554   .79156518  -.45106578
  .00042095  -.49491686  -.86894023          .00042095  -.49491686  -.86894023          .00031686  -.49497294  -.86890769
 -.91108286  -.35838706   .20368291         -.91108285  -.35838706   .20368291         -.91106343  -.35836411   .20380950

REFSMMAT ER  GIMBAL ANGLES   SDS CDU    AGC CDU    AGC SHAFT   SDS SHAFT              LØS DR    LØS ALT    LØS CR
 Y=   .000   AIG=   -.00       -.00       .01        -.03        10.99     R ERRØR      .00       .00       -.00
 Z=  -.000   AMG=    .00        .00       .00      AGC TRUN    SDS TRUN    V ERRØR      .00      -.00        .00
 X=   .000   AØG=    .01        .01       .00         .01       113.35

STATE VECT       RX          RY          RZ          RSS          VX         VY         VZ         RSS       TIME
 SDS CM      5725335.1   2912736.6  -1656192.7   6633737.3    -3911.85    5821.87   -3317.27    7758.94    1640.42
 AGC CM      5725335.1   2912736.6  -1656192.7   6633737.3    -3911.85    5821.87   -3317.27    7758.94
 DELTA              .0          .0          .0          .0         .00        .00        .00       -.00

 SDS LM      5726686.1   2948205.5  -1676134.6   6655544.8    -3941.82    5787.80   -3297.51    7740.17    1640.42
 AGC LM      5726686.1   2948205.5  -1676134.6   6655544.8    -3941.82    5787.80   -3297.51    7740.17
 DELTA              .0          .0          .0          .0         .00        .00        .00       -.00

PØSITIØN        LAT       LØNG       ALT        APØGEE     PERIGEE      AZIMUTH   ELEVATIØN    RANGE    RANGE RATE
 SDS CM        -14.55    -18.91   256909.9    282612.2    253710.6       -.37      32.24     40713.0      -40.35
 SDS LM        -14.68    -18.63   278740.9    282711.0    276459.4

FLAGWD 0 00000  FLAGWD 1 00000  FLAGWD 2 00000  FLAGWD 3 10000  FLAGWD 4 00000  FLAGWD 5 40200  FLAGWD 6 00000  FLAGWD 7 00100
FLAGWD 8 00000  FLAGWD 9 00000
RCSFLAGS 40011  RCSFLAGS 40011  RSBBQ     66102  RSBBQ+1   03304  CADRFLSH 56016  CADR+1    73174  CADR+2    10132  FAILREG  00000
FAILRG+1 00000  FAILRG+2 00000  PIPA X    77752  PIPA Y    77775  PIPA Z   77540  FLAG 10   00000  FLAG 11   00000  ØPTMØDES 00120
HØLDFLAG 77776  DAPDATR1 11103  DAPDATR2  11111  REDØCTR   00000  IMØDE 30 36001  IMØDE 33  26000  CHAN 11   01000  CHAN 12  00000
CHAN 13  00100  CHAN 14  00000  CHAN 30   37373  CHAN 31   37777  CHAN 32  77777  CHAN 33   67767

CDUX       .0000000      CDUY       .1098633E-01   CDUZ        .0000000       CDUT       -19.76440      ADØT/      -.3012188
ADØT/ØBP  -1.256434      ADØT/ØBY   .3407679       AK          .0000000       AK1        -57.75513      AK2         2.186279
THETAD X  -.5163574      THETAD Y  -64.36890       THETAD Z   10.53589        TIG         2372.820      BEST I      .0000000
BEST J     .1666667      MRKTIME1   404463.1       STAR X1     .7653723E-01   STAR Y1    -.2367206E-01  STAR Z1     .5000000
STAR RS1   .5063776      MRKTIME2   2368.290       STAR X2    -.1118258E-03   STAR Y2     1.219482      STAR Z2     .0000000
STAR RS2   1.219482      HAPØ       84000.00       HPER        .2709327E 08   RSP-RREC    575738.0      VGTIG X    -85.17761
VGTIG Y    409.6725      VGTIG Z    1078.389       ADØT/      -.3012188       ADØT/ØBP   -1.256434      ADØT/ØBY    .3407679
AK         .0000000      AK1       -57.75513       AK2         2.186279       THETAD X   -.5163574      THETAD Y   -64.36890
THETAD Z   10.53589      ØGC        .0000000       IGC         .0000000       MGC         .1301066      TEVENT      2372.820
LAUNCHAZ  -49.97680      LEMMASS    .0000000       CSMMASS     26520.00       ERRØRX     -.1428223      ERRØRY      58.55713
ERRØRZ    -2.208252      WBØD/ØCR  -.3902782       WBØD/ØCP   -3.980499       WBØD/ØCY    .5621408E-01  THETAD X    5.778809
THETAD Y  -111.4453      THETAD Z   15.16113       S/C MASS    1788.571       X CG        8.466005      Y CG        .3071489
Z CG       .5077438      JET FUEL   3.000366       SLØSH AR   -7.886368       IXX         31119.44      IYY         79900.74
IZZ        84710.12      IXY        2036.133       IXZ         246.1853       IYZ         1424.454      WX DEG/S   -.3497057E-02
WY DEG/S   .1049118E-01  WZ DEG/S   .3497057E-02   CØNFIG      1.000000       AXSCM/S2   -.9327403E-03  AYSCM/S2   -.1329155E-02
AZSCM/S2   .4605405E-02  XGYRØØUT   .0000000       YGYRØØUT    .0000000       ZGYRØØUT    .0000000      AXSMM/S2   -.9338093E-03
AYSMM/S2  -.1333495E-02  AZSMM/S2   .4614544E-02   ARSMM/S2    .4893284E-02
```

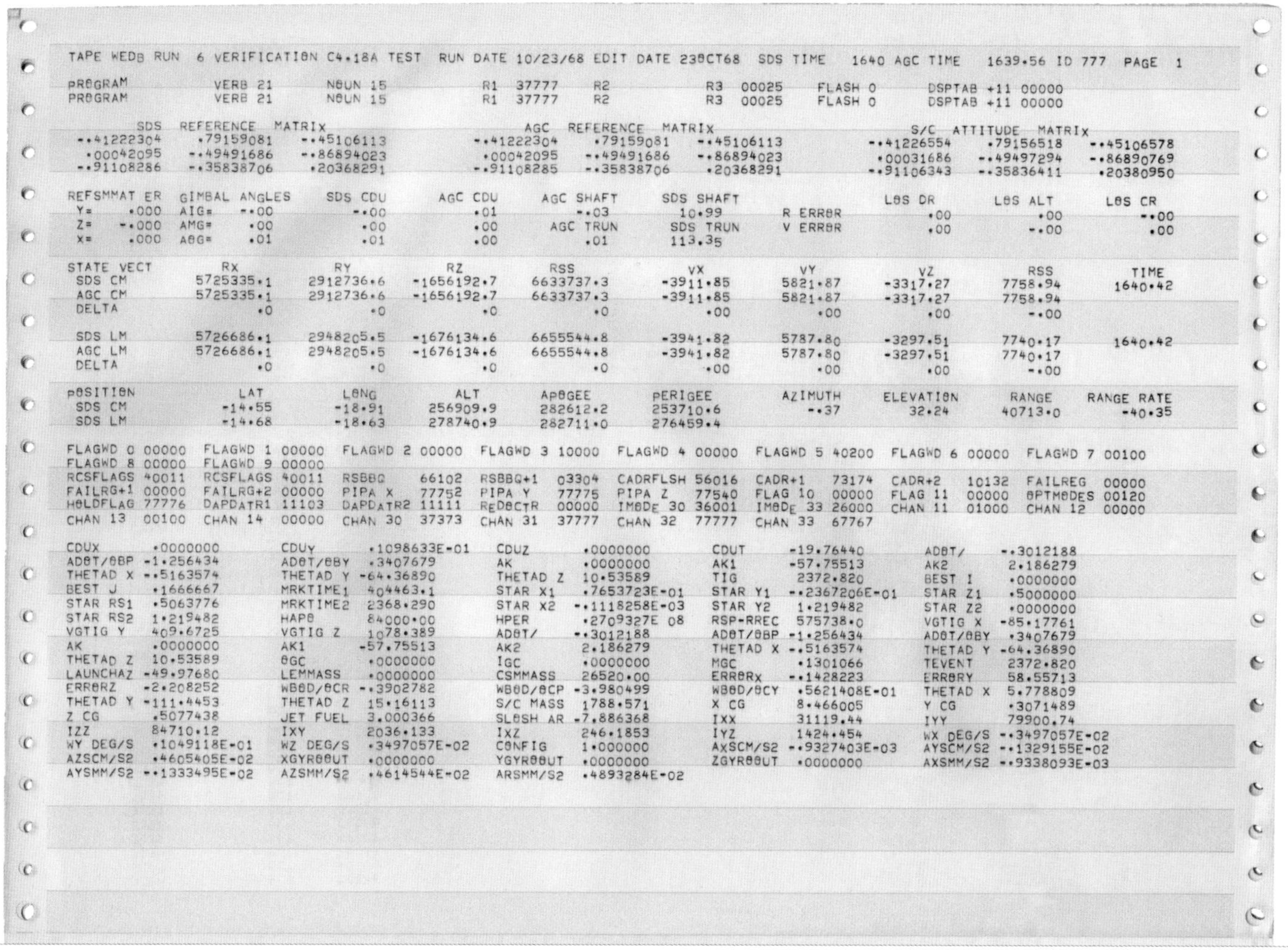

22.1 American computer engineer Frances 'Poppy' Northcutt, pictured here in 1969, was the first female engineer to work for NASA's mission control. She later became a lawyer focused on women's rights and civil rights issues.

22.2 Computer scientist and systems engineer Margaret H. Hamilton with Apollo computer software, 1969.

22.3 One of the many pages that make up Margaret Hamilton's program listings from the Apollo Guidance Computers, sourced from the Apollo Flight Guidance Computer Software Collection. This collection consists of reports, memoranda and related material documenting the Apollo flight guidance software developed by Margaret Hamilton's team at the Charles Stark Draper Laboratory (CSDL) in the late 1960s and early 1970s.

198 22.4  22.5

22.4, 22.5 The ILC seamstresses tracing patterns in nylon and stitching layers of aluminized plastic together, from a photo series by Ralph Morse published in *Life* magazine, 9 August 1968.

22.6 Hazel Fellows sewing an Apollo A7L spacesuit at ILC, 1968.

22.7, 22.8 Macro images of Buzz Aldrin's Apollo 11 A7L spacesuit, taken at the Smithsonian, Washington, DC, 2014, showing the weave of the fabric and the stitching in intimate detail. Instantly recognizable, it's one of the most celebrated engineered objects of the 20th century.

The astronauts' lives and the success of Apollo were hanging by their threads. Tolerances were small. Tolerance for errors even smaller. Eleanor Foraker was one of the supervisors who would regularly work eighty-hour weeks to meet the tight deadlines. One among her jobs was strict pin management: making sure all were accounted for so that none were left in the suits. Despite the pressures, there was a camaraderie and sense of pride at being part of something so significant. The women would share jokes with the astronauts who would come in for fittings. What a claim to fame for Iona Allen, who personally sewed Neil Armstrong's boots – boots that would make the most famous footprints in human history. In the years that followed it was revealed that some of the seamstresses had secretly written their names on the fabric hidden inside the suits' layers. Like everything made for Apollo, their work was destined for single use.

In Waltham, Massachusetts, women who had worked at the Waltham Watch Company were re-hired by the electronics company Raytheon to help build the Apollo Guidance Computer (AGC), which would manage the spacecrafts' guidance and navigation systems. Designed by MIT, the AGC used integrated circuits, a recent invention that revolutionized the electronics industry at a time when computers were the size of rooms. At Raytheon the skills of watchmaking and textile weaving were repurposed. From its earliest days, computing shared its roots with textile technology – Charles Babbage's Analytical Engine was inspired by the Jacquard loom in the early 1800s, which used punch cards to control the pattern of the weave. Building reliable computer memory in order to get to the Moon meant physically hardwiring information in the form of a copper wire weave called 'core rope memory'.

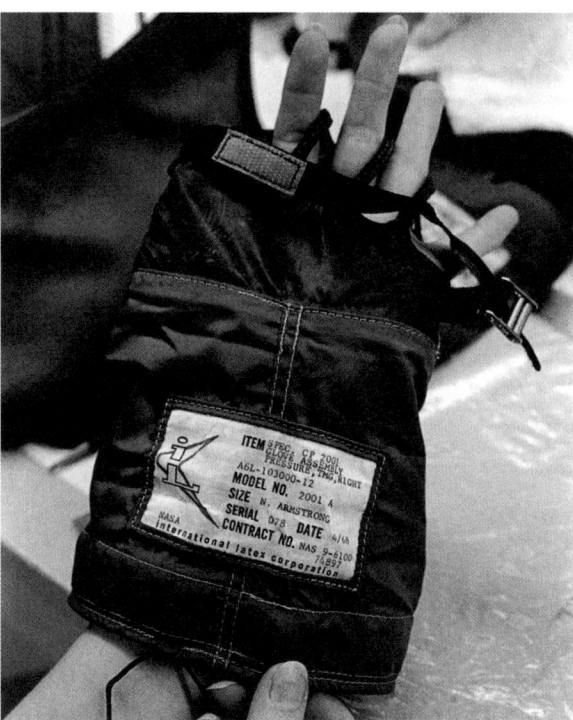

22.9 Supervisor Eleanor Foraker on the Apollo production floor, 1967.

22.10 Part of Neil Armstrong's right glove assembly for an A6L spacesuit, custom made by ILC for his role as backup commander of Apollo 8.

22.11 The gloves were made from plaster casts of astronauts' hands, each bearing the astronaut's name. Photo Ralph Morse, published in Life magazine, 9 August 1968.

22.12 The layers of the suit were laid out at left, from outside layers in front (right to left) to inner layers in back. Photo Ralph Morse, published in Life magazine, 9 August 1968.

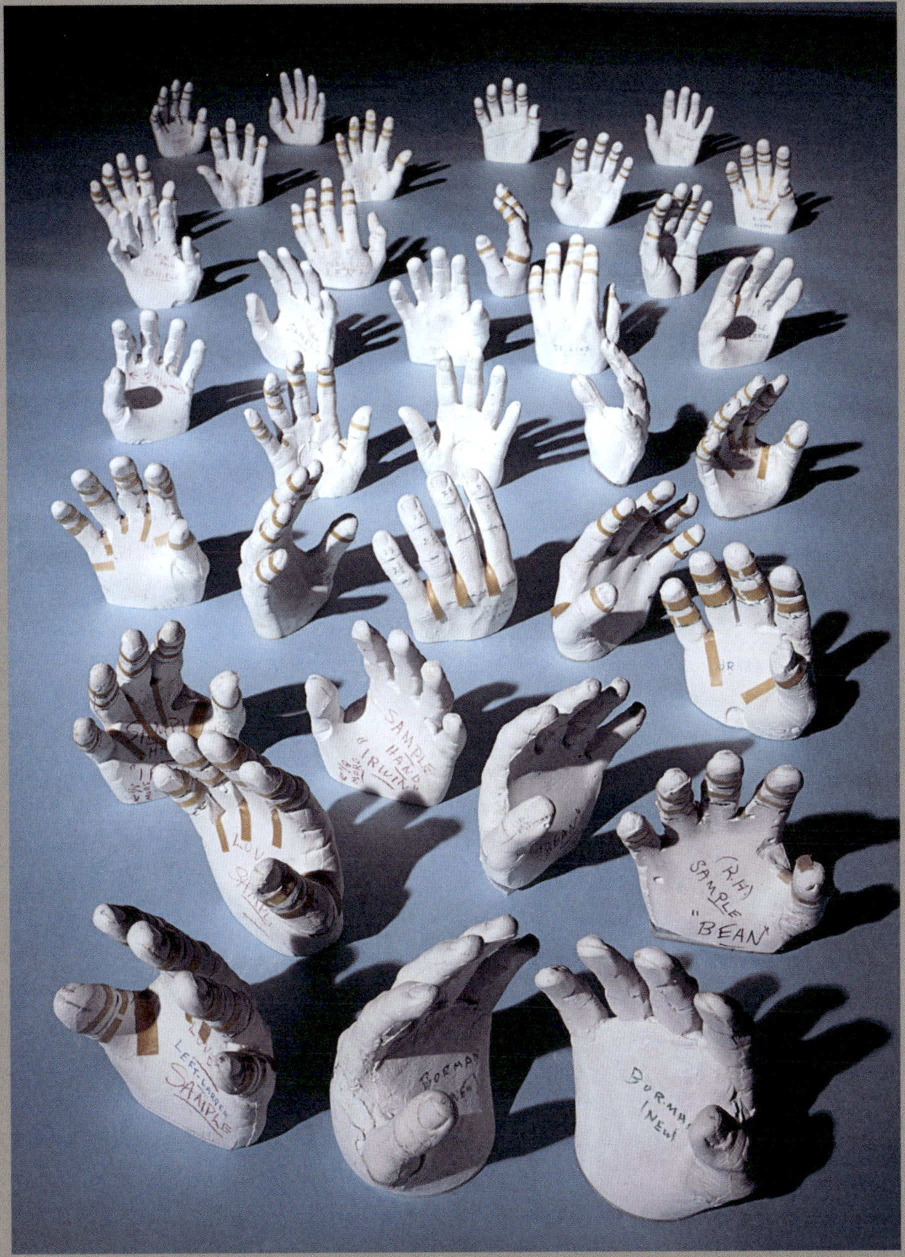

22.11　　　　　　　　　　　　　　　　　　　　22.12　　　　　　　　　　　　　　　　　　　201

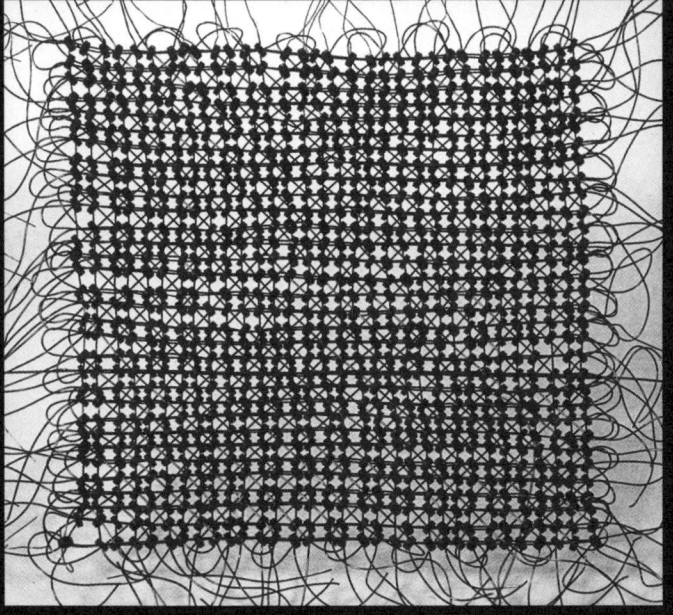

22.13–22.16 (CLOCKWISE TOP LEFT) Photographs from the DDC Magnetic Core Memory Archives at the Massachusetts Institute of Technology: memory plane (P74-27); 64 × 64 Module mat (F-2942); ferrite memory plane (P74-113); memory plane (P74-20).

22.17 Core rope memory works by magnetizing small rings with an electric current.

22.18, 22.19 A 1969 Raytheon press release featuring a 'space age needle worker', describing the core rope memory 'weaving' process used in the Apollo Guidance Computers.

22.20 Hands weaving core rope memory.

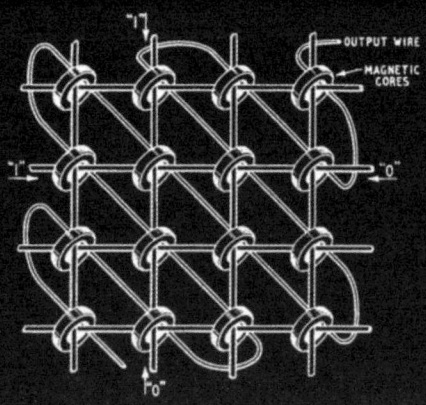

Using a special loom (also controlled with punch cards), the Waltham women would thread a long hollow needle with fine copper wire and pass it to and fro through and around a matrix of small magnetic ferrite 'core' rings. The path of the wire would denote a binary 0 or 1, depending on if the wire passed through or around the core. This process required dexterity, patience and precision, and resulted in a highly complex, tightly woven copper fabric. Here was physical written memory that couldn't be altered. Software woven as hardware, like a space-age Bayeux Tapestry. The technique became affectionately known as the 'Little Old Ladies' method.

The traditional artisanal techniques used in our early space missions have left an important cultural legacy and are still vital for the modern aerospace industry. Aerothreads is a women-owned company of artists and engineers based in Maryland who design and fabricate multi-layered insulation blankets for spacecraft. One of their employees, Morgan Betsill, worked on the Jet Propulsion Laboratory's Europa Clipper spacecraft, launched in 2024 to explore Jupiter's icy moon. Critical to that mission is the thermal blanket protecting the spacecraft's systems from the extreme environment of deep space. Understanding how to create fabrics suitable for the awkward geometries of a spacecraft, while also meeting the technical demands, requires the mind of an artist and engineer. These skills are a testament to how our most forward-thinking technology still relies on our most ancient. Techniques that are woven into our collective biological memory, that will serve us for all time, wherever we may go.

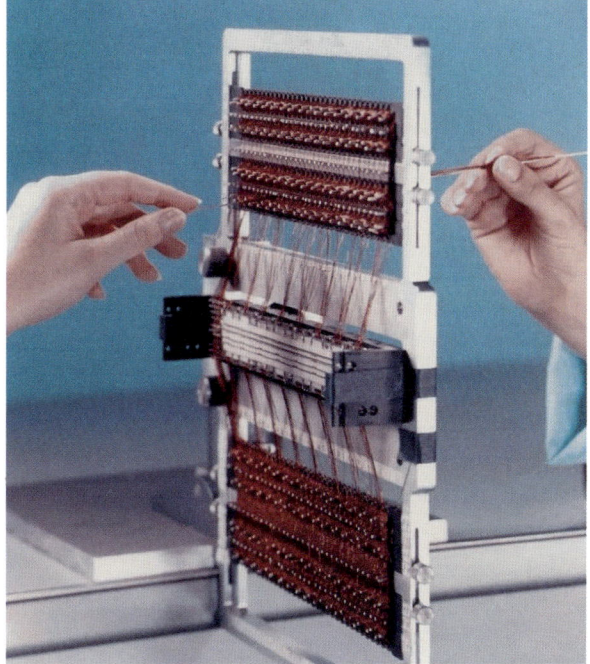

TREVOR BEATTIE

There was something about the Moon walkers' ungainly gait and oversized heads that reminded Trevor Beattie (b. 1958) of the marionette kids TV series *Thunderbirds* and *Fireball XL5*; shows that were populated with characters with exciting names like Colonel Steve Zodiac. For a boy growing up in Birmingham in the '60s and '70s, puppets were as close to space as it got, until the real-life Apollo adventures became the new obsession. Kids would rattle off the names of all the astronauts; they knew all the space facts and figures, and dutifully collected the Brooke Bond tea 'Race into Space' cards. But it was a school geography project that would prove to be the real catalyst for Trevor. His Silvine exercise book, carefully backed with brown paper, earnestly charts the history of the early space age with essays, timelines, clippings and portraits. The last page poses a philosophical question: 'What of the future?' Provided is a blank space to be filled in when the future finally arrives; a future which he wholeheartedly believed would include his own inevitable journey into space.

Mr Palmer (Class 2A, Geography) awarded him a disappointing 'B+' for his efforts, with a 'Good work' in red pen. Why hadn't he given him an A? Trevor wondered what his own space travel future might look like. He knew, if nothing else, the future would be expensive. It always is. In the Nicolas Roeg film *The Man who Fell to Earth* (1976), the alien Thomas Jerome Newton (David Bowie) has to raise tremendous funds to build a spaceship that will get him back to his home planet. By the 1990s, Trevor had generated his own considerable wealth, becoming one of the most important names in UK advertising and the creative mind behind some of the decade's biggest campaigns, notably the Eva Herzigová 'Hello Boys!' Wonderbra advert, and the marketing for Britain's latest political force, New Labour. By 2004, the planets were starting to align. The Ansari X Prize – a competition that awarded $10 million to the first private company to build a reusable spacecraft – had been claimed by Burt Rutan, and his design became the foundation for Richard Branson's Virgin Galactic, the first ever 'spaceline', offering suborbital flight for paying customers. Without hesitation, Trevor put up the cash and his name was on the waiting list. How long that wait would be, however, no one knew.

The Virgin Galactic spaceship echoes the experimental rocket planes of the early 1960s (namely the high-altitude X-15 that touched the

23.1 Trevor Beattie in seat harness during pre-flight training session, 2023.

23.2 The Galactic Golden Ticket.

23.3-23.6 Brooke Bond Picture Cards: THE RACE INTO SPACE, 1971.

Brooke Bond Picture Cards
THE RACE INTO SPACE
Man's first 50 steps into the universe
5p (14)

6. Vostok. First man to go into space was Russia's Yuri Gagarin. On Apr. 12, 1961, his 10,418 lb Vostok spacecraft made a single orbit of the Earth inclined at 65° to the Equator and ranging between 109 and 188 miles altitude. The spacecraft comprised a ball-like re-entry capsule, containing the cosmonaut on an ejection seat, and a service module with retro-rocket, air supply and gas-jet stabilization. The entire flight from lift-off to touchdown lasted 108 min. Five more cosmonauts later flew in Vostok spacecraft, including the first space woman Valentina Tereshkova. Gagarin was killed in a flying accident on Mar. 27, 1968.

Yuri Gagarin
Valentina Tereshkova

7. Russia's Baikonur cosmodrome is located east of the Northern end of the Aral Sea in Kazakhstan. It was from here that a Soviet heavyweight intercontinental ballistic missile sent Sputnik 1 into orbit in Oct. 1957. This 1½ stage rocket has since been developed to launch much heavier payloads—including the manned spacecraft Vostok, Voskhod and Soyuz—by adding new top stages and increasing take-off thrust.

There were six Vostok missions:			
SPACECRAFT	DATE	ORBITS	COSMONAUT
Vostok 1	12 Apr. 1961	1	Gagarin
Vostok 2	6 Aug. 1961	17	Titov
Vostok 3	11-15 Aug. 1962	64	Nikolayev
Vostok 4	12-15 Aug. 1962	48	Popovich
Vostok 5	14-19 June 1963	81	Bykovsky
Vostok 6	16-19 June 1963	48	Tereshkova

Ejection seat of the Vostok spacecraft.

7. Vostok on launch pad. The multi-stage rocket that lifted Yuri Gagarin into orbit had a two-stage central core plus four tapered strap-on boosters. Total thrust of all stages was 1,323,000 lb. Assembled horizontally on a transporter-erector, the combined rocket and spacecraft was taken out of the preparation building to the launch pad on a railed track. At the pad, the empty launcher was lifted into a vertical position and cradled by four hinged 'steady' arms. Servicing platforms also hinged into position around the rocket ready for final checkout and fuelling with liquid oxygen and kerosene.

How the seat could be ejected from the launch vehicle at an upward angle after the explosive separation of a circular hatch.

15. Voskhod. This manned spacecraft filled the gap between Vostok and Soyuz. With cosmonaut Vladimir Komarov in Voskhod 1 on Oct. 12, 1964 were a doctor Boris Yegorov and a scientist Konstantin Feoktistov. They completed 16 orbits in a flight lasting 24 hr 17 min 3 sec. Spacesuits were not worn and the capsule's parachute landing was cushioned by retro-rockets ignited just before ground contact. Voskhod 2, launched on Mar. 18, 1965, carried Pavel Belyaev and Alexei Leonov. Both wore spacesuits allowing Leonov to make history's first 'space walk' of about 10 minutes' duration from an extensible airlock. Their flight lasted 26 hr 02 min 17 sec. Cosmonaut Belyaev, aged 44, died in Moscow on Jan. 10, 1970, following an operation for a stomach ulcer, from peritonitis.

Alexei Leonov
Pavel Belyaev

15. During his pioneer 'space walk', Leonov received his oxygen from cylinders strapped to his back. The lifeline included cables for communications with co-pilot Belyaev (and ground command) and for relaying data from bio-medical sensors attached to his body. Leonov spent 23 min 41 sec outside the Voskhod 2 cabin, more than half that time in the airlock.

First Russian space walk

16. ...satellites were sent into Earth-orbit in 1965 to test mock-ups of the Apollo command and service modules and study the size and frequency of micro-meteorites. Each craft had concertina 'wings' spanning 100 ft—a sandwich of aluminium, Mylar and copper, carrying a mild electric current; when struck by a small particle a short-circuit was created for a fraction of a second, and the results telemetered to Earth. By mid-1966 the satellites had recorded more than 1,100 impacts, but none sufficiently damaging to deter the Apollo Moon flights.

17. Gemini 3 was the first of ten highly successful operational flights with the two-man American spacecraft. Flown by Virgil Grissom and John Young, it was launched by a Titan 2 rocket from Cape Kennedy on Mar. 23, 1965. It was placed in an initial orbit inclined at 33° to the Equator and varying in height between 100 and 139 miles. After a 3-orbit flight lasting 4 hr. 53 min., the spacecraft was recovered successfully in the Atlantic. The illustration depicts Gemini 4, with crew-member Ed White making the first American space walk, which was of 21 minutes duration.

8. Mercury. First American to orbit the Earth was Colonel John Glenn. Launched from Cape Canaveral by Mercury-Atlas MA-6 on Feb. 20, 1962, his Mercury capsule Friendship 7 completed three orbits in 4 hr 55 min. Two earlier manned Mercury flights—sub-orbital—were launched by Redstone rockets. Alan Shepard made a 15 min 22 sec ballistic lob over the Atlantic on May 5, 1961 and Virgil Grissom (later killed in the Apollo launch pad fire) made a similar flight on July 21, 1961. Both were successful but Grissom's capsule sank after the hatch blew out and it filled with water.

Virgil Grissom *John Glenn* *Alan Shepard*

Six missions were also flown by Mercury astronauts:			
SPACECRAFT	DATE	ORBITS	ASTRONAUT
Freedom 7	5 May 1961	Sub-orbital	Shepard
Liberty Bell 7	21 July 1961	Sub-orbital	Grissom
Friendship 7	20 Feb. 1962	3	Glenn
Aurora 7	24 May 1962	3	Carpenter
Sigma 7	3 Oct. 1962	6	Schirra
Faith 7	15-16 May 1963	22	Cooper

9. Few details are available of the large Cosmos satellites which eject capsules for recovery over the Soviet Union. The frequency with which launchings occur into low orbits suggest their main task is observation of the Western World and Communist China. Hundreds of Cosmos satellites, including recoverable and non-recoverable types, have been launched—many having standardised features to achieve 'production line' economy in manufacture.

9. Cosmos. Large quantities of these satellites have been launched. Some are used for scientific research, and others to test new spacecraft systems with civil and military applications. Cosmos 1, a non-recoverable scientific satellite, was launched on Mar. 16, 1962 from Kapustin Yar. Many of the larger spacecraft launched from Baikonur and Plesetsk apparently have reconnaissance duties; they eject capsules for recovery over Soviet territory usually within 8 to 12 days. Illustrated is one of the smaller non-recoverable scientific satellites.

10. Britain's Ariel satellites are designed purely for scientific research, and stem from a joint Anglo-American programme originated in Mar. 1959. In fact, Ariel 1 was the very first international collaborative space programme. Ariels 1 and 2 were built in America, but contained British scientific experiments. Ariel 3 was the first satellite to be built entirely in Britain, under the direction of the Ministry of Aviation (now the Ministry of Technology). Ariels 2 and 3 were launched on Mar. 27, 1964, and May 5, 1967, respectively. Development of a fourth satellite is under way.

10. Ariel 1. Britain's Ariel 1, built in America, was fitted with British scientific experiments, and was launched into a 754 by 242 mile orbit by an American Delta rocket from Cape Canaveral on Apr. 26, 1962. The 132 lb satellite carried seven experiments, from three British universities, designed to provide purely scientific information about the ionosphere, the complicated region of particle and magnetic fields which surrounds the Earth at an altitude of about 100 miles. Ariel 1 continued to transmit data until Nov. 1964, when it fell silent, but it was still in orbit at the beginning of 1970.

23.7

23.8

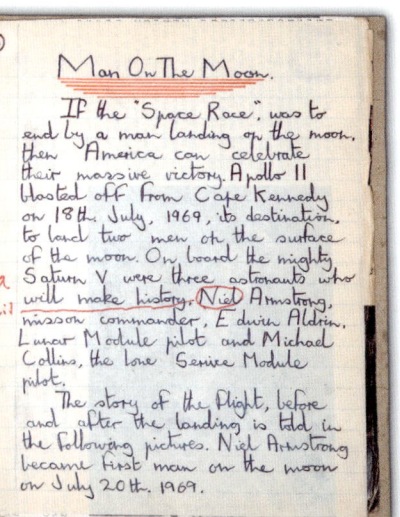

23.9

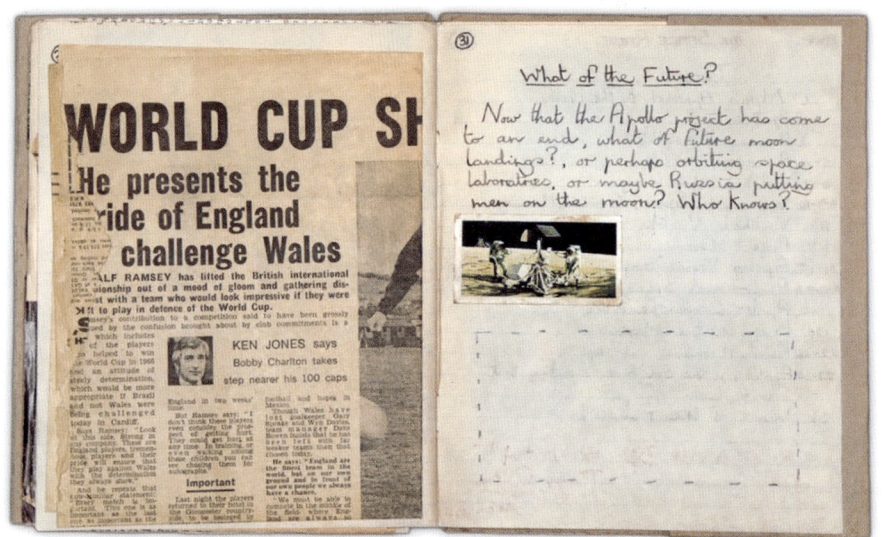

23.10

23.11

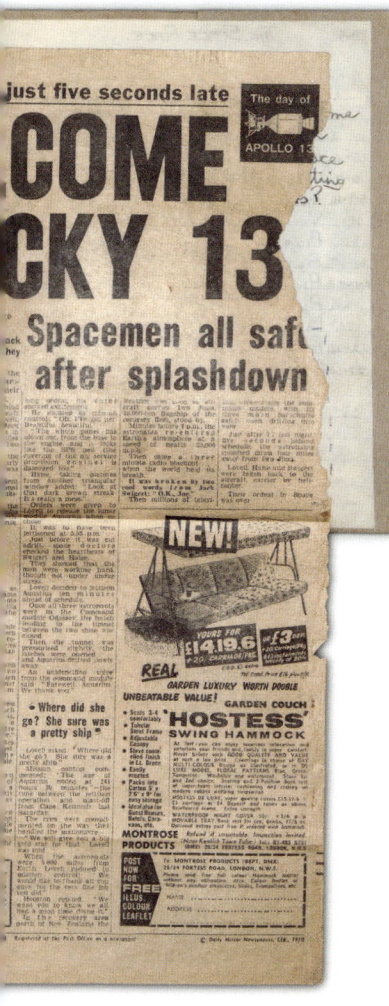

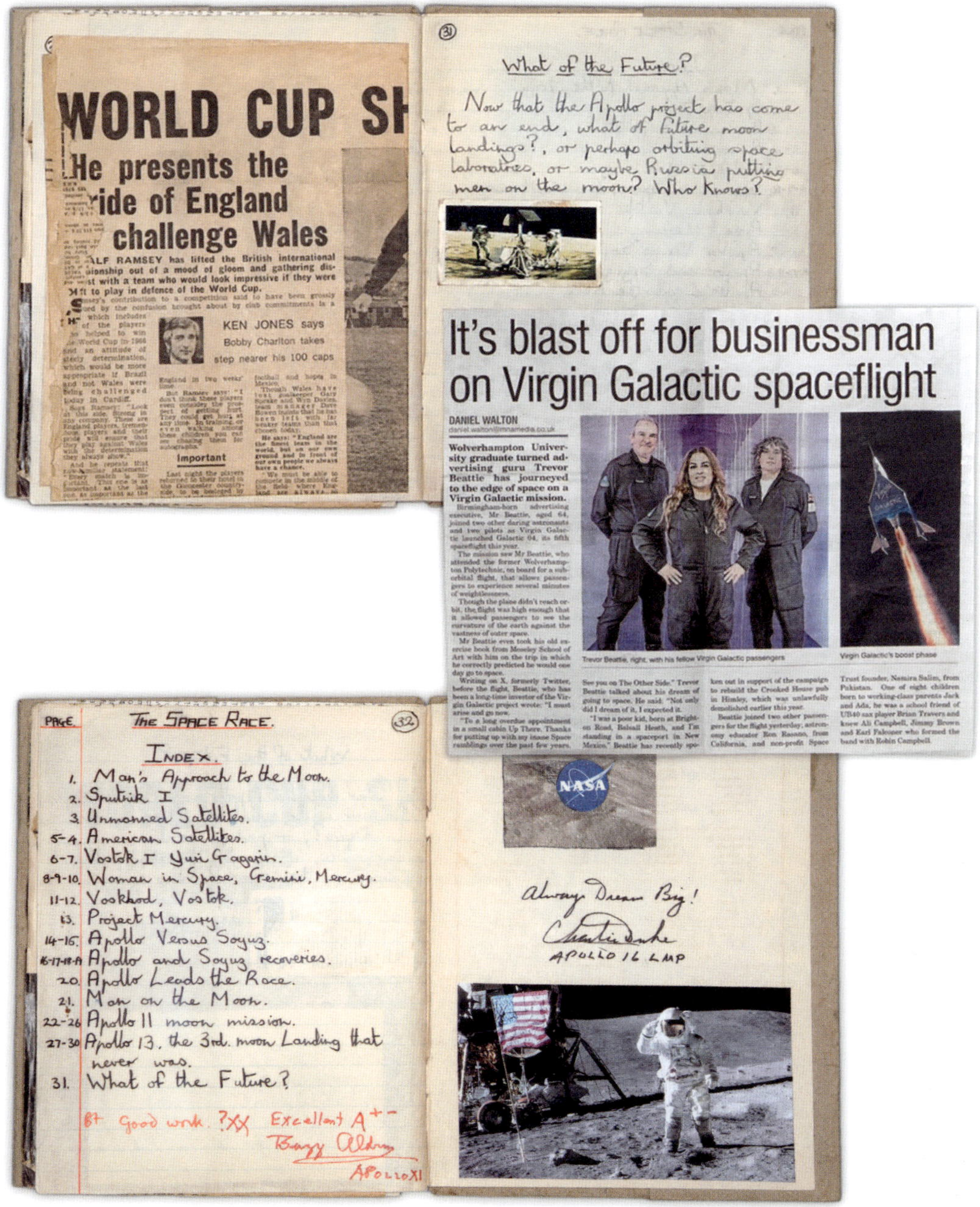

23.12 / 23.13

Pages from Trevor Beattie's Silvine exercise book:

23.7 Front cover: Trevor Beattie 2A: THE SPACE RACE.

23.8 Red-lined inside front cover, featuring legendary astronauts.

23.9 'MAN ON THE MOON' with misspelling of 'Neil', corrected by Buzz Aldrin.

23.10 'WHAT OF THE FUTURE?', with rectangular space left blank.

23.11 'WELCOME HOME LUCKY 13', *Daily Mirror* cutting, 18 April 1970.

23.12 'WHAT OF THE FUTURE?', with 'IT'S BLAST OFF FOR BUSINESSMAN' newspaper cutting added to blank space.

23.13 Back page of Beattie's SPACE RACE project: marked 'B+ GOOD WORK' by Beattie's teacher, re-graded as 'EXCELLENT A+-' by Buzz Aldrin.

edge of space, flown by silver-suited legends like Scott Crossfield, Joe Engle and Neil Armstrong). When it drops from its carrier aircraft, the rocket engine ignites and you are greeted by a tremendous kick of g-force in your back, pushing you upwards with a violent rattle as you ascend into the blackness. As the spaceship reaches the top of its parabolic flight path, the engine stops and you become weightless, like a marionette dangling from its strings. For a few bone-shaking moments in October 2023, eighteen years after he'd booked his seat, Trevor Beattie became Colonel Steve Zodiac. In an hour, he was back on solid ground.

Tucked away under the pilot's seat was Trevor's geography project, which had since been re-marked by Buzz Aldrin, no less, who upgraded it from 'B+' to an 'Excellent A+-'. The Apollo 16 Lunar Module pilot Charlie Duke had given Trevor some advice on how best to spend his time on such a short trip: 'Look out the goddamn window!' Most people will tell you that space begins at 100 kilometres (62 miles); Trevor will tell you that space begins when the Earth is at its most intense blue, the sky is at its deepest black and you're in floods of tears. Outside, the Earth glows as if it possesses some kind of internal light, a beatific 'earthshine' that has only bathed the faces of a privileged few hundred people. From his pocket, he pulls out another personal memento – a cheque for $2.23 that was written by Orville Wright a hundred years previously. He folds it up into a paper plane and launches it across the cabin. Legacy, present and future held in a perfect moment.

Trevor doesn't want to be called an astronaut. Space traveller is fine, but 'astronaut' is a moniker reserved for those who have been in orbit. The blank space at the end of his geography homework is now filled with the *Wolverhampton Express and Star* newspaper reports of his adventure. But why had Buzz Aldrin included a minus in his new grade? 'I'd spelled "Neil" wrong,' said Trevor. 'I'd never noticed, even after all these years.' Homework completed. Handed in. Better late than never. Not perfect, but nothing of real value ever is.

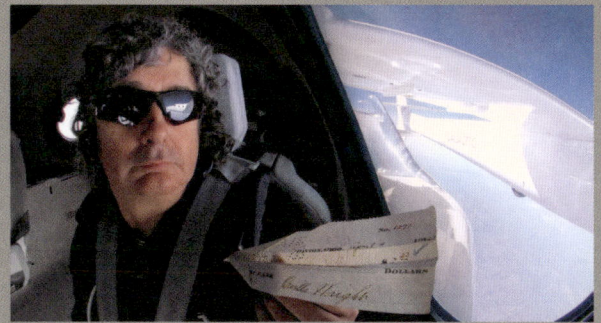

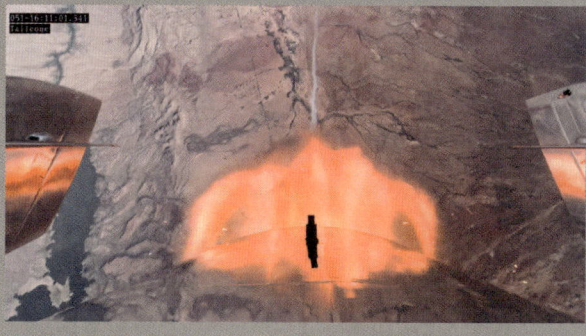

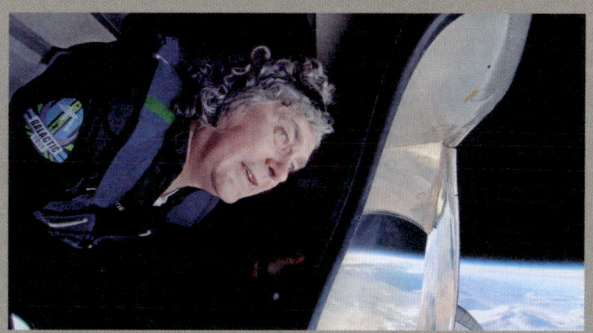

23.14, 23.15 Trevor Beattie's metallic Galactic 'Astronaut Wings' badge and Mission Patch.

23.16, 23.17 The Orville Wright cheque (unfolded and folded) that travelled with Beattie into space.

23.18–23.25 Ticket to ride. Beattie's journey into space courtesy of Virgin Galactic.

ACT 5.

VITAM QUAERITUR

'THE SEARCH FOR LIFE'
WHERE IS EVERYBODY?

OLIVER POSTGATE

The voice of Oliver Postgate (1925–2008) is one of the most cherished sounds of the British cultural landscape. Postgate was responsible for animated television classics of the 1960s and '70s such as *Bagpuss* and *Ivor the Engine*. Together with puppeteer Peter Firmin, he worked knee-deep in celluloid and cardboard in Firmin's disused pigsty at his farm in deepest rural Kent. Here a creative alchemy transformed odds and ends into beautiful stop-motion short films, motivated not by lofty ideals of high art but by the urgent need to feed the voracious television machine, ever hungry for more. Among their most celebrated characters were the inquisitive pink woollen Clangers (knitted by Peter's wife Joan), who lived on their isolated, moonlike micro-planet, happily minding their own business, occasionally disturbed by the activities and detritus of bothersome Earth.

The Clangers was of its time. The first episode aired just a few months after the Moon landing in 1969, a period of excitement and adventure as well as political tension and uncertainty. Each episode begins, appropriately, with the Earth hanging in the blackness of space, as per the Apollo 8 *Earthrise* photographs (figs 20.17–20.18) that had become such a potent symbol of the moment. Just a small camera move away, carried by Vernon Elliott's enchanting and melancholic musical score, we traverse past a few jewel-like stars to arrive at the Clangers' strange home, a place that despite its barren, cratered surface represents the type of rural idyllic escape so familiar in children's literature. Their own swanee-whistled language was instantly understandable by young children. For grown-ups, Postgate's typed shooting scripts are a Rosetta stone – the Clangers' dialogue was written in English before it was replaced by whistles. Each episode is introduced by Postgate's monologue, steering us through cautionary musings on the nature of alien life, space junk ('expensive rubbish'), environmentalism and troublesome earthly politics. It is prose as elegant and poignant as Carl Sagan's famous ode to the Earth, his *Pale Blue Dot* essay, which was still two decades away.

24.1 Oliver Postgate at work on Firmin's farm, Kent. To his left is Bagpuss, meticulously supervising.

24.2 Clanger puppet exhibited at the V&A Museum of Childhood, London, in 2016.

24.3, 24.4 Knitting patterns for *The Clangers*, as created by Joan Firmin.

24.5-24.8 (OVERLEAF) Pages from *The Clangers' Bible*, the booklet first presented to the BBC to pitch the idea.

24.9-24.12 (OVERLEAF) Storyboard and proposal pages from *The Clangers' Bible*, where Postgate first detailed the Clangers' musical language.

'... this calm serene orb sailing majestically among the myriad stars of the firmament. Perhaps this star too is home for somebody. Can we imagine the sort of people who might live on a star like this? Let us go very close. Let us look and listen very carefully. And perhaps we shall see. And hear.'

THE CLANGERS (1969)

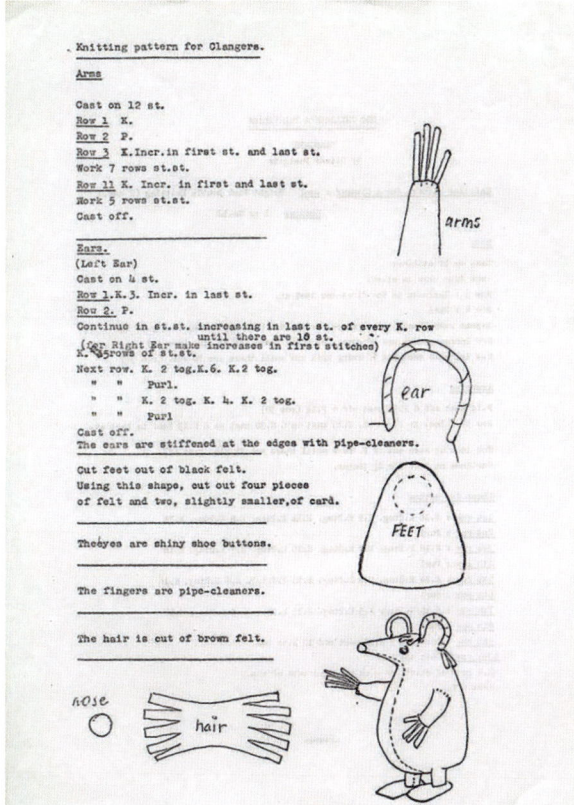

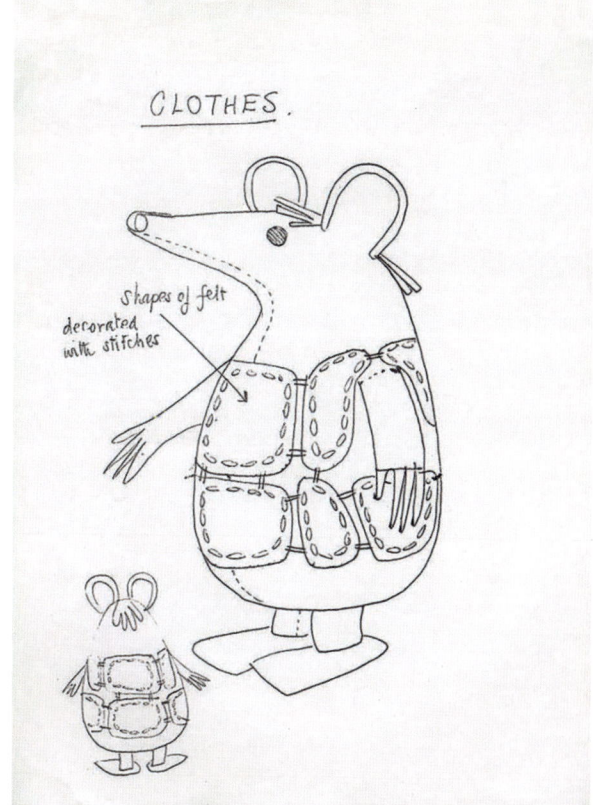

1

SMALLFILMS

First draft of proposed film Series for B.B.C.

This is colour and to be shown first on Sunday evening and then, later, in Watch with Mother.
If necessary we would make two sound-tracks for the two types of audience. This would allow us to avoid being too obvious in the Sunday evening treatment.

Apparently aerial view of earth.
(Super Space-type non-pan sup   music.)

see sputnik pass
see sattelite
see space probe
see tin bath or such.
see rowing-boat
  fly past in orbit.

pann up.
mix to sky with
odd-looking worlds

track through worlds
(slow fade music)

orb across
Track in on it

mix in track to
studio set
examine surface of planet.
see dustbin-type lids.

Nar. This is the twentieth Century. After a mere two thousand years of Civilisation man's scientific genius has at last brought him to the threshold of interplanetary travel.
Now that the Sputniks and the sattelites and the space probes and many other of the strange and complex works of man are hurtling in orbit around our cloudy planet, it is fitting that we should turn our eyes away from this little earth of ours and travel, in our imaginations across the boundless silent stretches of outer space towards other, stranger worlds.
Is it not possible that somewhere, in some far galaxy, in some unknown star there lives maybe a race of living beings whose standards of civilisation are far in advance of ours?
This serene orb, sailing majestically amid the myriad stars of the firmament. How far from the mad bustle and noise of our world.
How calm and silent. Who can say what god-like creatures may inhabit such a place?

3

(We have our first sight of the inside of the planet which is where the Clangers live. It may be a labyrinth of translucent caves with houses carved in the expanded polystyrene rock. There may be forests of copper trees with copper in leaves. The Clangers are metal-workers and most of their equipment is made from copper sheet and rock. They have a scource of power which we do not have which is that the berries of the some of the trees have the faculty of generating steam-like pressure when enclosed.
This means that they have some rather basic cars and power-tools.)
We follow the people down the corridor to where some other Clangers are waiting. They stop and, pointing back, tell what has happened.
The aeronaut Clanger comes disconsolately down the slope, dangling his helmet.
(He is, incidentally our father-figure. The story now becomes less general.)
He sits in his chair and looks sad. Mrs. Clanger dabs his brow and murmurs words of comfort. She suggests that he might like a bowl of soup. He nods. Small Clanger(our hero) is despatched to fetch a bowl of soup. He takes the copper can and gets on his bike thing Tiny Clanger(our baby heroine) is distraught at his departure and demands to go as well. She jumps on the back of the bike and they ride away to the forest where the soup-wells are.

The soup-wells are little pointed hills or stalagmites. Small Clanger and Tiny Clanger sniff about like Bisto Kids until they smell the flavour soup they are after. Small Clanger unscrews one of the hills and looks in. There is a shout of rage from inside and he replaces it hastily. He unscrews another hill and speaks civilly to the soup-keeper inside. He passes in the jug to be filled. Tiny Clanger leans too far over and falls in. There is a great noising inside the well and she is pitched out again rather soupy. The soup-keeper hands out the can of soup. Tiny Clanger is told to pick a flower from the copper tree. She does this and tosses it into the well. The soup-keeper is heard to receive it with delighted sounds and he chews it up with relish. xxxxxxxxxxxxxxxxxxxxxx xxxxxxxxxxxxxxxx The Clangers turn to go.
The Soupkeeper calls to them(to shut the door)
Small Clanger runs back and screws on the hilltop. They ride home. Deliver the soup to Mum.
Pop drinks the soup. Scratches himself. He settles down for a nap, puts a circular newspaper over his face and snores.

> 'Now that the Sputniks and the satellites and the space probes and many other of the strange and complex works of man are hurtling in orbit around our cloudy planet, it is fitting that we should turn our eyes away from the little earth of ours and travel, in our imaginations across the boundless silent stretches of outer space towards other, stranger worlds.'
>
> THE CLANGERS' BIBLE

Throughout the episodes there's a light dusting of Postgate's own political background. He was a lifelong pacifist and a conscientious objector during the Second World War. His father and grandfather were both prominent socialists. He was educated at Dartington Hall in Totnes, famous for its progressive 'no rules' approach, which he notes as being one of the unhappiest periods of his life. The Clangers' politics are a reflection of Postgate's own navigation of the world. His love of invention, his reactions against the frenetic pace of modernity and waste, and the absurdities of war. The Clangers were living in a socialist utopia on their own tiny island in space, occasionally disturbed by the values of an outside world made manifest through the assortment of alien objects that would wash up on their shore – from a noisy television set to an intrusive robotic probe and a net of gold coins. There are some more overtly political motifs too, such as the visiting astronaut who plants a flag that features both the American Stars and Stripes and the Soviet Union's Hammer and Sickle. It is soon repurposed by the Clangers as a new tablecloth. 'It's nice to have visitors …' narrates Postgate as the trespassing astronaut leaves, 'But sometimes it's nicer to see them go.' For a generation, the Clangers' small adventures were our first introduction to space and politics. Like the space race on Earth, the Clangers had their own exploratory space programme: in an episode titled 'The Intruder', Major Clanger sits on his rocket, ready for his journey to our planet. But through a telescope, the Clangers see the reality of the Earth beneath its occluded clouds. We see the vast monolithic skyscrapers of a city. Soulless. Inhuman. Confusing. The mission is cancelled at once. The Earth is too muddled a place for a Clanger. The rocket ship is packed away, safely stored underground by the soup wells to be repurposed for something more useful.

24.13–24.15 Stills from 'The Tablecloth', *The Clangers*, S2E1.

24.16 Still from 'The Intruder', *The Clangers*, S1E5.

24.16

CARL SAGAN

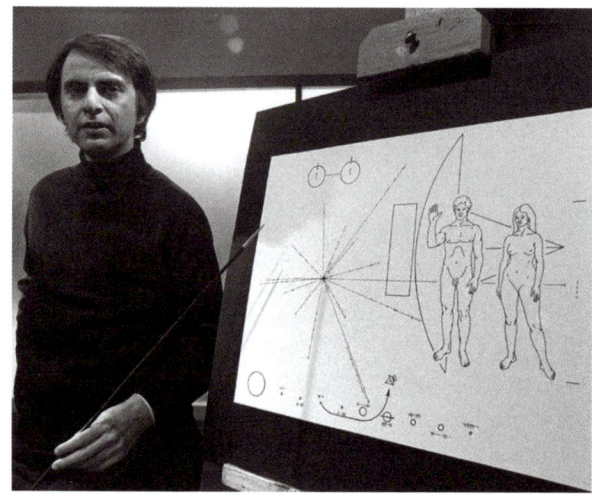

1977 was an important year for humans. We saw a universe filled with exotic life, courtesy of *Close Encounters of the Third Kind* and *Star Wars*. The husband-and-wife designers Charles and Ray Eames expanded our minds with their short film *Powers of Ten*, taking us to the very edge of the universe and back the other way into the nucleus of an atom. It was also a big year for music. Elvis had left the building for the last time. Fleetwood Mac's *Rumours* and the Sex Pistols' *Never Mind the Bollocks* were released, coinciding pleasingly with the centenary of Thomas Edison's invention of the phonograph. But another double album that year was destined to outshine – and outlive – them all; one that would become synonymous with its curator, the young, enigmatic Cornell University astronomy professor Carl Sagan (1934–1996), whose clarity of thought, melodious baritone voice and infectious, poetic wonder about science would inspire generations to follow.

This particular release was the 'Sounds of Earth' Voyager Golden Records, described as the Earth's greatest hits. The discs were attached to the two Voyager spacecraft, built in haste by the Jet Propulsion Laboratory to take advantage of a once-in-a-175-year window of planetary alignment. They launched in the summer of 1977; their mission: a grand scientific tour of the mysterious outer planets, of Jupiter, Saturn, Neptune and Uranus, and their moons. Decades later, blind and frail, the spacecraft sail on, still collecting data beyond our solar system. Even after we lose contact with them they will continue, our most distant emissaries, a billion years of sailing ahead of them into the calm waters of the interstellar void – the benign, silent darkness between the stars. Artefacts from a forgotten planet, carrying with them a gift for the once-imagined extraterrestrial audience who might chance upon them in a universe of soul-crushing indifference.

Sagan's passion for time-capsuling, as well as science, was ignited as a young child at the 1939 New York World Fair's 'World of Tomorrow' exhibition. Westinghouse Electric Corporation

had buried a torpedo-shaped time capsule under Flushing Meadows containing all sorts of 1930s ephemera, to be opened in the year 6939. It's not hard to see why he was so enthralled – the time capsule as an idea is a declaration of hope and optimism, an acknowledgment that some kind of tomorrow, perhaps different from today, will eventually come.

The Voyager Records were not the first cosmic time capsules to be sent into space by Sagan. In 1972 the Pioneer 10 and 11 spacecraft would become our first interstellar probes whose trajectories would see them flung outside our solar system. SETI (Search for Extraterrestrial Intelligence) founder and astronomer Frank Drake discussed the significance of this event with Sagan, and between them it was decided to construct an engraved metal plaque that could be attached to mark the remote possibility that an extraterrestrial civilization might chance upon it. What should it say? Science would be the language of choice, given that we share a common mathematics, physics and chemistry wherever we are in the universe. Sagan suggested a pictorial map of our solar system. Drake suggested a pulsar map as a location marker. A diagram of the energy states of a hydrogen atom, the most common element in the universe, was included. Artist and screenwriter Linda Salzman, then Sagan's wife, drew two naked figures: a man and a woman standing next to each other.

25.1 Carl Sagan on the TV programme *Camera Three*, 28 January 1974, with the Pioneer 10 interstellar plaque illustrations.

25.2 Stills from Charles and Ray Eames, *Powers of Ten* (1977), a masterpiece about scale and perspective. The audience travels outwards to the edge of the universe before returning to the proton in the nucleus of a carbon atom, in the hand of a picnicker in Chicago. It's an idea that has since been copied many times, but never bettered.

25.3, 25.4 Pages from G. E. Pendray, *The Story of the Westinghouse Time Capsule* (1939).

25.5 Officials lowering the Westinghouse time capsule into ground, New York World's Fair, 1939.

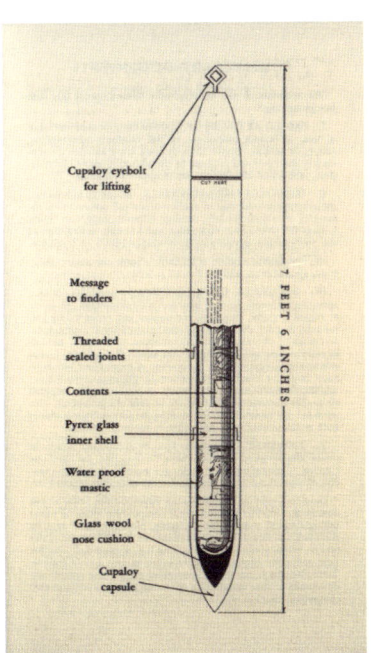
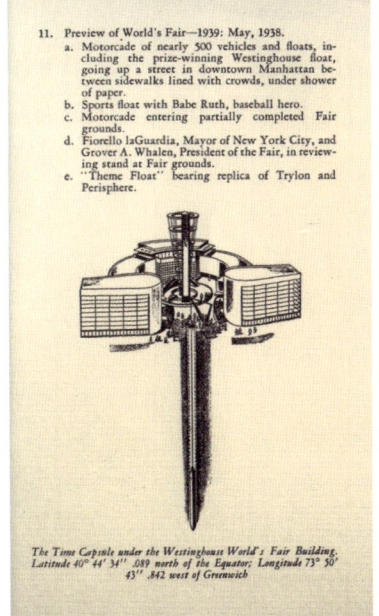

25.6 'The Sounds of Earth' Voyager Golden Record, 1977.

25.7 The Voyager Record cover, with instructions on how to play it (see also fig. 25.9).

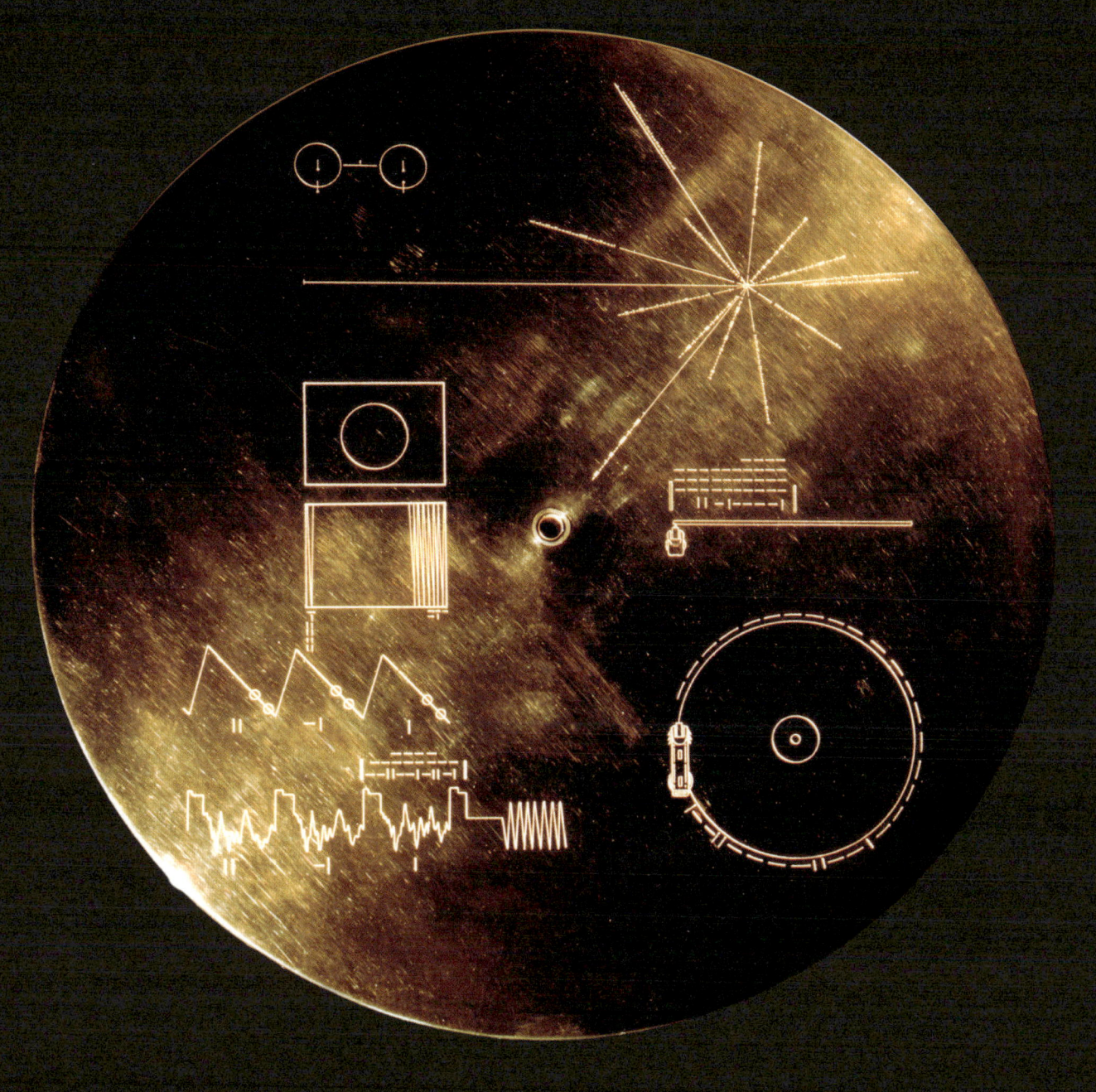

'I had monuments made of bronze, lapis lazuli, alabaster ... and white limestone ... and inscriptions of baked clay ... I deposited them in the foundations and left them for future times.'

ESARHADDON, KING OF ASSYRIA, QUOTED IN *MURMURS OF EARTH* (1978)

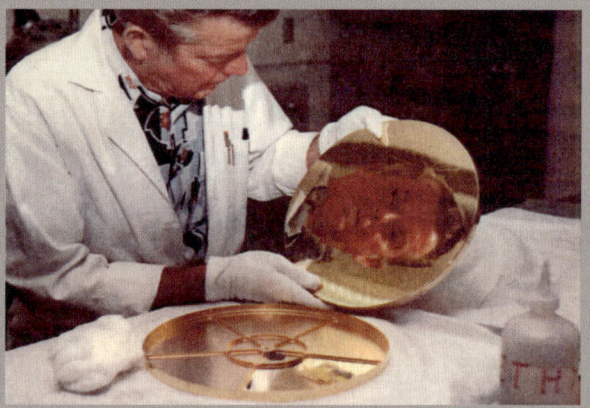
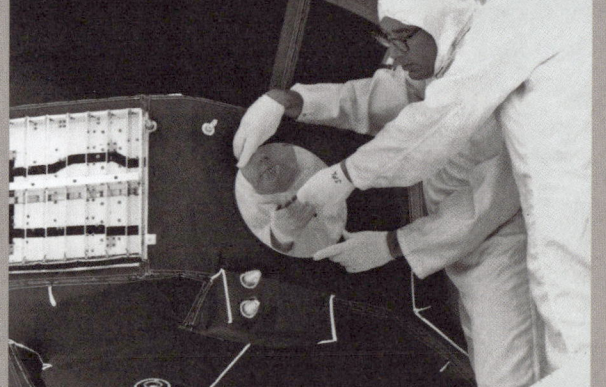

25.8 Production of the Voyager Golden Records in 1977.

25.9 Explanation of the Voyager Records' diagrammatic instructions for use.

The nudity raised eyebrows and questions down here on Earth. Why was the man's hand raised and not the woman's? Why was she standing slightly behind him? Would an alien even know what the open-palm gesture meant? What if they have tentacles? If nothing else, it showed that universal agreement on how we represent ourselves as a collective species is political – and impossible.

If science seemed a credible *lingua cosmica*, then music was a reasonable choice as a medium that better reflected our emotional and cultural selves. Where language is inadequate, music expresses the inexpressible. The Voyager Records were made of gold-plated copper, with an aluminium cover that featured coded instructions on how to play them. A stylus and cartridge were also included for the convenience of any extraterrestrials unfamiliar with 1970s analogue music technology. But there was a problem: Sagan and his team had just six weeks to choose the content, compile all the recordings and images, and sort out the paperwork and various legal affairs.* Within the groove of the record is coded over a hundred images, greetings in fifty-five languages and a *Sounds of Earth* audio essay, including the brain waves of Ann Druyan, one of the key creative directors of the project, recorded during a meditation session. There is also ninety minutes of music from around the world. The physician Lewis Thomas had suggested that the

---

* There's nothing like an unrealistic deadline to focus the mind.

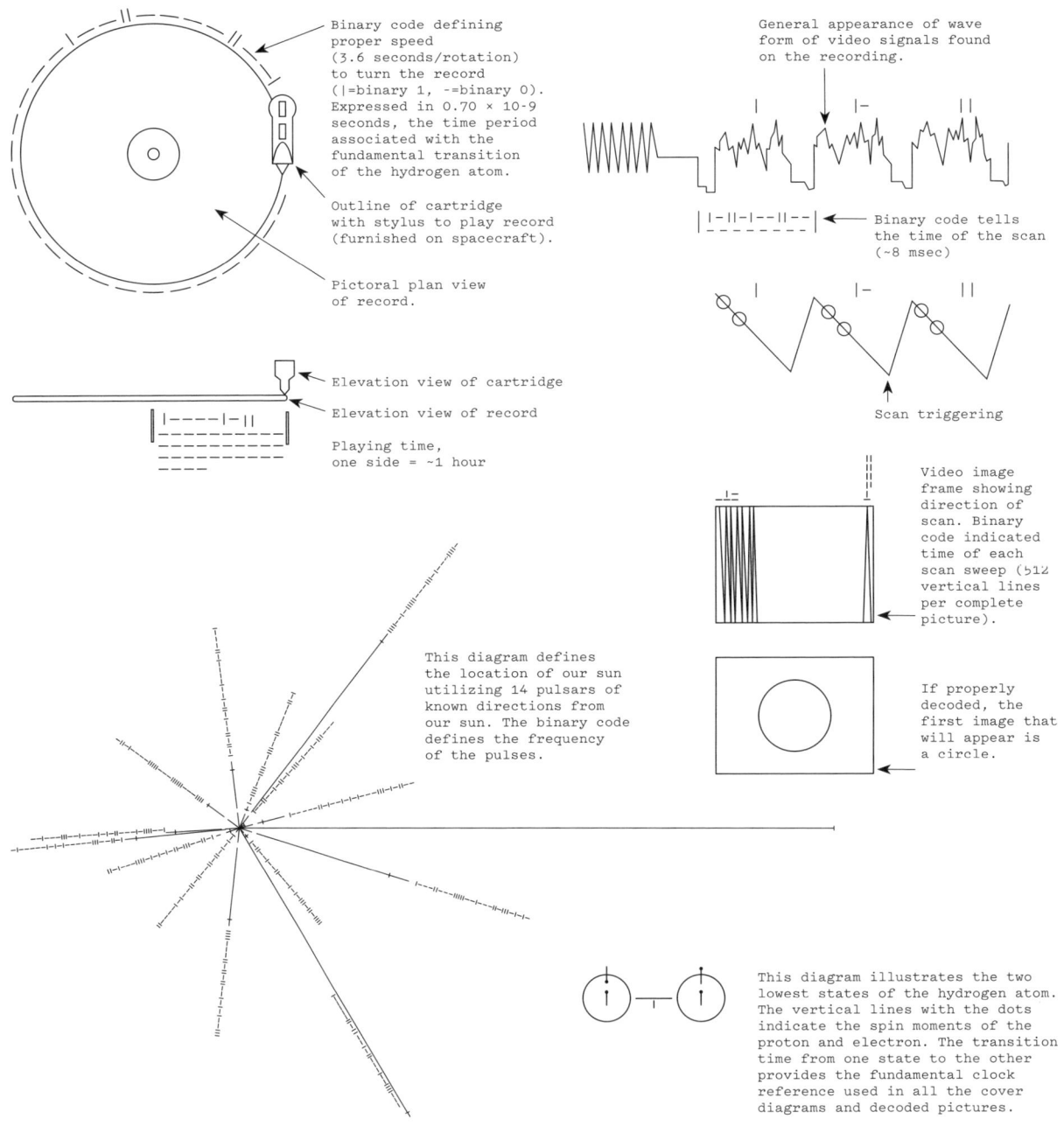

record should just be the music of J. S. Bach, but later conceded, 'that would be boasting'. Alan Lomax, an ethnomusicologist who studied the relationship between music and culture, was brought in to advise on the selection. The music chosen was representative of as many cultures and time periods as possible, everything from Chuck Berry to an Indian raga to Azerbaijani bagpipe music. But Bach is by far the most represented composer, with three out of the twenty-seven pieces. The music section ends with Beethoven's Cavatina from String Quartet No. 13 in B flat major, Opus 130. Perhaps no other piece of music has ever captured the essence of being human quite so profoundly.

Those fraught six weeks in 1977 honed in on a more fundamental question: what is it we'd like to say about ourselves? The real audience for the Golden Records is, of course, the one back home. Those of us pondering our place in the universe, wondering who else might be out there. Those spellbound by the Voyager 1 photograph taken from above the plane of the solar system, the *Pale Blue Dot* image of the Earth as a tiny speck. It's an image that will always be accompanied by Carl Sagan's voice in our minds: 'a mote of dust suspended in a sunbeam'.

Sagan once described the project as 'hopelessly quixotic'. If nothing else, its gesture, like any message in a bottle cast into an ocean, is one of hope.

Hope is all we have.

25.10 Message from US President Jimmy Carter. This document was stored as an image on the Golden Record.

25.11-25.29 Images from both NASA and the United Nations featured on the Golden Record.

```
                    STATEMENT

     This Voyager spacecraft was constructed by the
United States of America.  We are a community of 240
million human beings among the more than 4 billion who
inhabit the planet Earth.  We human beings are still
divided into nation states, but these states are rapidly
becoming a single global civilization.

     We cast this message into the cosmos.  It is likely
to survive a billion years into our future, when our
civilization is profoundly altered and the surface of
the Earth may be vastly changed.  Of the 200 billion
stars in the Milky Way galaxy, some -- perhaps many --
may have inhabited planets and spacefaring civilizations.
If one such civilization intercepts Voyager and can
understand these recorded contents, here is our message:

        This is a present from a small distant world,
        a token of our sounds, our science, our images,
        our music, our thoughts and our feelings.  We
        are attempting to survive our time so we may
        live into yours.  We hope someday, having solved
        the problems we face, to join a community of
        galactic civilizations.  This record represents
        our hope and our determination, and our good
        will in a vast and awesome universe.


                                    Jimmy Carter
                              President of the United States
                                    of America


THE WHITE HOUSE,
  June 16, 1977
```

25.11–25.29

JOCELYN BELL BURNELL

Jocelyn Bell Burnell didn't win the 1974 Nobel Prize for Physics. The honour was bestowed on the radio astronomers Martin Ryle and Antony Hewish (her supervisor), even though it was she who had turned over the pebble and made the crucial discovery. Perhaps it was a blessing in disguise. The infamous oversight resulted in countless other prizes and honours she might not have received otherwise, but it also drove her to become one of the most vocal champions of equality in science.

Born in 1943 in Northern Ireland, Susan Jocelyn Bell grew up in a Quaker household. Her faith has been a constant companion throughout her life. Her father was an architect and would call upon her inquiring mind and forensic attention to detail to help with his site surveys. From a very early age she felt the nagging injustice of the way women and girls were valued in society. Brothers were seen as more important than sisters; for girls, science classes at school meant 'domestic science' only. At the University of Glasgow she was the only female physics honours student in her year, and her entrance into the large, echoing lecture theatres was met with wolf-whistles and the stamping of feet and desks.

Astronomy was her calling – radio astronomy in particular, which was emerging as the new and exciting field for astronomers in the 1950s and '60s. Radio waves emitted from sources in space provided a much broader range of frequencies than visible light, in turn giving scientists a much

26.1 Jocelyn Bell Burnell with the Cambridge University radio telescope that discovered the first pulsar.

26.2 The first chart recording of the detected signal from pulsar CP1919 on 6 August 1967. The pulsed nature was not suspected at this time, however; the first observation of pulses was on 28 November 1967 (see **26.3**).

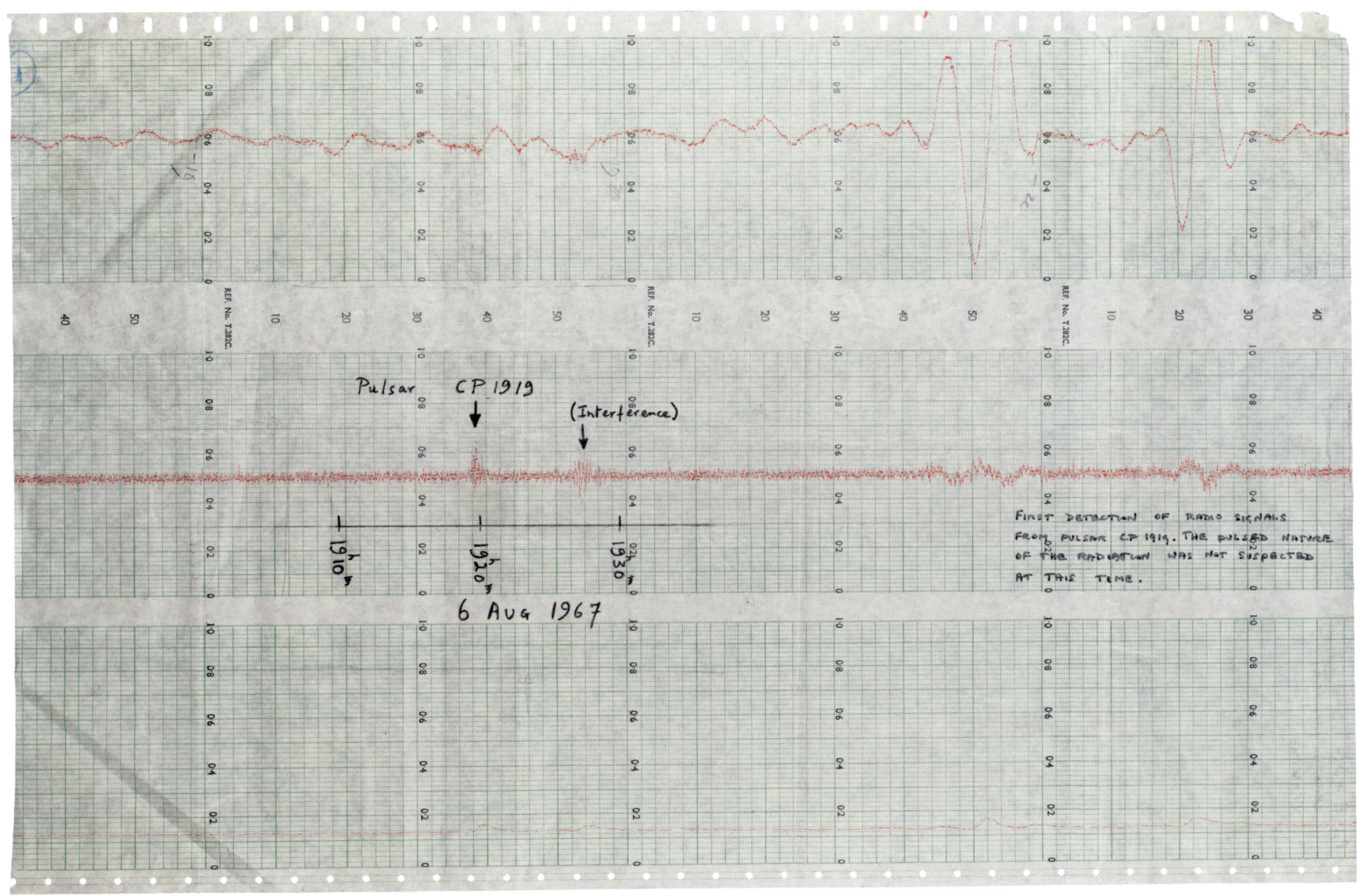

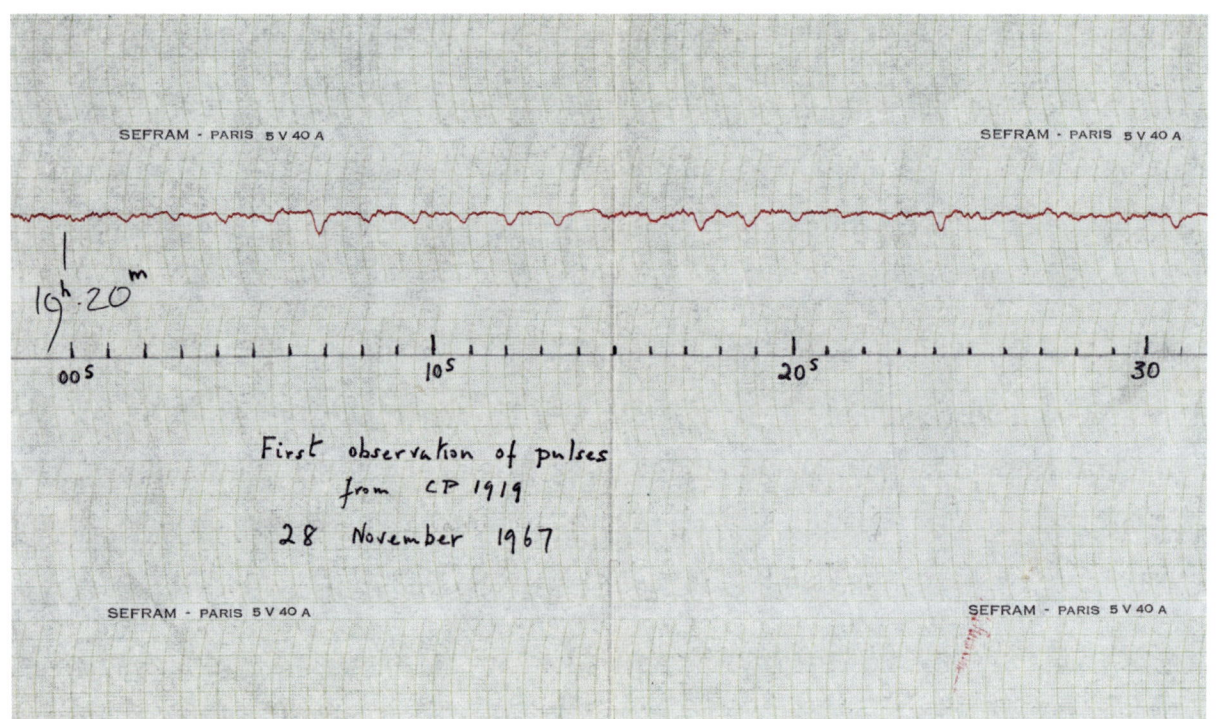
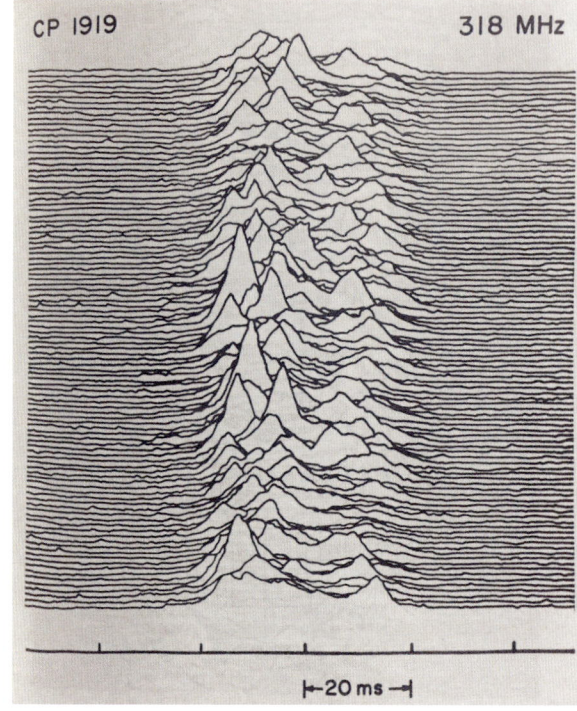

broader view of the universe.* She continued her studies as a PhD student at the Mullard Radio Astronomy Observatory at Cambridge, where she helped build the telescope which would make her famous. Not that it looked anything like a traditional telescope. It was a great agricultural-looking web of wire and wooden posts, referred to as 'four-and-a-half acres of washing lines'. The purpose of this giant antennae was to look for quasars: bright galactic nuclei that were the hot topic in radio astronomy. The data from these observations would be recorded as a line of squiggly peaks and troughs on long reams of paper, some 30 metres (100 feet) of paper every 24 hours that needed to be carefully analysed by eye. Looking at the long strip chart recorder one day, she noticed a small smudge in the readout she couldn't explain. A fault in the equipment that she had helped build? Heaven forbid. Radio interference? It didn't look like it. She showed it to her supervisor, who dismissed it as a piece of 'scruff', but her curiosity and persistence paid off.

On closer inspection, the strange artefact showed strong peaks at regular and rapid intervals, like the readout of a heartbeat on an electrocardiogram. The other possible explanation for such a regular and stable pulse was that it was intelligent in origin – perhaps the signal broadcast of some extraterrestrial civilization. And so the mysterious radio source was nicknamed 'LGM-1' (Little Green Men), but after three more of these sources were discovered, this hypothesis was dismissed. What she'd actually discovered was an entirely new and exotic astronomical phenomenon: pulsars,

---

* Radio astronomy also comes with the added bonus of not having to work exclusively at night!

incredibly dense spinning remnants of collapsed stars that emit regular pulses of radiation, like a celestial lighthouse sweeping its beam across space. CP1919, as it was first named, was one of the great astronomical discoveries of the age. Another beautiful stone turned over in our pursuit of knowledge.

Honour and recognition eventually came her way, if not by the Nobel Prize committee. After a lifetime in science, Jocelyn Bell Burnell's greatest legacy will be something that outshines her famous discovery. She will forever be known as one of science's great defenders, and a great practitioner of the scientific method itself: to live for the questions, because discovery is a journey without end.

26.3 Cambridge pulsar at right ascension 19 hours 19 minutes (CP1919), proof that the universe reveals itself to those who look closer.

26.4 Many consecutive pulses from CP1919. Time increases towards the right and towards the top of the figure. From Harold D. Craft Jr, 'Radio Observations of the Pulse Profiles and Dispersion Measures of Twelve Pulsars', September 1970.

26.5 An illustration of the enigmatic spectrogram from CP1919 was spotted by musician Bernard Sumner while he was in the Manchester Central Library café eating a sandwich. This cosmic transmission from a strange and distant object became the cover art for one of the finest records of all time, Joy Division's 1979 debut album, Unknown Pleasures.

26.5

JILL TARTER AND THE ORIGINS OF SETI

Are we alone? Humans hold a deep conviction that there must be life out there. Not just microbial life but intelligent life. Life that can think and feel the way we do; life that can invent telescopes and spaceships and iPhones. For most of our history, these deep convictions have been the only thing we've had. But convictions, however deep, don't give you an answer. For forty years, the American astronomer Jill Tarter (b. 1944) has been at the vanguard of science's attempts to answer this question using astronomy to hunt for evidence of extraterrestrial technology out there among the billions of whispering stars.

In 1984 she co-founded the SETI Institute, and she was the driving force behind the dedicated SETI listening post, the Allen Telescope Array (ATA) in northern California, a facility for hunting for signals of intelligent life beyond Earth. She has been a key part of SETI's large-scale searches, as well as being its most celebrated public advocate. She also has the honour of being the inspiration for Carl Sagan's 1985 book *Contact*, which follows the story of Ellie Arroway (played by Jodie Foster in the 1997 film adaption), a SETI astronomer battling the statistical (and political) odds to look for an alien signal among the galaxy's radio noise. Tarter's own story touches on similar themes. As a young girl she would walk along the beaches of Manasota Key in Florida with her father, looking up at the night sky and wondering if someone out there might be looking back. The difference between Jill Tarter and Ellie Arroway is, of course, that Arroway eventually finds what she's looking for. Jill is still looking.

The search for intelligent life beyond Earth has a long and colourful history that straddles pseudoscience, alternative religions, philosophy and media hype. A great example can be seen in the September 1919 edition of *Popular Science*, which considers how we might communicate with our Martian neighbours. Its cover features a glorious image of a giant moveable signalling mirror, and inside, methods of alien communication are discussed by the popular scientists and alien advocates of the day: Percival Lowell, William Pickering and the French astronomer Camille Flammarion. It was a fantasy that foreshadowed the actual 'greetings' sent in 1974 from the Arecibo Observatory in Puerto Rico (featured in fig. 27.1) – not to Mars but out to the M13 globular star cluster, 25,000 light years away. We haven't heard anything back, but every few years we get excited about possible candidates. In 1967 it was the LGM (Little Green Men)

27.1 Dr Jill Tarter on a platform over 150 metres (500 feet) above the radio telescope dish at the Arecibo complex, where she directs the search for signs of civilization in outer space.

27.2 Film still from *Contact* (1997), featuring Jodie Foster as Ellie Arroway.

27.3 'Earth flashes a message to Mars', from the September 1919 issue of *Popular Science*.

'If it's just us, it seems like an awful waste of space.'
CONTACT (1985)

signal discovered by astrophysicist Jocelyn Bell Burnell, which turned out to be the first recorded pulsar. A decade later came the strange anomaly logged by Ohio State University's 'Big Ear' radio telescope that prompted astronomer Jerry Ehman to scribble *Wow!* in the margins of his computer printout. Or the 'Oumuamua cigar-shaped interstellar object that glided across our solar system in 2017, which some less rigorous commentators proclaimed might be an alien technological artefact. As Carl Sagan reminds us: 'Where we have strong emotions, we are liable to fool ourselves.'

The origins of SETI as a formal scientific discipline can be traced back to the summer of 1950. A group of well-known physicists at Los Alamos, New Mexico, were musing about the flying saucer news that had made the headlines the year before, when Enrico Fermi blurted out, 'yes, but where is everybody?!' They all knew exactly what he meant. If intelligent life is common, why the deafening silence? In 1959, around the time Tarter was still in high school, the physicists Philip Morrison and Giuseppe Cocconi published a landmark paper in the prestigious science journal *Nature*, 'Searching for Interstellar Communication', which suggested the possibility of radio astronomy as a method of interstellar contact. A year later, Frank Drake conducted Project Ozma, the first modern radio SETI search at the Green Bank Observatory in West Virginia. The year after that, the first SETI conference was held, and Drake's now-famous equation was introduced. It was a way to guesstimate the number of advanced civilizations that might be in our galaxy – a broad framework for organizing our ignorance that multiplies seven variables assumed necessary for intelligent life to evolve.

Tarter originally studied engineering at Cornell University, but moved into astronomy as a graduate student. Her initiation into the world of SETI happened in 1971, when a report from the NASA-sponsored engineering-design and fact-finding study Project Cyclops landed on her desk. It was the first official paper on the subject, and it opens with a quotation from Frank Drake:

At this very minute, with almost absolute certainty, radio waves sent forth by other intelligent civilizations are falling on the earth. A telescope

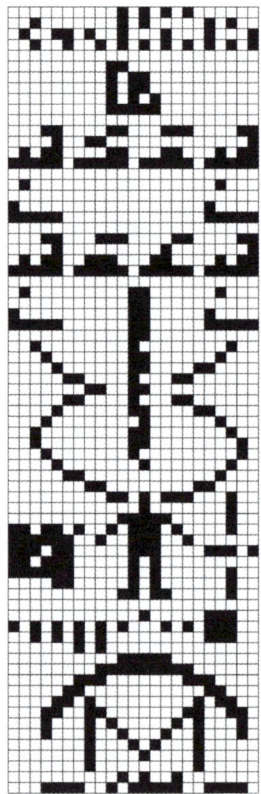
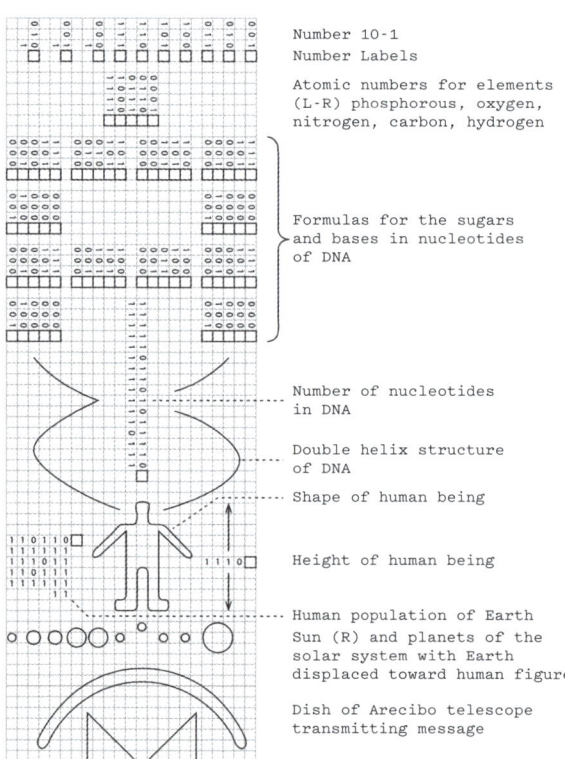

27.4 Arecibo message, computer artwork.

27.5, 27.6 The original Arecibo message (uncoloured, left) and decoded key with labels (right). This message was broadcast on 16 November 1974 by the Arecibo radio telescope in Puerto Rico, USA. The message was aimed at the M13 globular star cluster (25,000 light years distant). The broadcast lasted less than three minutes and consisted of 1,679 binary digits. The seven parts of the message were (from top): the numbers 1-10, elemental atomic numbers for DNA, nucleotide formulae for DNA, double helix of DNA, a human figure, the solar system and the transmitting telescope.

27.7 WOW! signal original printout, 15 August 1977, from the North American Astrophysical Observatory Records collection.

can be built that, pointed in the right place, and tuned to the right frequency, could discover these waves. Someday, from somewhere out among the stars, will come the answers to many of the oldest, most important, and most exciting questions mankind has asked.

She was hooked. Suddenly, here she was, at the right time and in the right place, equipped with computers, telescopes, scientists and funding – although the latter would become SETI's other great uncertainty.

Her fifty-year tenure has seen three major developments in the field. First, the discovery of exoplanets. While they were always assumed, it's only since the 1990s that they have been directly observed. We now know of thousands. Second, we know that the environmental parameters of life are far broader than we thought, opening the possibility for simple life to be thriving in some of the most seemingly inhospitable corners of our solar system. Third, technology is going through a rapid uptick. The stone-age radio astronomy of the mid-twentieth century has been transformed by the digital revolution, and will be revolutionized further with AI.

The reality of searching for evidence of alien techno-signatures is akin to looking for a needle in a haystack with no certainty that there's a needle in the haystack to begin with. However good your eyesight, there are no guarantees of finding anything, and just as significantly, disproving the hypothesis is impossible. Does that mean it's a pointless exercise? Whether or not a signal is found, the act of looking directs us to all kinds of interesting and healthy questions about life, the universe and everything in between. These are the lines of inquiry that motivate us to undertake scientific research in the first place. Engaging in SETI is a very human thing to do. In the words of Cocconi and Morrison, 'the probability of success is difficult to estimate; but if we never search, the chance of success is zero'.

Success is a relative term. Seek and ye shall discover something.

27.8 The Allen Telescope Array (ATA), University of California, Berkeley. The ATA, an array of 350 radio telescope antennae, is used simultaneously for the SETI project and for radio astronomy research.

27.9 Cover of *Project Cyclops: A Design Study of a System for Detecting Extraterrestrial Intelligent Life* (1971).

27.10, 27.11 Artist's concept of ground-level view of Cyclops system antennae, showing the central control and processing building, from *Project Cyclops* (1971).

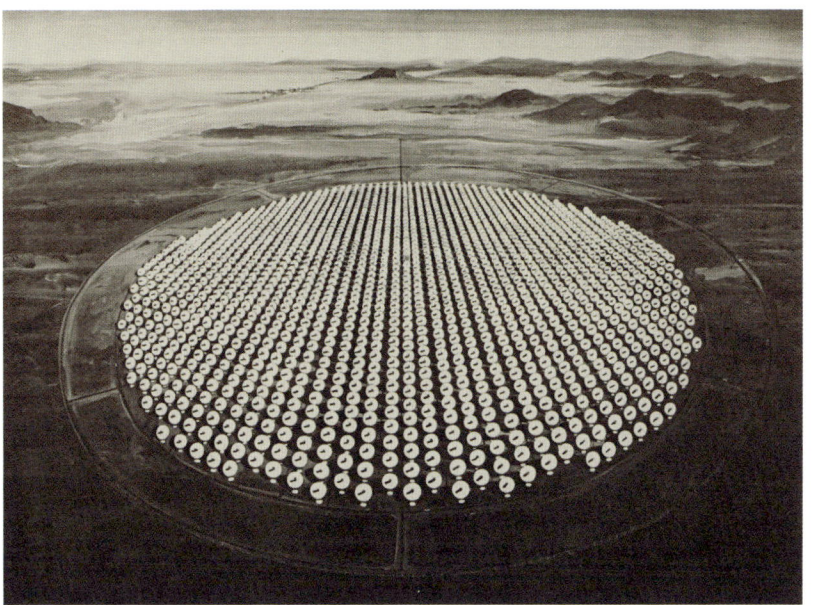
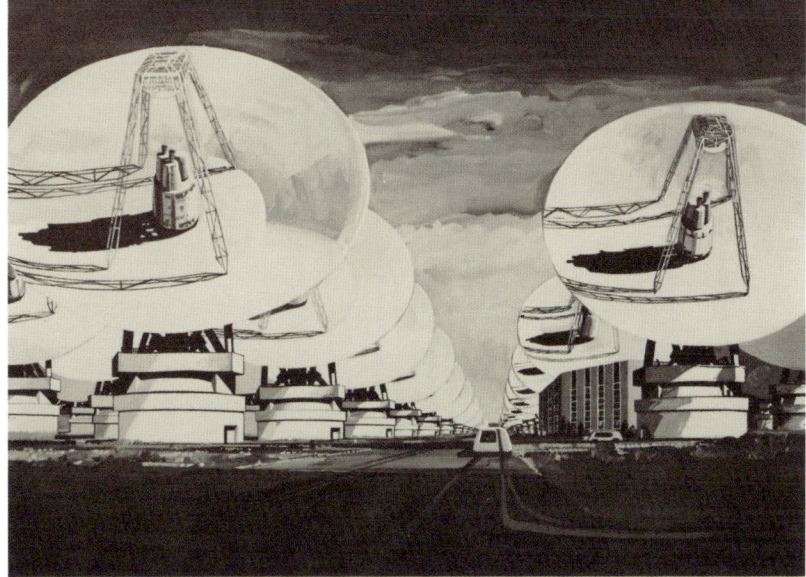

TOMOHIRO NISHIKADO

A phalanx of aliens make their way down the screen accompanied by a descending four-note chromatic loop, increasing in tempo as they approach. From behind your fortifications you dart out left and right, firing your blippy laser gun, picking them off one by one. Can you wipe them out before they touch land?

Nobody on Earth had experienced anything like it. This was not just a new video game but a new genre: the arcade shoot 'em up ('shmup') was born. Video game companies, vast and cool and unsympathetic, regarded our loose change with envious eyes, and slowly they drew their plans against us.

H. G. Wells's *War of the Worlds* has a lot to answer for. The story has persisted through both decades and mediums, from Orson Welles's infamous radio adaptation in the 1930s to its incarnation as the most famous video game of all time. Tomohiro Nishikado (b. 1944) is the man behind *Space Invaders*, which he let loose into the world in 1978. It was part of a wave of post-war Japanese invention that gave the world karaoke, the Sony Walkman and later *Hello Kitty* and *Pokémon*, anime and emojis. Video games, still in their infancy in the 1970s, were central to this new revolution.

In the late 1960s, Nishikado embarked on a career in audio electronics but switched his attention to video game development, working for the video games company Taito. Rival company Atari had launched *Breakout* in 1976 – a classic bat-and-ball game, the aim of which is to clear all the coloured blocks in order to progress to the next level. Its 'all clear' scoring concept was a huge inspiration for Nishikado, and a challenge was set for him to come up with a game to rival *Breakout*'s success. *Space Invaders* was the result. It was the very definition of 'game changing'. He chose aliens rather than people as the target – a more palatable enemy at which to aim. As we've seen, aliens had a powerful grip on the public imagination in the late 1970s: *Close Encounters of the Third Kind* and *Star Wars* were released within

28.1 Tomohiro Nishikado.

28.2 *Space Invaders* flyer, 1978.

Space Invaders flowcharts, drawn by Tomohiro Nishikado:

28.3 Flowchart showing the process of detecting the type of Invader laser beam (RAY A, B or C).

28.4 There are three types of the Invader laser beam, which are named RAY A, RAY B and RAY C. For Ray B and RAY C, the timing of appearance is preset, and this table determines that which Invader (each one is numbered in the array of the Invaders) fires RAY B/RAY C. Ray A, in the meantime, is fired from the Invader closer to the player's laser cannon.

28.5 Flowchart showing the process of detecting whether or not an Invader laser beam hits the player's laser cannon.

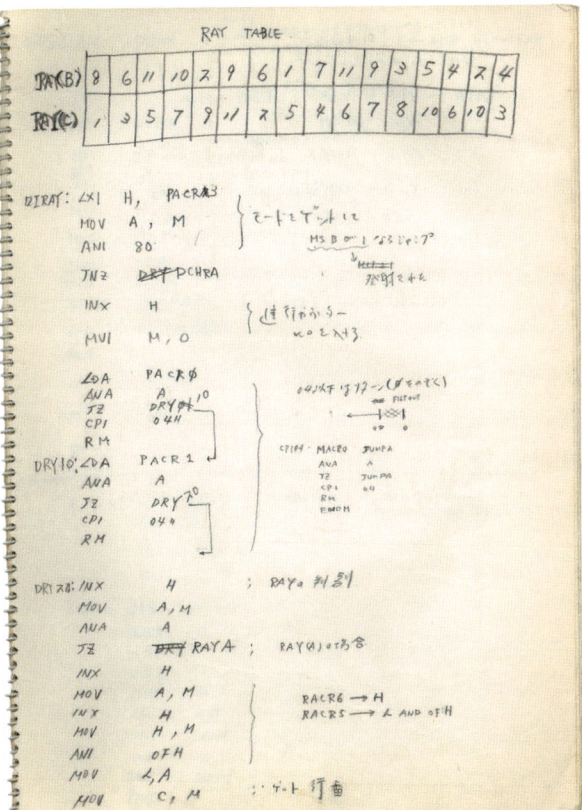

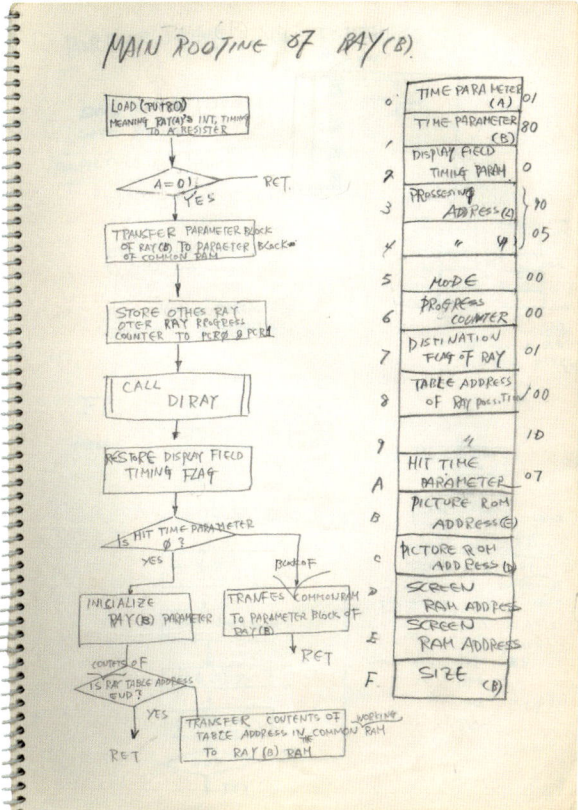

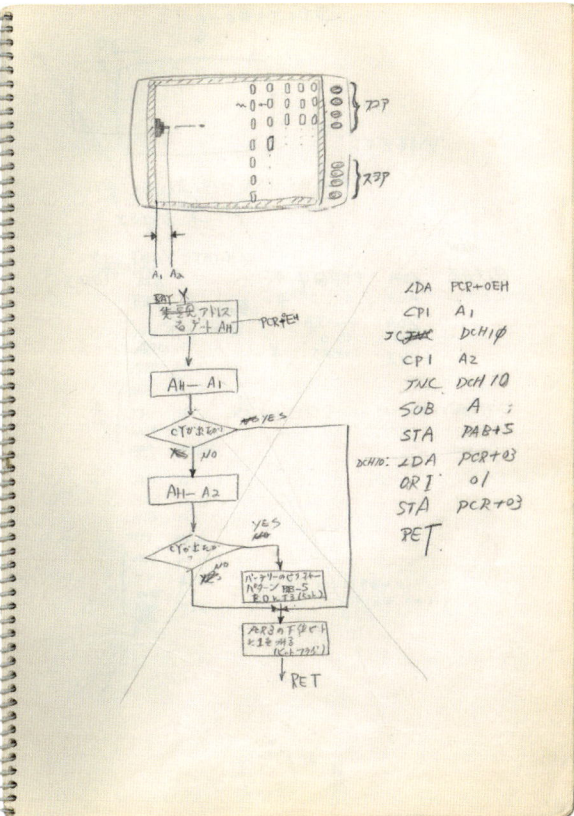

28.6, 28.7 Illustrations by Henrique Alvim Corrêa for the 1906 Belgian edition of H. G. Wells's *War of the Worlds*.

28.8, 28.9 Nishikado's sketches for the *Space Invaders* characters, turning tentacles into pixels.

28.6

28.7

キャラクター（イメージ）

※モンスター

※ UFO

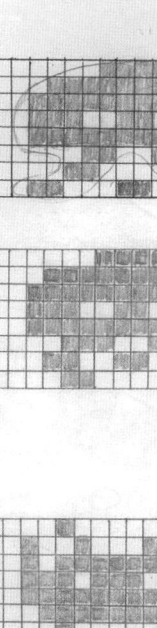
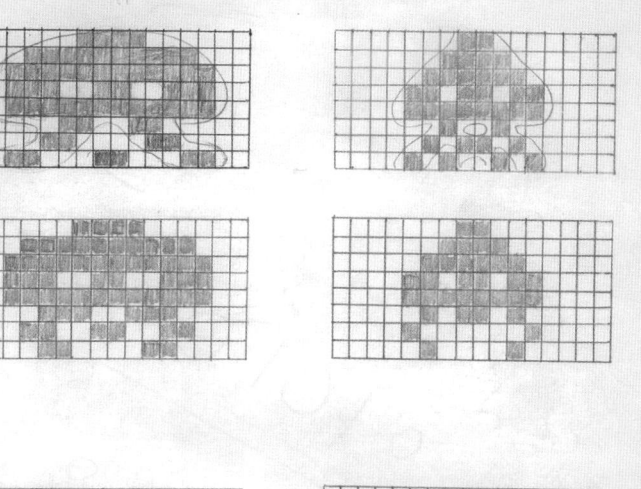
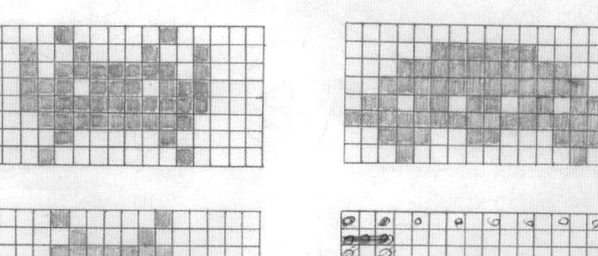

months of each other in 1977, and Ridley Scott's *Alien* followed a couple of years later. Nishikado's own brand of aquatic-looking characters were inspired by artist Warwick Goble's tentacled invaders that illustrate Wells's classic story from the 1890s. In order to devise a game as technically sophisticated as *Space Invaders*, he was forced to make his own microprocessor hardware; the computers of the day simply weren't up to the job. But try as he might to create a smooth descent for his aliens, he was stuck with their famous jerky movements. Technical limitations became one of the secrets of the game's success.

Space Invaders had it all – a simple story, dramatic tension, the skill of the game play and a hi-score list to beat that made players come back for more. It was quickly licensed around the world, creating a billion-dollar phenomenon, and like many new, disruptive technologies, moral panic followed in its wake. Legend has it that it spawned a 100-yen coin shortage in Japan. In the UK, tabloid newspapers ran fear-inducing stories of kids stealing money from their parents to fuel their *Space Invaders* addiction: 'SPACE INVADERS DRIVE A 13 YEAR OLD BOY TO CRIME!' What detrimental and corrupting effect was it having on the youth? Urgent questions were raised in Parliament to try and control this new video game menace.

Today the Invader character (👾) isn't just symbolic of the game, it's a symbol of gaming itself. It's as familiar an emblem of Japanese culture as Katsushika Hokusai's *Great Wave off Kanagawa* (1830/32). It has broken out of the

28.10

confines of the video screen and morphed into a new analogue art form in its own right, inspiring a clandestine French street artist working under the pseudonym 'Invader', who recreates 8-bit 'Rubikscubist' nostalgia in urban landscapes with his coloured square tiles.

The aliens that we once battled have become cherished allies. Unlike Wells's Martians, who perished on Earth, Nishikado's Invaders managed to conquer not just our attention but our planet, and have decided to hang around and make it their permanent home. They're more than welcome. We've got more dangerous things to worry about now.

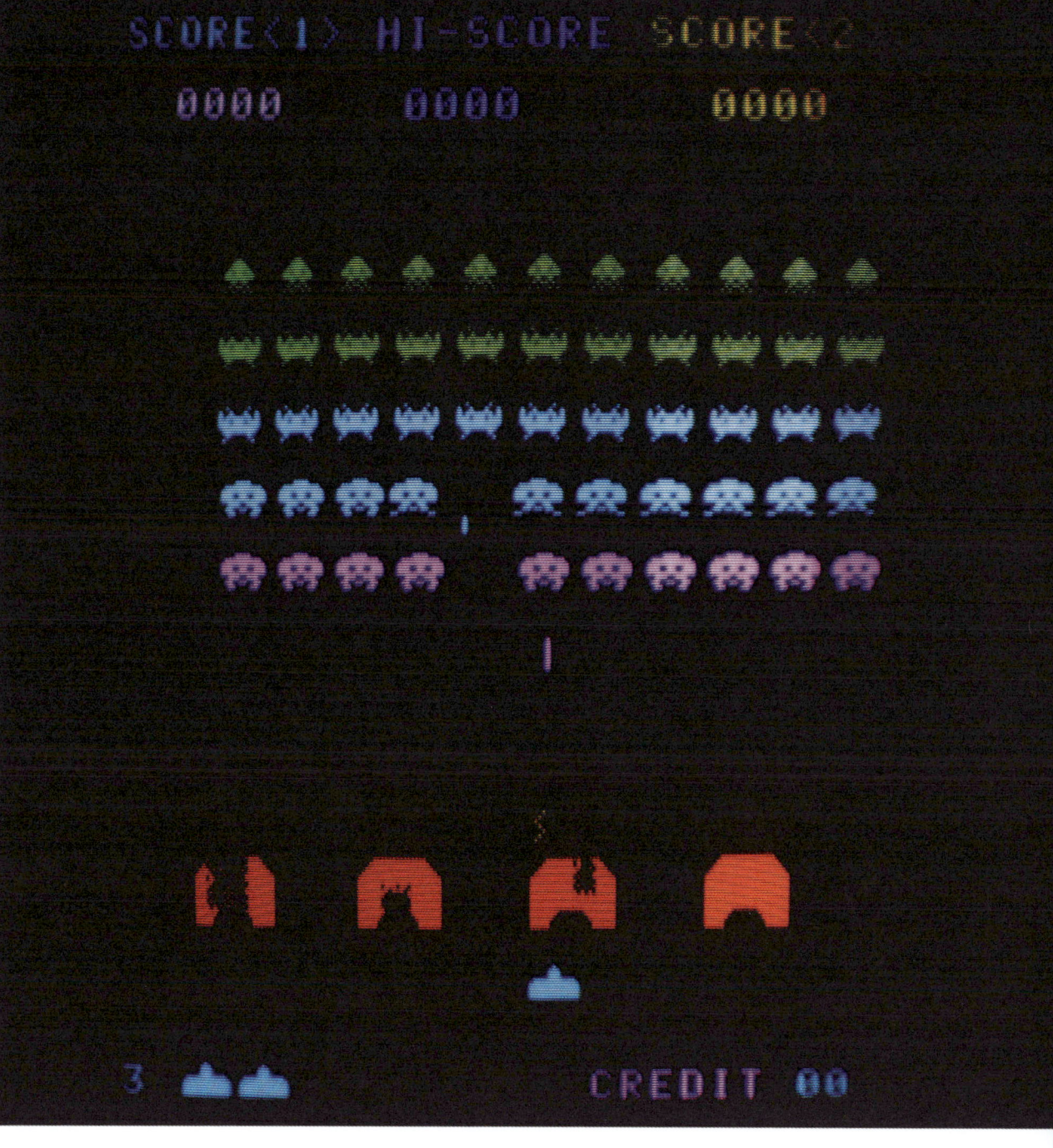

28.10 Young men playing with *Space Invaders* table-type cabinets in an Invader House.

28.11 *Space Invaders* in-game screen, 1978.

Invaders by the French artist Invader.

28.12 *PA_320*, Paris, 2000.
28.13 *PA_800*, Paris, 2009.
28.14 *LDN_122*, London, 2011.
28.15 *LDN_110*, London, 2009.

28.16 Detail of the *Invasion Map of Paris*, 2024.

28.16

ACT 6.

TEMPORIS PEREGRINI

'WANDERERS IN TIME'
RETRO-FUTURISTS, ARCHAEOLOGISTS AND THOSE PUSHING TO NEW HORIZONS

RON JONES

On 28 January 1986 the world watched in horror as the Space Shuttle Challenger broke up as it climbed into space from Cape Canaveral in Florida. For Ron Jones (b. 1953) and many of the young aerospace engineers working at the Vandenberg US Air Force base, more than 3,000 kilometres (2,000 miles) away in California, it signalled the end of the road. At the time Vandenberg was under major refurbishments, preparing to support its own military payload shuttle launches. The Challenger accident put an end to those plans and threw the future of human space flight in America into question. Unemployed three months later, Ron found himself in Hawaii with some time to think. His Vandenberg experience had taught him that few people understood how the West Coast launch complex fitted into the broader scope of national space activities. What was needed was a realistic assessment of where we were going and how we were going to get there. Sitting on the beach, Ron conceived a framework for what would become the most comprehensive vision for human expansion into space ever attempted.

As teenager growing up in Silicon Valley in the 1970s, Ron was fascinated with the futuristic space architecture proposed by writers like the physicist and space visionary Gerard K. O'Neill – human communities living in rotating orbital cylinders made from materials mined from the Moon and asteroids (fig. 32.2) – and the NASA Ames/Stanford Torus study, which proposed habitation in a giant wheel-shaped space station. These wild ideas were brought to life in vivid detail by a new generation of space artists like Rick Guidice (fig. 32.3) and Don Davis. It was hard not to get excited about the future, but

29.1 An overall view of Space Launch Complex Six at Vandenberg. The structures are (L-R): the payload changeout room, the shuttle assembly building, the access tower and launch mount, and the mobile service tower.

29.2 The earliest version of Ron Jones's Integrated Space Plan, then called the Western Space Program, made in 1989.

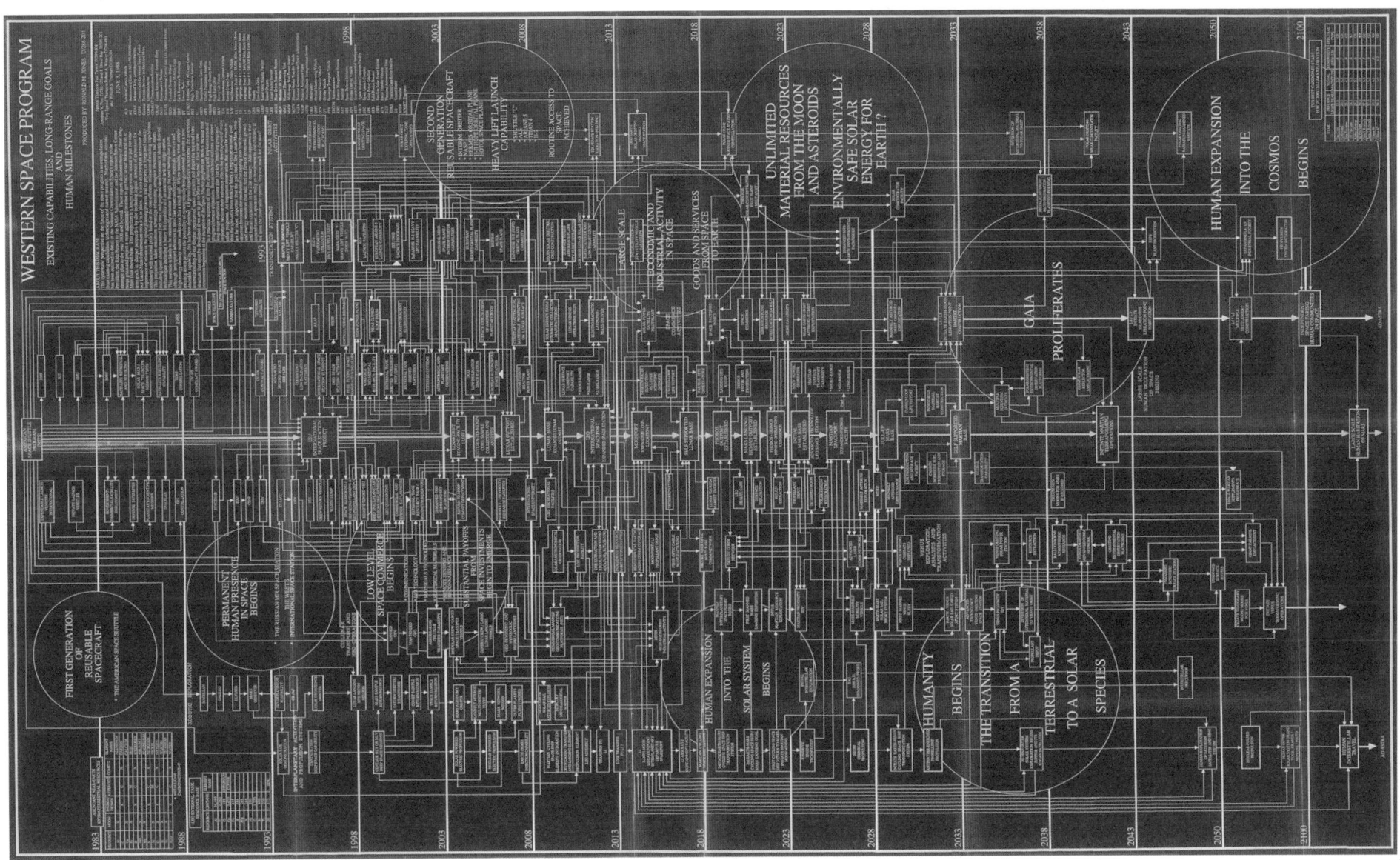

29.3

reality meant frustration with the lack of clear goals in the American space effort post-Apollo. The new semi-reusable Space Shuttle at least hinted at a new era of cheaper access to space, and a departure from single-use launch vehicles. Unfortunately, for all its triumphs, it was neither of these things.

Upon his return from Hawaii, Ron was hired by the aerospace giant Rockwell International to work on the Challenger's replacement, Endeavour. In his spare time, he began converting his ideas into a visual graphic, hand-drawn on a large pad of engineering graph paper, which he then redesigned as a flow chart on an early Apple Macintosh. It wasn't long until his project came to the attention of Rockwell's Advanced Projects Group, a team working specifically on lunar and Mars mission trade studies in support of President George Bush's Space Exploration Initiative. As part of the Advanced Projects Group, he had access to an array of exotic space exploration studies and plans that never materialized, dating back to the early days of Wernher von Braun. Here seemed an opportunity to revisit some of these ideas and to really interrogate the questions: what would it actually take for humans to expand into space? What technologies might we need to develop? How long will it take? How might it all fit together?

Over time the Integrated Space Plan grew into a vast, interconnected circuit board of ideas. Down the sides, time is marked in five-year intervals, beginning in 1988 and continuing more than a hundred years into the future. The central spine represents important milestones – from the Space Shuttle and various associated technologies to the first international space station, to permanent human habitation on Mars and beyond. The large circles contain the 'big picture' implications for humanity, which become increasingly esoteric and speculative. Right at the bottom we are left with the words 'Ad Astra' (to the stars); a cosmic 'to be continued'. Rockwell International recognized the visual appeal of Ron's wild, gigantic flow chart and began using it as a marketing tool, producing and distributing many hundreds of copies that would soon grace office walls around the world, a curious piece of art and a springboard for radical ideas.

Despite being called the 'Integrated Space Plan', it wasn't really a plan at all. It was a call to action to the many isolated departments of the American space programme that lacked unified long-term thinking. Over the years it's been revisited, redesigned and updated, but somehow its early iterations remain the most potent – an artefact from simpler times, a thought from the margin that sprang to life and spread like the roots of a strange and beautiful tree. A moment when we planned for a daring future, with all possible outcomes.

29.3 Illustration by Don Davis of a toroidal colony, exterior view, 1976.

29.4 Detail from the Rockwell Integrated Space Plan, v. 4, 1997, showing the incremental timescale from 1988 to 2100.

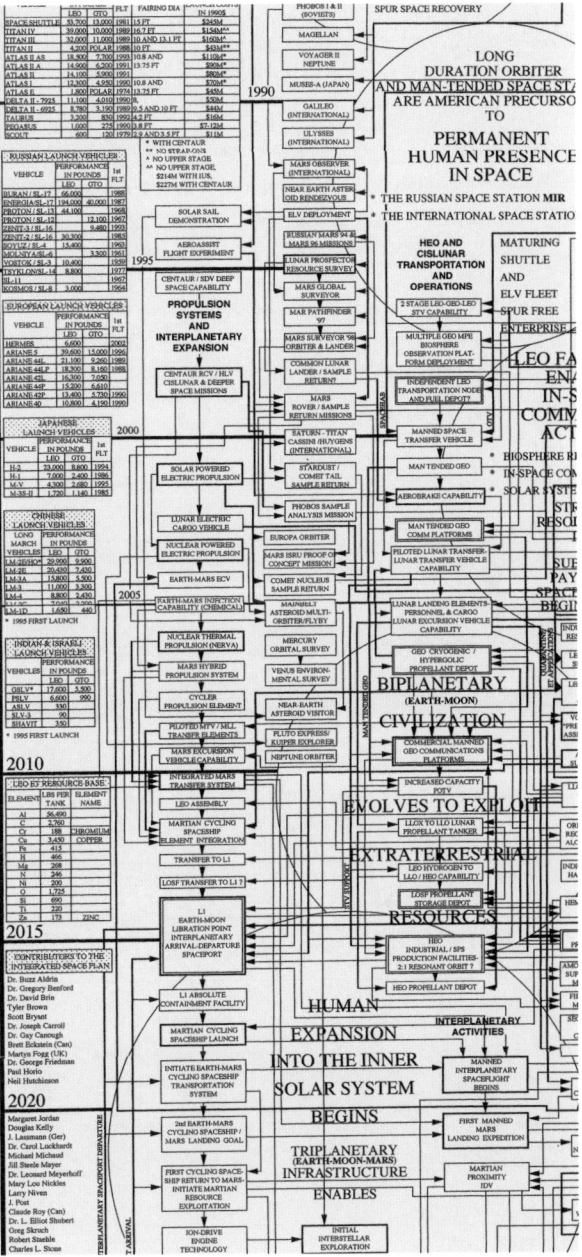

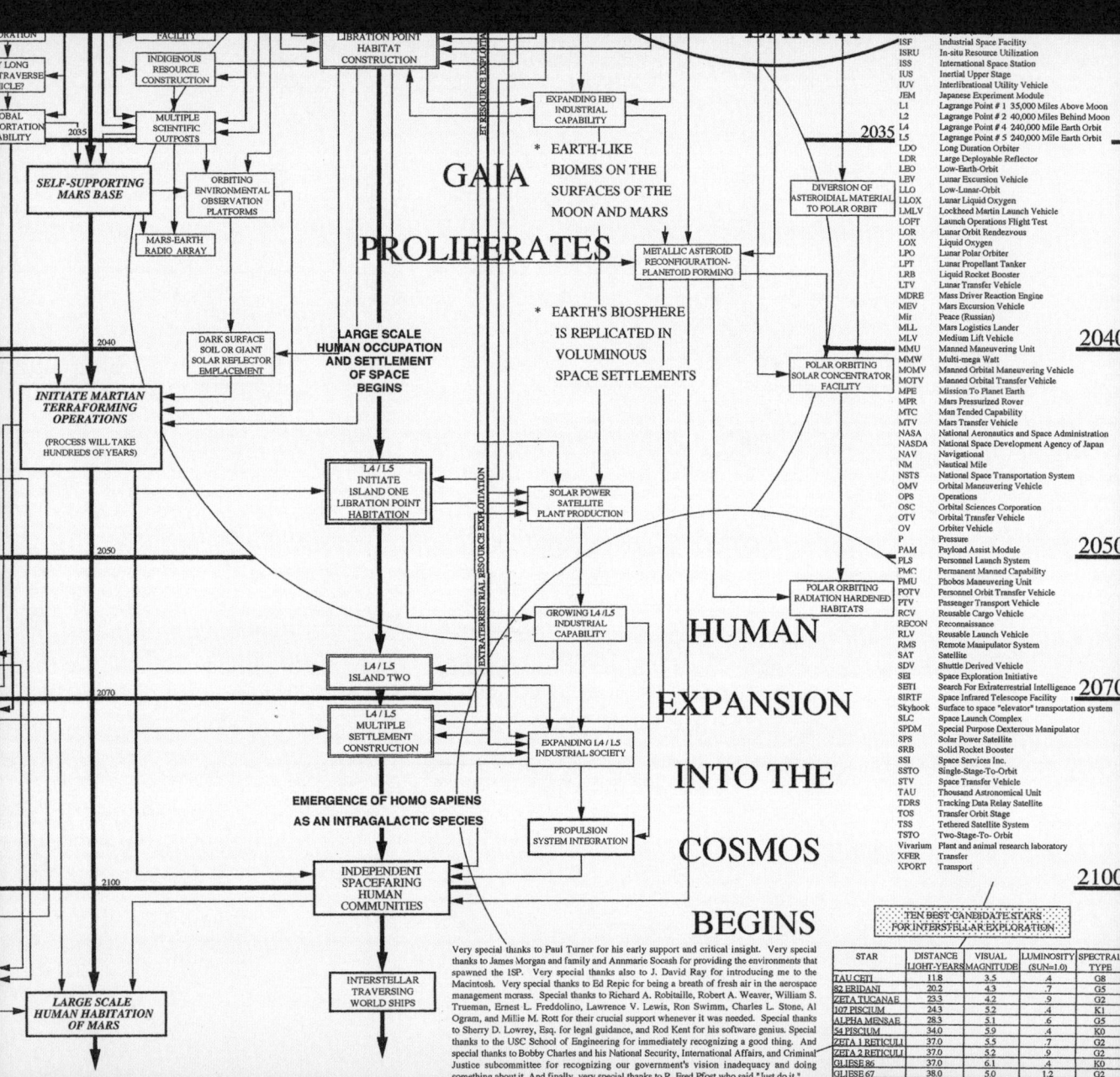

'The Integrated Space Plan is an outstanding example of the type of consolidated and far-reaching planning that must be done if we are to expand human presence outward into our solar neighbourhood.'

BUZZ ALDRIN (B. 1930), ASTRONAUT VETERAN OF APOLLO 11 AND GEMINI 12

29.5 Detail from the Rockwell Integrated Space Plan, v. 4, 1997. These large circles are key to understanding what the underlying architecture of the ISP is all about. These larger circles are intersecting nodes connecting multiple elements of the ISP, showing the major aspirational milestones of human space exploration.

29.6 Don Davis, *Space Colony: Torus Wheel*, 1975. Large assemblies can be put together in space. On this part of the rim, panels of a colony are being fitted by small vehicles called ANTS (Assembly Non-Tethered Ships

FREEMAN DYSON

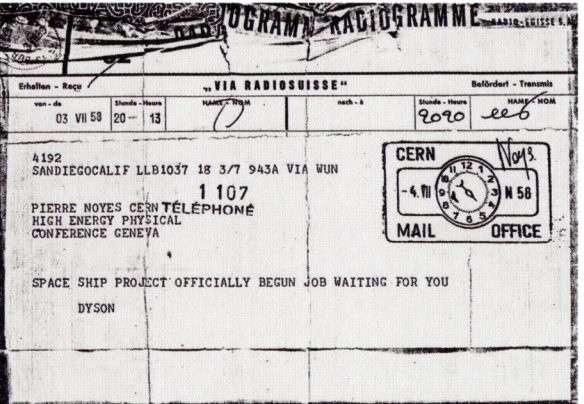

'Daddy builds spaceships!' the young Esther and George Dyson were told in 1958 as their family moved from the East Coast out to La Jolla, California. Back then, the cars, buildings and pretty much everything else in southern California looked like spaceships. The Soviet launch of Sputnik the year before had sent America into existential crisis, and physicist Freeman Dyson (1923–2020) was moving to California to begin work at General Atomics on the highly classified Project Orion. This government study looked to bring space travel in lockstep with the other great Cold War technology of that time, atomic energy. It had been just over a decade since the first nuclear weapons ended the Second World War, and thoughts were now turning to harnessing their vast release of energy as a way of reaching the stars. Project Orion was a roller coaster of an idea: a 4,000-tonne crewed spaceship would be constructed, powered by thousands of small nuclear bombs. One by one the bombs would be ejected and detonated, and the rapid succession of explosions would propel the craft in the opposite direction like a giant atomic machine gun. The 'opposite direction' in this case was towards Mars, Saturn, Jupiter and their respective moons, and who knows where else. After all, this was being dreamt up before America had even put a foot in space. In theory the physics was sound. Plans were drawn up. Small prototypes were made and tested using conventional explosives. But despite having some of the world's finest minds working on the project, the engineering practicalities and the unresolved issue of radioactive fallout, not to mention the complicated politics, led to its demise. It was all just too much for 1958, and by 1965, as the more modest Saturn V rocket was preparing to take us to the Moon, Daddy's spaceship was declared dead.

Although a naturalized American, Freeman Dyson was born and raised in England. Since his youth, space travel and problem-solving had never been far from his mind. At the age of five he calculated the number of atoms in the Sun. By nine, he had written a Jules Verne-style

30.1 Portrait of Freeman Dyson, photographed at his desk by Francis Bello in 1956.

30.2 Telegram from Freeman Dyson to Pierre Noyes, 3 July 1958.

30.3 Dyson (far right) at General Atomics' Point Loma test site with a high-explosive-driven tethered model of the Project Orion spaceship design, summer 1959.

30.4 Original concept of the Orion ship, reproduced on p. 70 of Orion, drawing by Michael Treshow, GAMD-3597, 25 October 1962.

30.5 Cover of General Atomics' Project Orion report, 1959.

'It's better to be wrong than to be vague.'

THE SCIENTIST AS REBEL
(2006)

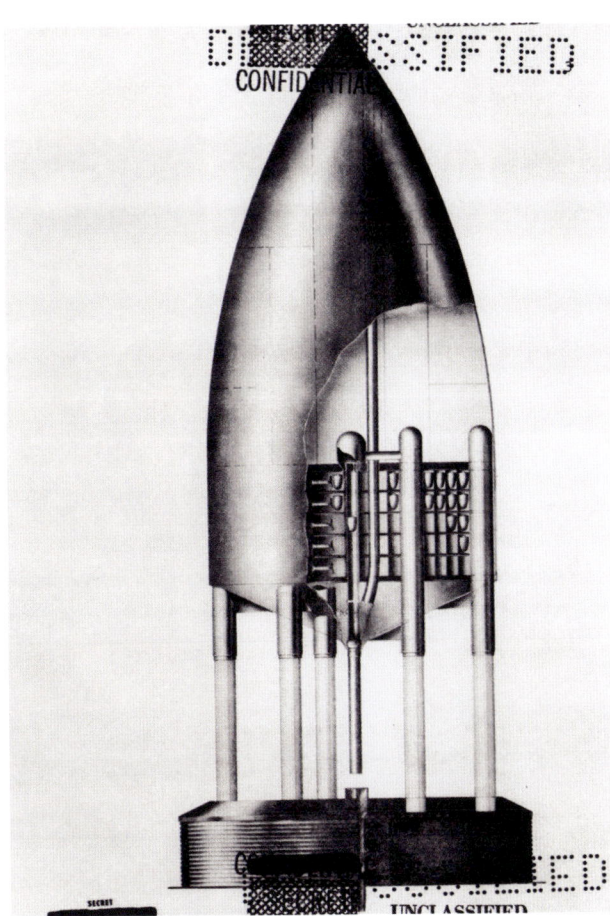

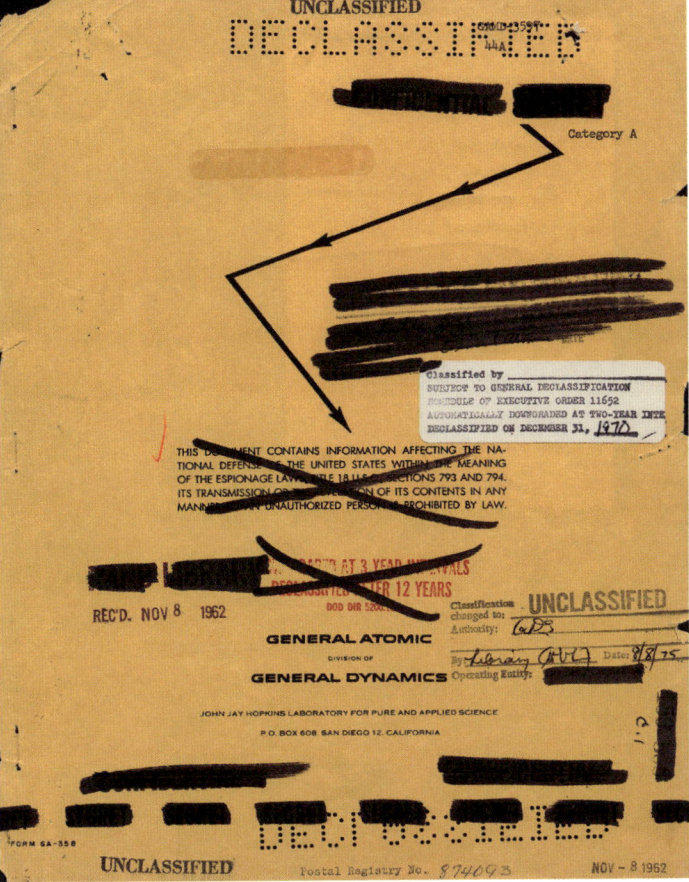

30.6-30.9 Freeman J. Dyson, 'A Space Traveler's Manifesto', 5 July 1958: 'Either through inadvertence, or by a deliberate act of wisdom, the American government has announced to the public that we are working on the design of a space-ship to be driven by atomic bombs ... It is my belief that this scheme alone, of the many space-ship schemes which are under consideration, can lead to a ship adequate to the real magnitude of the task of exploring the Solar System.'

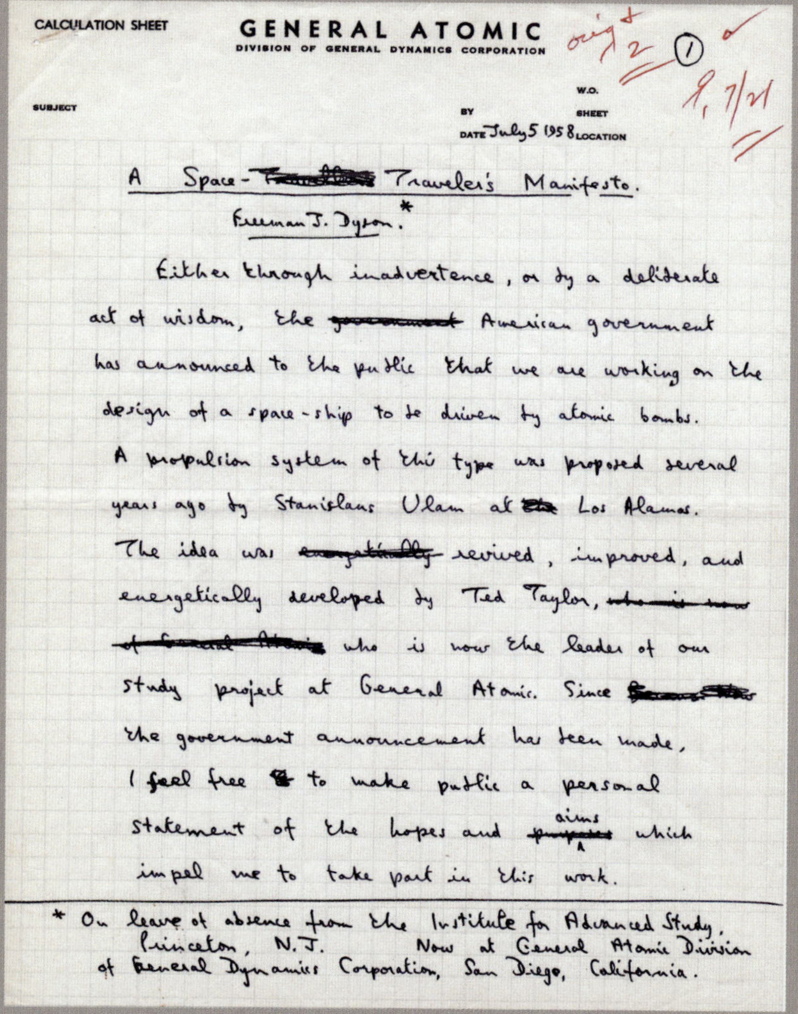

CALCULATION SHEET
GENERAL ATOMIC
DIVISION OF GENERAL DYNAMICS CORPORATION

orig + 2 ① 9, 7/21

SUBJECT
BY
DATE July 5 1958
W.O.
SHEET
LOCATION

A Space-Traveler's Manifesto.

Freeman J. Dyson.*

Either through inadvertence, or by a deliberate act of wisdom, the American government has announced to the public that we are working on the design of a space-ship to be driven by atomic bombs. A propulsion system of this type was proposed several years ago by Stanislaus Ulam at Los Alamos. The idea was revived, improved, and energetically developed by Ted Taylor, ~~who is now~~ who is now the leader of our study project at General Atomic. Since the government announcement has been made, I feel free to make public a personal statement of the hopes and aims which impel me to take part in this work.

* On leave of absence from the Institute for Advanced Study, Princeton, N.J. Now at General Atomic Division of General Dynamics Corporation, San Diego, California.

30.6

CALCULATION SHEET
GENERAL ATOMIC
DIVISION OF GENERAL DYNAMICS CORPORATION

②

SUBJECT
BY
DATE
W.O.
SHEET
LOCATION

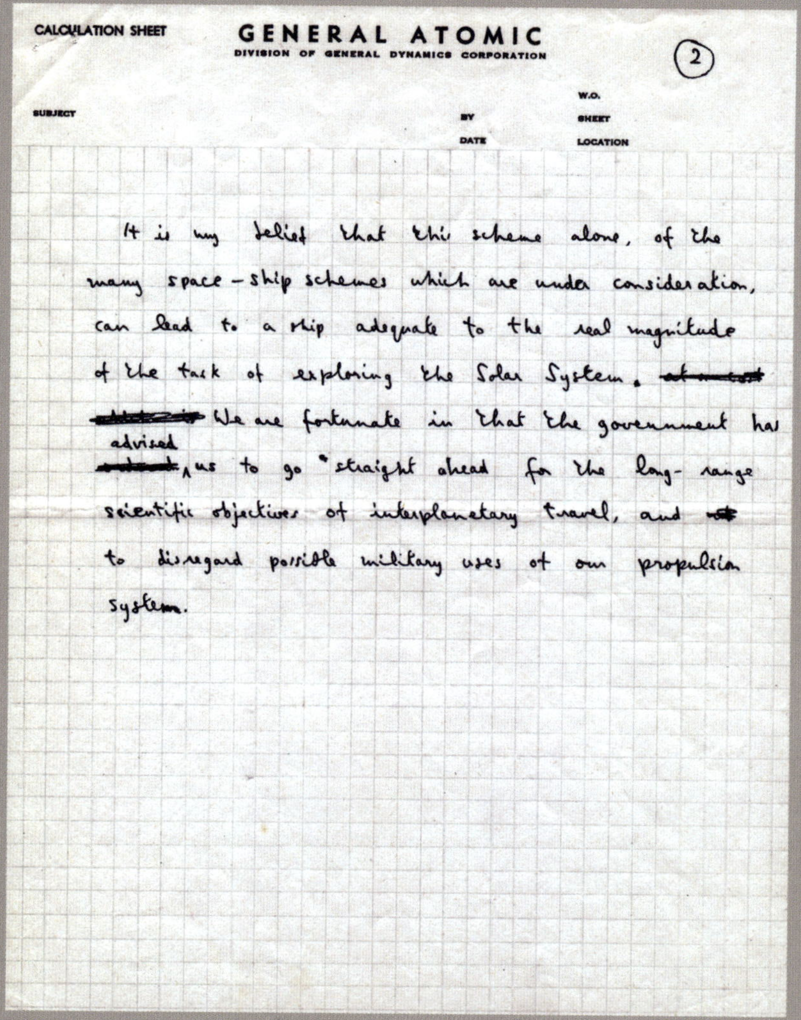

It is my belief that this scheme alone, of the many space-ship schemes which are under consideration, can lead to a ship adequate to the real magnitude of the task of exploring the Solar System. We are fortunate in that the government has advised us to go "straight ahead for the long-range scientific objectives of interplanetary travel, and to disregard possible military uses of our propulsion system.

30.7

'My purpose and my belief, is that the bombs which killed and maimed at Hiroshima and Nagasaki shall one day open the skies to man.'
'A SPACE TRAVELER'S MANIFESTO' (1958)

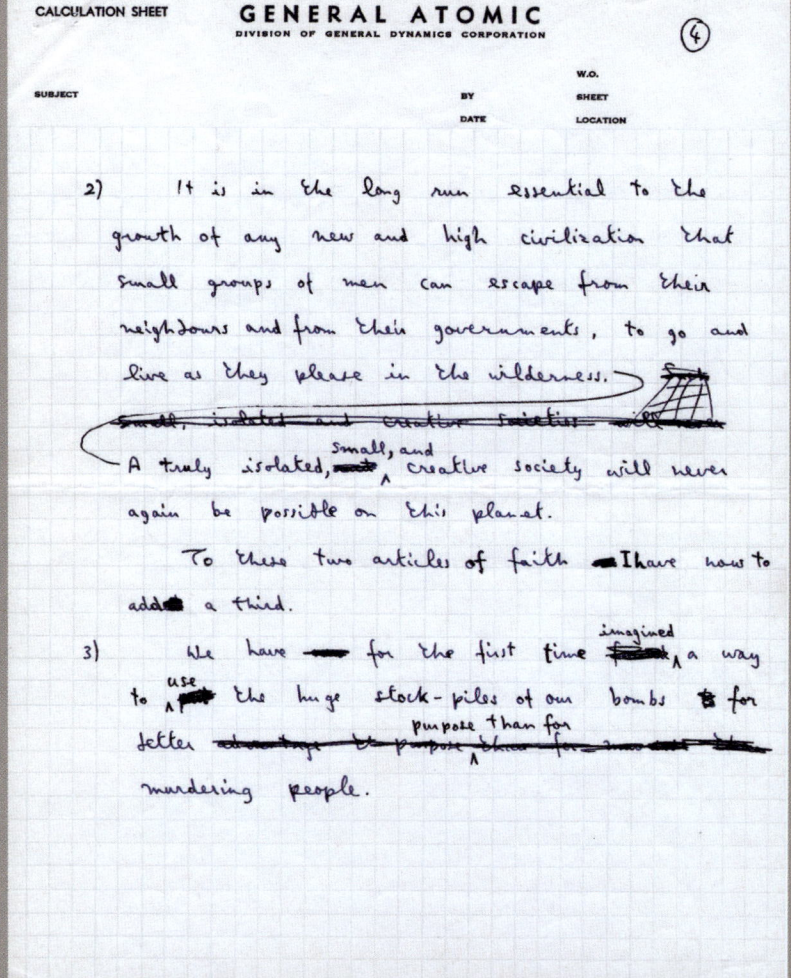

As problems in pure physics, our problems are certainly less interesting than those which were faced by the builders of the first nuclear reactor or the first atomic bomb. I work here not for the beauty of the work itself, but because I feel an overwhelming desire that this project should succeed.

From my childhood it has been my conviction that men would reach the planets in my lifetime, and that I should help in the enterprise. If I try to rationalize this conviction, I suppose it rests on two beliefs, one scientific and one political.

1) There are more things in heaven and earth than are dreamed of in our present-day science. And we shall only find out what they are if we go out and look for them.

2) It is in the long run essential to the growth of any new and high civilization that small groups of men can escape from their neighbours and from their governments, to go and live as they please in the wilderness. A truly isolated, small, and creative society will never again be possible on this planet.

To these two articles of faith I have now to add a third.

3) We have for the first time imagined a way to use the huge stock-piles of our bombs for better purpose than for murdering people.

voyage extraordinaire story called 'Sir Phillip Roberts's Erolunar Collision', about an expedition to witness a collision between the asteroid Eros and the Moon. Writing and calculating were Freeman Dyson's two great talents. As a physicist, his academic home was the Institute for Advanced Study at Princeton, New Jersey (also home to another European immigrant, Albert Einstein), where his work tidying up quantum electrodynamics alongside the great Richard Feynman became his most celebrated accomplishment.

But for those with an interest in space history (i.e. you), Freeman Dyson will forever be remembered as a Futurist prophet who enjoyed tinkering with many highly speculative ideas around space travel, interstellar communication and the proliferation of life throughout the cosmos. He imagined a 'greening of the galaxy' – everything from genetically modified potatoes that can grow easily on Mars to his famous 'Dyson trees', engineered to grow on comets or frozen Kuiper belt objects such as Pluto that lurk way out at the frontier of our solar system. What might such strange cosmic island habitats look like? The low gravity could mean the proliferation of vast, expansive biological structures. And where there are trees, other life forms might follow. It's a concept, if nothing else, that has taken root in the fertile minds of science fiction writers and artists.

Dyson was also interested in ways of detecting advanced extraterrestrial civilizations, beyond the radio astronomy favoured by SETI.* He reasoned that advanced civilizations would require vast amounts of energy (much more than just a planet could provide), so it seemed plausible that some kind of massive engineered structure could be built around a star in order to harvest its energy directly. These stellar 'Dyson spheres', as they became known, were a concept first imagined by the English science fiction writer Olaf Stapledon in his strange cosmic travelogue of 1937, *Star Maker* (time charts from which are illustrated fig. 4.9). In 1960 the Russian physicist Nikolai Kardashev took this idea even further and created a hypothetical framework for categorizing a civilization's technological sophistication by the amount of energy it's able to harness. A type 1 civilization (let's call ourselves a 0.5 – we're not there yet) should be able to harness and store all of its planet's energy. A type 2 civilization should be able to harvest its parent star's energy, using technology like a Dyson sphere. A type 3 civilization would have become so sophisticated it could capture an entire galaxy's energy. Periodically the astronomical world gets excited by possible results of detecting alien megastructures. So far, they have only been disappointed.

Freeman Dyson was an outlier. A provocateur, controversial in his opinions about other matters of science.† He saw life, the universe and everything as an ongoing set of calculations. Puzzles waiting to be solved. Most of all, he opened the minds of ordinary people to the infinite possibilities of the nature of the cosmos. However wild (or wrong) they might be.

---

* Search for Extraterrestrial Intelligence – see Chapter 27.

† Notably on matters of climate science.

30.10 Olaf Stapledon, *Star Maker* (1937).

30.11 Freeman Dyson's simple sketch of a white dwarf binary star being used to accelerate an object to 2,000 km/s without the need for any other propellant. Might advanced civilizations use these stars as a galactic transport system for crossing vast distances?

30.12 Jon Lomberg, *Comet Trees*, 1986, inspired by the 'Dyson tree' concept. The illustrations of such esoteric objects (however beautiful they might be) bear little resemblance to Dyson's actual idea. Humans like looking at beautiful things as much as we like thinking about interesting things.

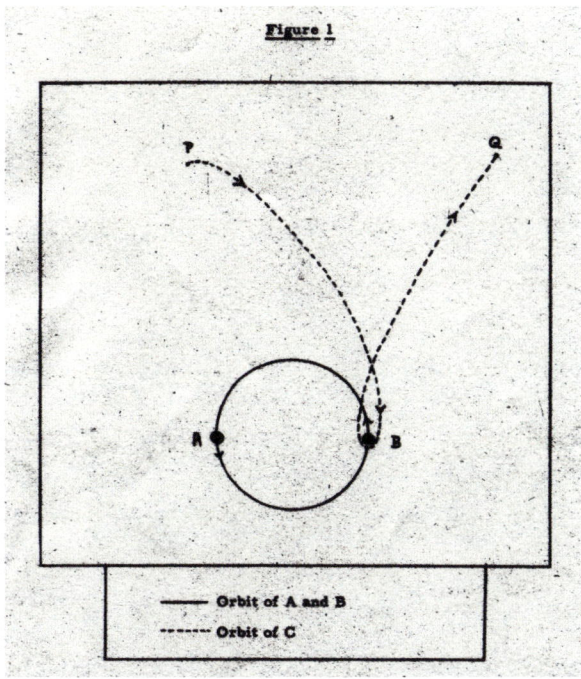

'In all my writing the aim is to open windows, to let the experts inside the temple of science see out and to let the ordinary citizens outside see in.'

FROM EROS TO GAIA (1992)

30.13 Schematic impression of a Dyson sphere by Jon Lomberg: *Dyson Lattice* from *Encyclopedia Galactica*, 1975.

30.14 Lomberg's *Starship Augury* (1994) depicts a ship constructed using auger seashell DNA. The artist was inspired by Dyson's notion that advanced technology would use DNA-like coding to grow rather than build spaceships.

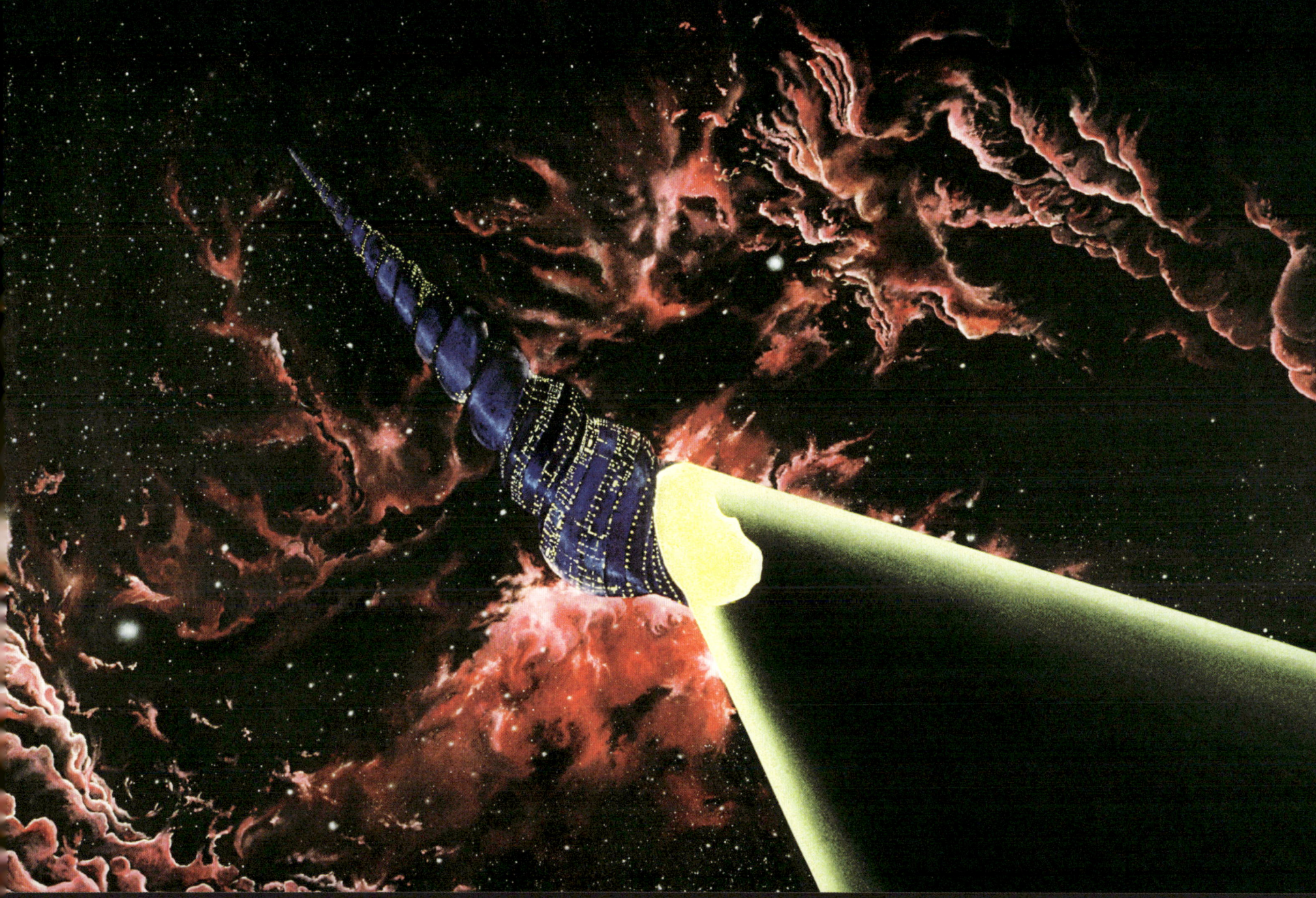

PAUL VAN HOEYDONCK

31.

There's a lot of our stuff littering the Moon's surface – a glut of recent lunar landers; the two strange Soviet Lunokhod rovers from the early 1970s that look like the skeletons of some prehistoric alien animals. Then there's all the Apollo detritus: the descent stages of the six lunar modules; rovers and retro-reflectors; tools and flags and bibles. Bits of discarded spacesuit shed like snakeskins. There are astronauts' personal items too: family photographs, golf balls, a hammer and a feather. And look! A 9-centimetre (3½-inch) abstract figurine carved from a single block of aluminium, lying in the lunar dust at Hadley–Apennine, the Apollo 15 landing site, next to a small memorial plaque with the names of fourteen astronauts who had lost their lives. The artwork, known as the *Fallen Astronaut*, was placed there by Commander David Scott in 1971. While nobody was watching, he quietly pulled the figurine out of his spacesuit pocket and laid it on the ground in a dignified gesture of remembrance. No mention of it was made in the moment. Its existence was only revealed during the post-flight press conference, but no details were given of the statue's origin.

The work was conceived by Paul Van Hoeydonck (1925–2025), a Belgian artist who, with the Waddell Gallery in New York, had come up with the wild idea of putting a piece of his futuristic space art on the Moon. What more prestigious exhibition space was there to display your art than the lunar surface? After much behind-the-scenes wrangling, the Apollo 15 crew were persuaded to take it along. But there was a problem: it was never intended as a memorial to fallen comrades. Quite the opposite. Paul had envisaged it as a bold, forward-facing statement that represented mankind's transhumanist future, inspired by the famous black monolith in Stanley Kubrick and Arthur C. Clarke's *2001: A Space Odyssey* (1968).

Van Hoeydonck was deeply troubled by his artistic vision being changed so fundamentally, and became increasingly angry at his forced anonymity. Like any artist, he wanted recognition for his work. For Commander Scott, however, this simple gesture was always intended to fly under the radar, nervous as he was about the possible conflicts of interests that involving a non-sanctioned commercial artist might bring. Apollo had to be squeaky clean. To this day, Scott insists the opposing narrative was 'fabricated by NASA personnel'. Eventually in 1972, Van Hoeydonck broke his silence and went public in an interview with American journalist Walter Cronkite.

As the story gained public interest, two copies of the work were commissioned – one for the Smithsonian in Washington, DC, and one for the king of Belgium. The artist and the gallery also arranged the production of 905 limited edition copies to be sold at $750 a piece, but they were stopped by a furious NASA, deeply sensitive about profiteering from the Apollo missions, which were paid for with public money.

Art is always political. Even more so when nervous government agencies, reputations and money are involved.

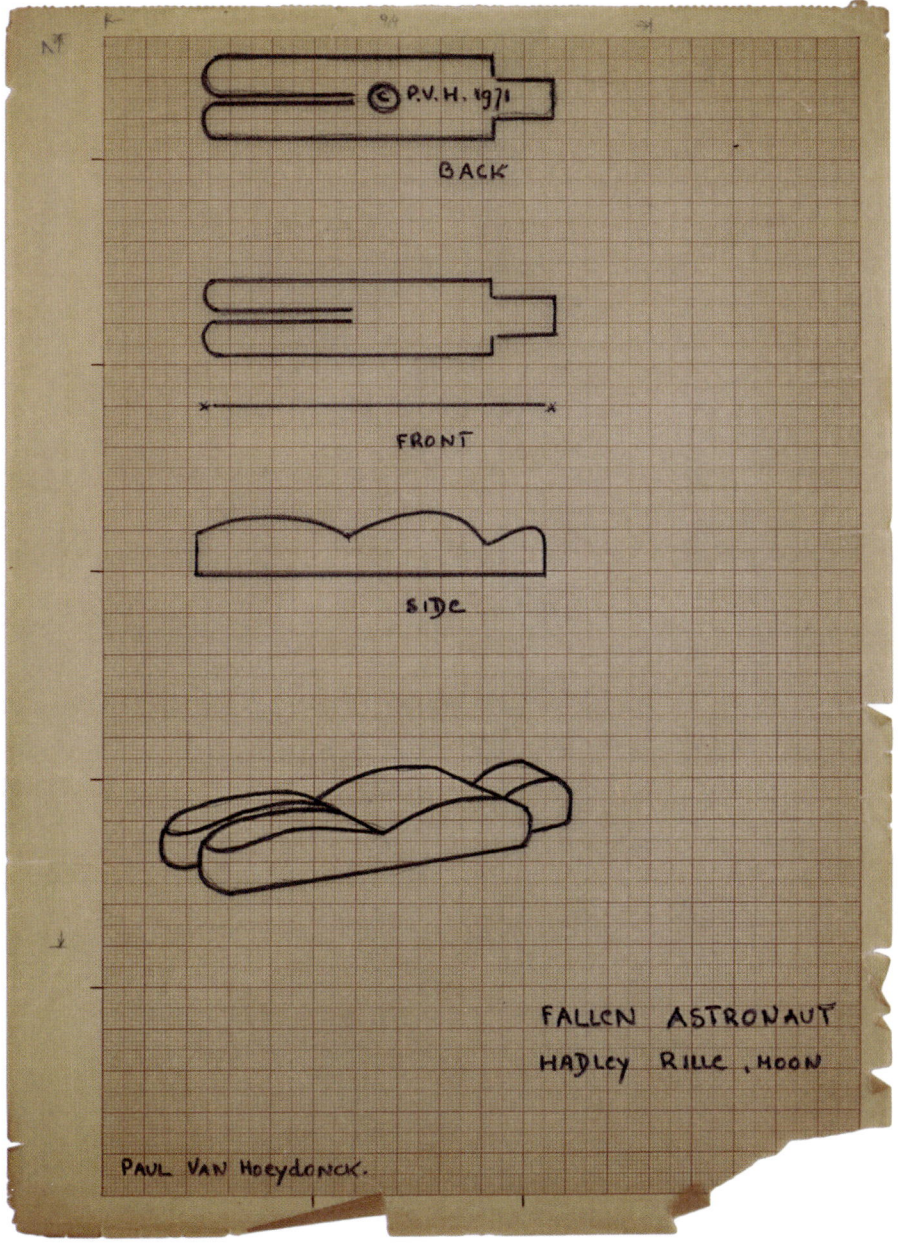

31.1 Paul Van Hoeydonck poses with a replica of Fallen Astronaut in front of the Empire State Building in New York in 1971.

31.2 Preparatory drawing of Fallen Astronaut.

31.3 Astronaut David Scott on slope of Hadley Delta during Apollo 15 Extravehicular Activity (EVA), using a 70mm camera.

31.4, 31.5 Front and profile views of *Fallen Astronaut*.

31.6 A close-up view of a commemorative plaque left by Scott at the Hadley-Apennine landing site, in memory of 14 NASA astronauts and USSR cosmonauts, now deceased: Charles A. Bassett II, Pavel I. Belyayev, Roger B. Chaffee, Georgi Dobrovolski, Theodore C. Freeman, Yuri A. Gagarin, Edward G. Givens Jr, Virgil I. Grissom, Vladimir Komarov, Viktor Patsayev, Elliot M. See Jr, Vladislav Volkov, Edward H. White II and Clifton C. Williams Jr.

> 'The most romantic place on Earth is Cape Kennedy.'
> PAUL VAN HOEYDONCK, 1965

31.4 31.5

31.6

BEZOS, MUSK AND THE HIGH FRONTIER

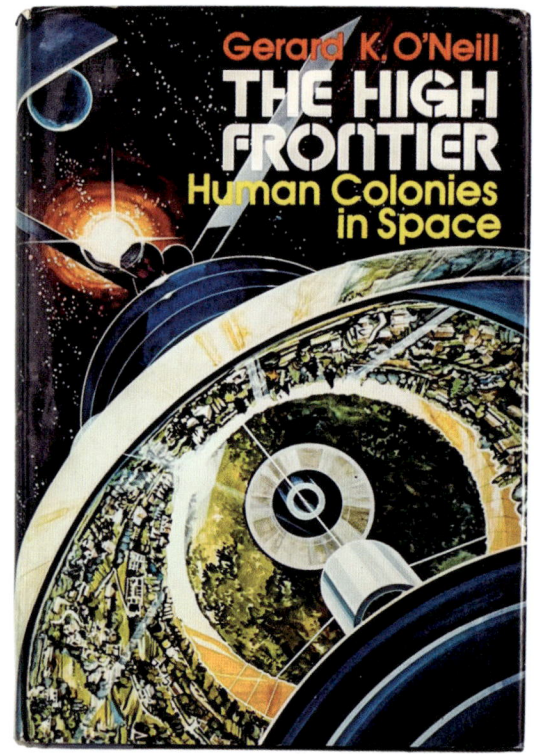

There's an old joke: how do you become a millionaire? Take a billion dollars and start a rocket company. The often-trotted-out maxim 'space is hard' can be seen in the embers of the failed rocket start-ups that have come and gone. Companies like the now defunct Rotary Rocket Company, whose prototype 'Roton', with its bizarre helicopter descent mechanism, stands as a monument at the Mojave Space Port – a warning to those who would attempt what many have tried and failed. At the turn of the twenty-first century, a rapidly reusable rocket (the Holy Grail) didn't exist, and no private company had ever delivered a payload into orbit (the other Holy Grail). Today the two richest men in the world, Jeff Bezos and Elon Musk, are leading the charge in doing just that. Rivals with the same aims of extending human presence beyond the Earth and exploiting the new space goldrush.

Way back in 2000, Elon Musk (b. 1971) found himself with $180 million burning a hole in his pocket after his ousting as the CEO of PayPal. His to-do list was short and ambitious: make human life on Earth sustainable, and make humans a multiplanetary species. Like the rocketeers who had opened up space in the twentieth century, his drive was kindled by science fiction: Isaac Asimov, Robert Heinlein and in particular Douglas Adams's prescient *Hitch Hiker's Guide to the Galaxy* (1979). Perhaps Elon found a kinship with the chaotic Zaphod Beeblebrox, President of the Galaxy, with an ego so huge he needed a second head surgically attached to contain it. Space exploration looked very different then: America had the Space Shuttle, but it didn't have anywhere to go beyond the International Space Station. Where was the excitement, the adventure, the wild things that the Apollo generation had been promised? He decided to invest half of his fortune in doing something about it, imbedding himself with other New Space advocates, like Peter Diamandis, the founder of the X Prize, who in the mid-1990s offered a huge cash incentive to anyone who could launch a reusable spacecraft above the magical 100-kilometre (62-mile) line that marks the edge of space. Peter was one of the first in a long list of well-meaning people who urged Musk not to start a space company. Unless he wanted to go broke. Or go mad. The keys to space were still firmly in the pockets of governments and their closed-shop contractors, such as Boeing and Lockheed Martin. But not for much longer.

At the same time, Amazon's Jeff Bezos (b. 1964) was founding his own space exploration company, Blue Origin. Like Musk, Bezos was motivated by cultural markers of the 1960s and '70s: Apollo, *Star Trek* and most significantly physicist Gerard K. O'Neill's wild visions of giant, habitable space colonies. Both knew that when it came to bold, visionary space projects, governments were dragging their feet – they were risk averse, bureaucratic and hobbled by short political cycles, which led to short-term thinking. Rocket technology was prohibitively expensive and burdened by over-regulation. Private companies with vast personal wealth would have the freedom to create a viable sustainable space economy that would then open the door to innovation, enterprise and ultimately exploration. Bezos also knew that infinite growth on a finite Earth is impossible. The need for energy and raw materials to satiate our unquenchable thirst for comfort and convenience meant eventually having to expand off-planet. The turn of the century was the moment that rampant capitalism and radical environmentalism, rather than raw geopolitics, became the primary force behind space exploration. This is the stuff that's beyond the reach of everyday politics.

Start-ups building viable rockets and rocket engines from scratch were unheard of in the 2000s. For SpaceX, based out of Hawthorne in Los Angeles, this meant manufacturing their own components in situ, or where necessary, purchasing off-the-shelf components. With Musk came a personally vetted band of young,

32.1 Jeff Bezos at Launch Site One in West Texas.

32.2 Cover of Gerard K. O'Neill, *The High Frontier* (1976).

32.3 NASA artist Rick Guidice's cutaway view of a Bernal Sphere, an orb-like, slowly rotating space colony, which featured on the cover of *The High Frontier* in 1976.

32.4 Blue Origin's New Shepard lifts off for its historic first human flight, NS-16, 20 July 2021.

32.5 Graphic of SpaceX's Falcon 9 rocket, the launch vehicle which has become the workhorse of the space industry, transporting crews to the International Space Station as well as other payloads to orbit. Its ability to land back on Earth has revolutionized launch capabilities by reducing the cost of sending mass into space.

32.6 Graphic of SpaceX's Crew Dragon spacecraft, used for all crewed missions to and from the ISS as well as private space missions such as Polaris Dawn, which saw Jared Isaacman and Sarah Gillis perform the first commercial space walk.

32.7 A graphic overview of SpaceX mission operations.

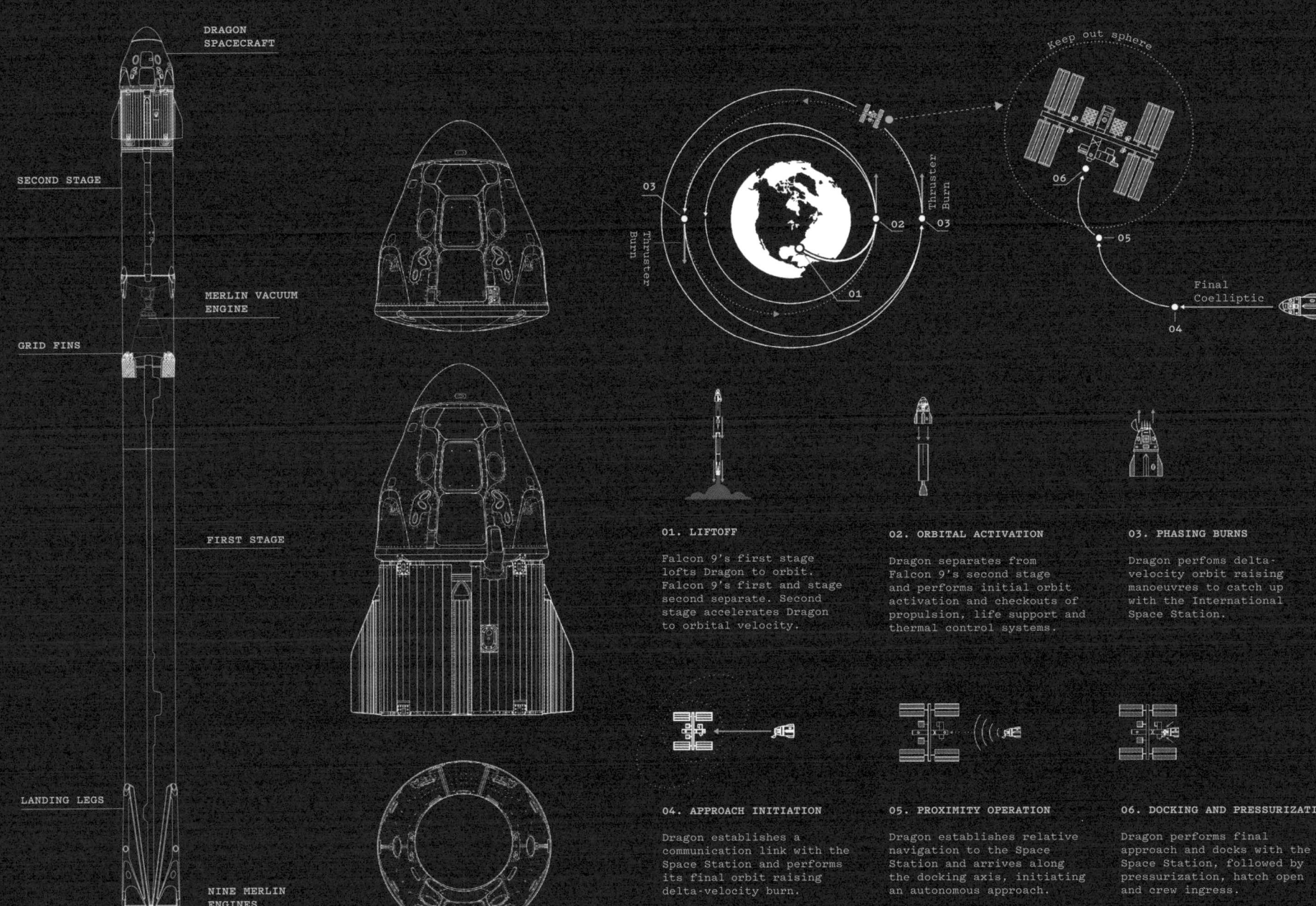

high-flying can-do engineers and managers including Gwynne Shotwell, the secret weapon who would sell this new vision of space exploration to the world. By 2003 they had their first vehicle, the Falcon 1 (named after the Millennium Falcon from *Star Wars*). A launch site was chosen on the Kwajalein Atoll in the Marshall Islands. Building a launch facility from scratch on an island the size of a couple of football fields in the middle of the Pacific Ocean was a logistical nightmare. But it had three advantages: first, its remoteness meant that a wayward experimental rocket wouldn't hit anyone. Second, they could pretty much be left to their own devices, and third was the island's position near the equator. The angular momentum of the Earth's rotation from that latitude would give the small, underpowered rocket an extra boost as it flew eastwards, translating into more mass in orbit. On a tiny isolated island in a string of tiny isolated islands, like breadcrumbs on a map, the team worked at the very edge of what was possible, under the darkest of star-filled skies. March 2006 saw the first launch. As Falcon 1 rose above the atoll, its Merlin engine caught fire, and thirty seconds later it fell back to Earth. The second attempt a year later was closer to success, but was foiled by a weight-versus-risk decision not to add slosh baffles* into the second-stage fuel tank. The third launch in 2008 was thwarted by human error and monumental bad luck – at the moment of stage separation, the first stage, rather than falling away, bounced up and bumped into the second stage, sending it off-course. They had just enough spare parts in the factory for one final shot. To show the world that SpaceX could do what no other private company had done. On 28 September 2008 the Falcon 1 rose into the sky flawlessly, and this time it kept going. Upper-stage and fairing separation nominal. The mass simulator payload inserted gently into its orbit. Days from bankruptcy, SpaceX had done the impossible.

---

* Slosh baffles = internal structures designed to stabilize the motion of the liquid fuel in flight.

Meanwhile Bezos's company Blue Origin had been at work, with little to no fanfare. Founded in 2000, in west Texas, its coat of arms is a pair of tortoises looking to the heavens over the motto 'Gradatim Ferociter' (Step by Step, Ferociously). The company's first small step was to exploit the exclusive market for suborbital space tourism, building a rocket and crew capsule that would take paying passengers on a parabolic ride above the space demarcation line for a few minutes of weightlessness and a spectacular view of the Earth. With only Richard Branson's Virgin Galactic as competition, this was a niche market. But it was exciting, aspirational and an opportunity to generate income and experience for new projects. The first iteration was Bezos's demo vehicle, named after Robert H. Goddard, that flew just under 30 metres (100 feet) in altitude. This became New Shepard (named after the first American in space, Alan Shepard), which has now been taking the wealthy and deserving into suborbital space since 2021, with passengers including eighty-two-year-old Wally Funk (one

---

32.8 The sleek SpaceX (IVA) spacesuit worn by all astronauts who fly on the Crew Dragon spacecraft. The original suit concept was designed by Hollywood costume designer Jose Fernandez at the request of Elon Musk, who said they needed to look 'badass'. Since the very beginning, spacesuit design and fashion have walked side by side.

32.9 Barbara Diener, *First SpaceX Dragon Capsule to go into Orbit*, 2021.

'Where there's a will there's a way. Where there's no will there's no way.'

ROBERT ZUBRIN (B. 1952), PRESIDENT OF THE MARS SOCIETY

of the Mercury 13 women who completed the astronaut selection tests with the first group of American astronauts), ninety-year-old Captain Kirk actor William Shatner and of course pop star Katy Perry. It's an idea that's been long imagined. In the 1940s the visionary British Interplanetary Society had proposed putting a human on board a V-2 missile and doing exactly the same thing. But most important for both companies' strategies was their *iterative* approach to engineering – accepting risk, building prototypes, testing them to destruction and learning the lessons moving forward – something that is impossible for large government-funded space projects.

Everything has changed in twenty years. SpaceX now launches 90 per cent of all mass that's carried into orbit, as well as being the primary ferry service for astronauts to and from the International Space Station. Blue Origin is now catching up, moving beyond its space tourism beginnings into orbital delivery of payloads. The next chapter has already begun, with spectacular test flights of two new giant heavy-lift vehicles, Space X's behemoth Starship and Blue Origin's New Glenn (after John Glenn (fig. 18.8), the first American astronaut in orbit), as well as the development of new lunar landers and other space architecture for our return to the Moon and onwards to Mars – and beyond.

Why should we care? The twentieth century demonstrated that space travel is possible. But the twenty-first is showing that it's essential and profitable. We are now completely dependent on our space activities, from communication to navigation to Earth observation, science, finance and defence. This is just the beginning. Soon we'll have the ability manufacture huge structures in orbit; we'll be able to harvest solar power and to make better materials and pharmaceuticals.

The Moon will take on new economic importance. For Bezos, this is one step in a lifelong vision to move all polluting industry and infrastructure away from Earth, returning it to a pre-industrial form. And of course, to explore. To boldly go where no billionaire has gone before. Because it's there. Bezos and Musk, with their vast wealth, are spearheading this new future. By building a new road to space, they're encouraging others to drive along it. A new gold rush to the High Frontier, and delivery on the promise of space that was made almost a century before.

32.10 Falcon 9 rocket debris on Tresco Island, near Scilly, 2015.

32.11 'Megaroc' (1946) - the forerunner to Blue Origin's New Shepard. A single-occupancy suborbital rocket proposed by R. A. Smith of the British Interplanetary Society. The design was based on a modified V-2, but sadly the idea was rejected.

32.12 New Glenn's fairing with the NG-1 mission patch, 10 January 2025.

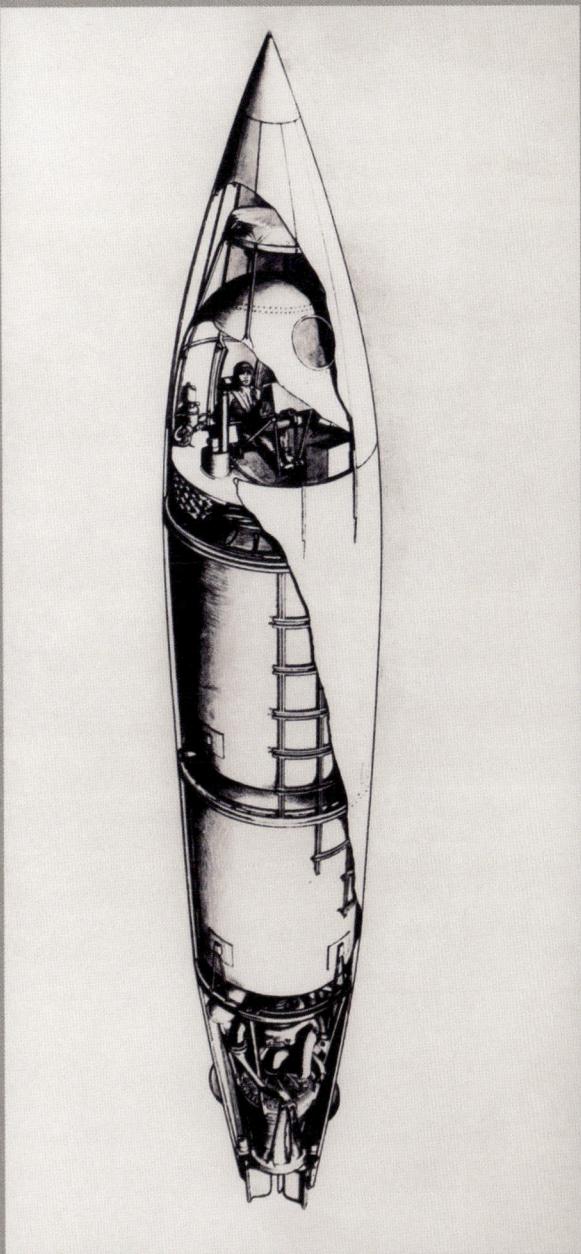

32.11

32.12

ALICE GORMAN

33.1 Alice 'Dr Space Junk' Gorman pictured here with a model of Sputnik, holding *Soviet Union*, no. 136 (1961).

33.2, 33.3 Sample locations from the Sampling Quadrangle Assemblages Research Experiment (SQuARE). **33.2** shows Square 03 in the starboard Maintenance Work Area of the International Space Station; **33.3** shows Square 05 on the aft wall of the Node 3 module.

Everything we do leaves an imprint, and every imprint tells a story. The brief time we've been exploring space is no exception. Yet our associations with the word 'archaeology' are so rooted in the distant past it seems odd to couple it with a word as forward-facing as 'space'. The Australian archaeologist Alice Gorman (b. 1964), also known as Dr Space Junk, has built her career at this new frontier, looking at the cultural landscape of our space activities. It is an area of study that satisfies our divided brain – science and technology on one side, and social science, history, politics and poetry on the other. It's understandable that Alice wanted to be both an astrophysicist *and* an archaeologist. In 2001, while working in Queensland on an Indigenous heritage project, she found herself in the Outback at night, looking up instead of down. She watched the smooth pass of a satellite in front of a backdrop

'So many myths and legends are centred on the Moon. But now it's a human landscape. Tranquility Base where those astronauts first set foot on the Moon is an archaeological site. They've left artefacts there. They've left footprints.'

ALICE GORMAN, 'SPACE ARCHAEOLOGY' AT TEDXSYDNEY, 2013

of fixed stars, and realized she could make a career connecting the two hemispheres of Earth and sky.

Space offers a unique challenge for archaeologists, who are used to getting their hands dirty. Recently the International Space Station (ISS) has become a living archaeological project in an effort to help us understand how humans adapt to such a unique environment. For over twenty years, it has grown into its own off-world culture; a diverse mini community free from the effects of gravity. Gorman and her co-chief investigator Justin Walsh have studied it in a variety of ways. For example using the trove of historic astronauts' photographs to catalogue the shifting displays of Russian space heroes (such as Gagarin, Tsiolkovsky and Korolev) and religious icons that adorn the Russian Zvezda module. What might these changes tell us? It turns out astronauts make good archaeologists too. The Sampling Quadrangle Assemblages Research Experiment (SQuARE) asked crews to take daily photographs of 1-metre-square (3-foot) grids aboard the station and track the types of objects that fall into those areas – a traditional archaeological sampling technique. From these pictures, we can track how physical space inside the station is used, how it changes over time and what that might mean. It's the first archaeological dig to take place in space.

Outside the ISS, low Earth orbit has become increasingly crowded, with thousands of pieces of orbital debris posing a serious threat to the working satellites. But one person's space junk is another's space treasure. Among the flotsam and jetsam remain some historic gems for Gorman to study. Early satellites like America's Vanguard 1,

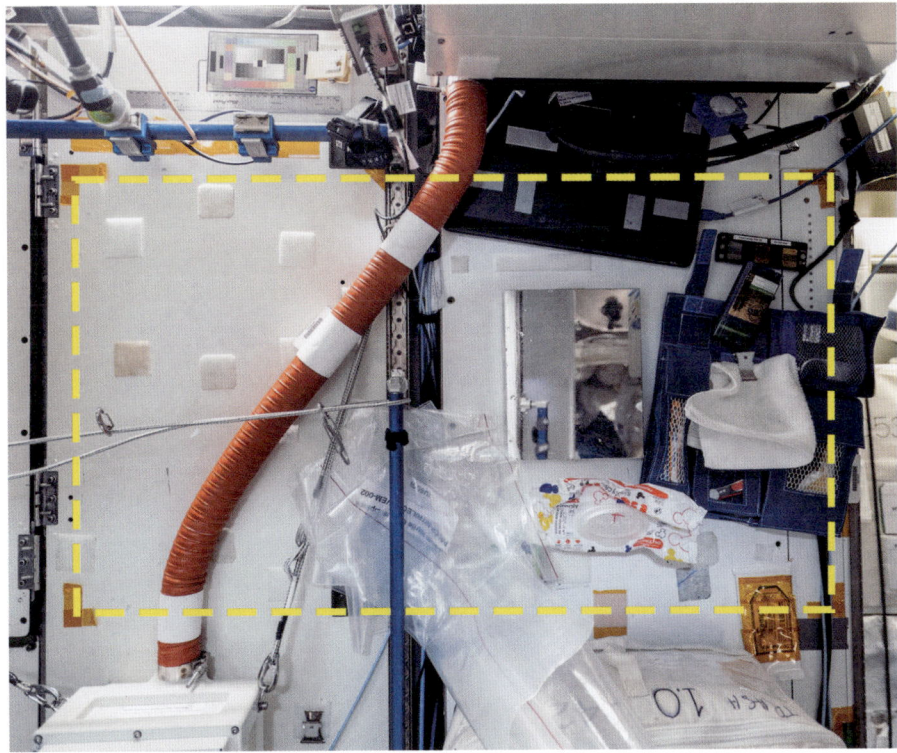

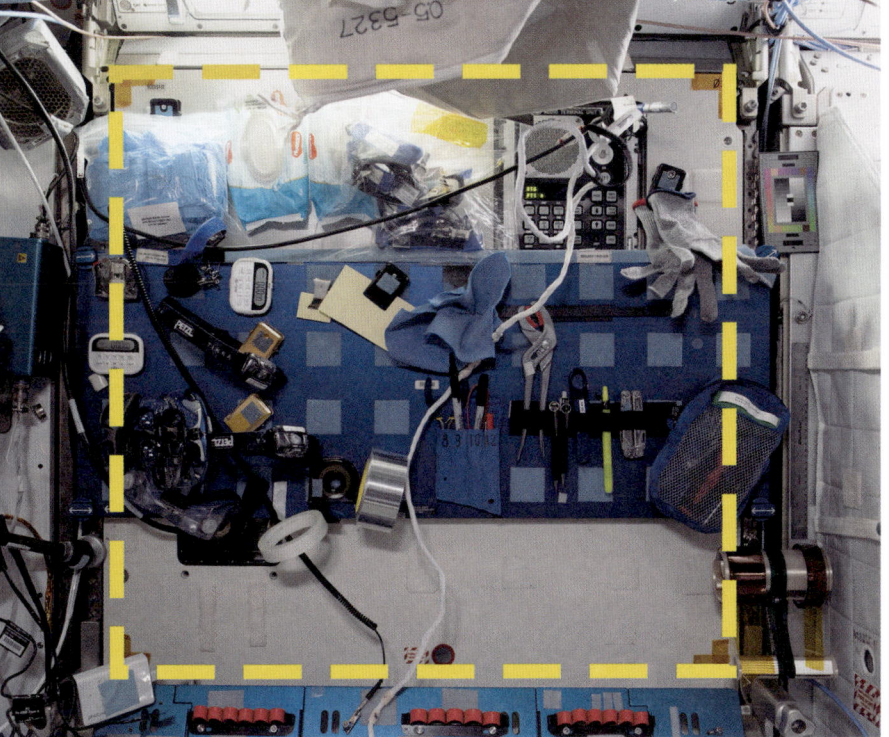

33.4-33.6 An artist's impression of the vast amount of space debris now orbiting Earth. The main concentration is in Low Earth orbit (with an altitude of 2,000 kilometres/1,242 miles), but it can also be found in geosynchronous orbit (around 36,000 kilometres/22,000 miles), whereby a satellite remains fixed over the same geographical location. Space debris has archaeological importance, but it becomes increasing problematic when considering the safety of future space hardware.

33.7 Dr Mark Burchell (top) and Dr Simon Green study coloured light micrographs of impact craters caused by space junk on a solar panel of the Hubble Space Telescope. The solar panels were constantly bombarded with tiny high-velocity particles. While some of these were natural small meteoroids, there were impacts from debris left in orbit from previous missions.

now the oldest object in orbit, still haunt this new continent and have become important artefacts that mark the early history of the space race.

Glowing objects falling from the sky have always inspired our most vivid imaginings, superstitions and portents. While most of what we send into space remains out of reach, sometimes fragments of de-orbited spacecraft come back down to Earth and end up landing in populated areas. In Argentina in 1991, a washing-machine-sized piece of the early Russian Salyut 7 space station crashed through the roof of a woman's house as she was doing the ironing. In Manitowoc, Wisconsin, a piece of Sputnik 4 embedded itself into North 8th Street one morning in 1962. A replica of the fragment sits in a nearby museum like a holy relic, now part of a local folklore that's spawned an annual community festival, 'Sputnikfest'. The de-orbiting of the first American space station, Skylab, became a national media event as debris rained down over Western Australia. 'I survived Skylab' hats and T-shirts were printed, and Skylab fragments became highly collectable, much like the craze for Egyptian artefacts in the nineteenth century. As a publicity stunt, NASA was issued with a $400 fine by the shire of Esperance for littering. They never paid.

Our space stuff has now reached other planets – an important milestone in the evolution of our species. The Soviet Venera landers still sit on Venus's hellishly hot surface. What might they look like now? On Mars there's an ever-growing variety of human-made objects, from the Viking landers of the 1970s to the scattered bones of more recent rovers and their associated debris. The Perseverance rover's parachute, whose pattern contains a secret message; protective covers; old netting; a helicopter drone; bits of broken wheel; and Curiosity's footprints – the holes in the wide wheels of the Curiosity rover that spell out 'JPL' (for Jet Propulsion Laboratory) in Morse code as it slowly trundles across the Martian surface: ·–––/·–––·/·–··. A pleasing reminder of our extraterrestrial presence, lest we forget. Out beyond the planets are probes looping around the solar system and beyond. If the Earth ever gets destroyed by Vogons to make way for a hyperspace bypass, the two Voyagers with their Golden Record time capsules

33.8 Poster for Skylab Protective Helmets. The helmets were designed and produced by William Foard, based on a concept by Jeff Hall. Photo courtesy Foard Media, Los Angeles.

33.9 Paperweight made from debris that fell to Earth as the US Skylab space station re-entered the atmosphere in July 1979. This piece of debris was recovered at Rawlinna, Australia.

33.10 This image from the Mars Hand Lens Imager (MAHLI) camera on NASA's Mars rover Curiosity shows a small bright object, just over a centimetre (½ inch) long, on the ground at the 'Rocknest' site. The rover team has assessed this object as debris from the spacecraft, possibly from the events of landing on Mars.

33.8

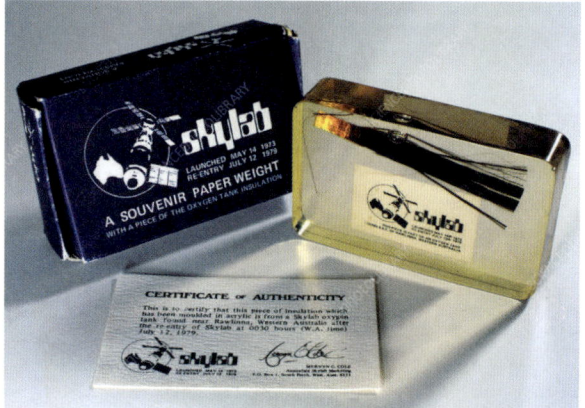

33.9

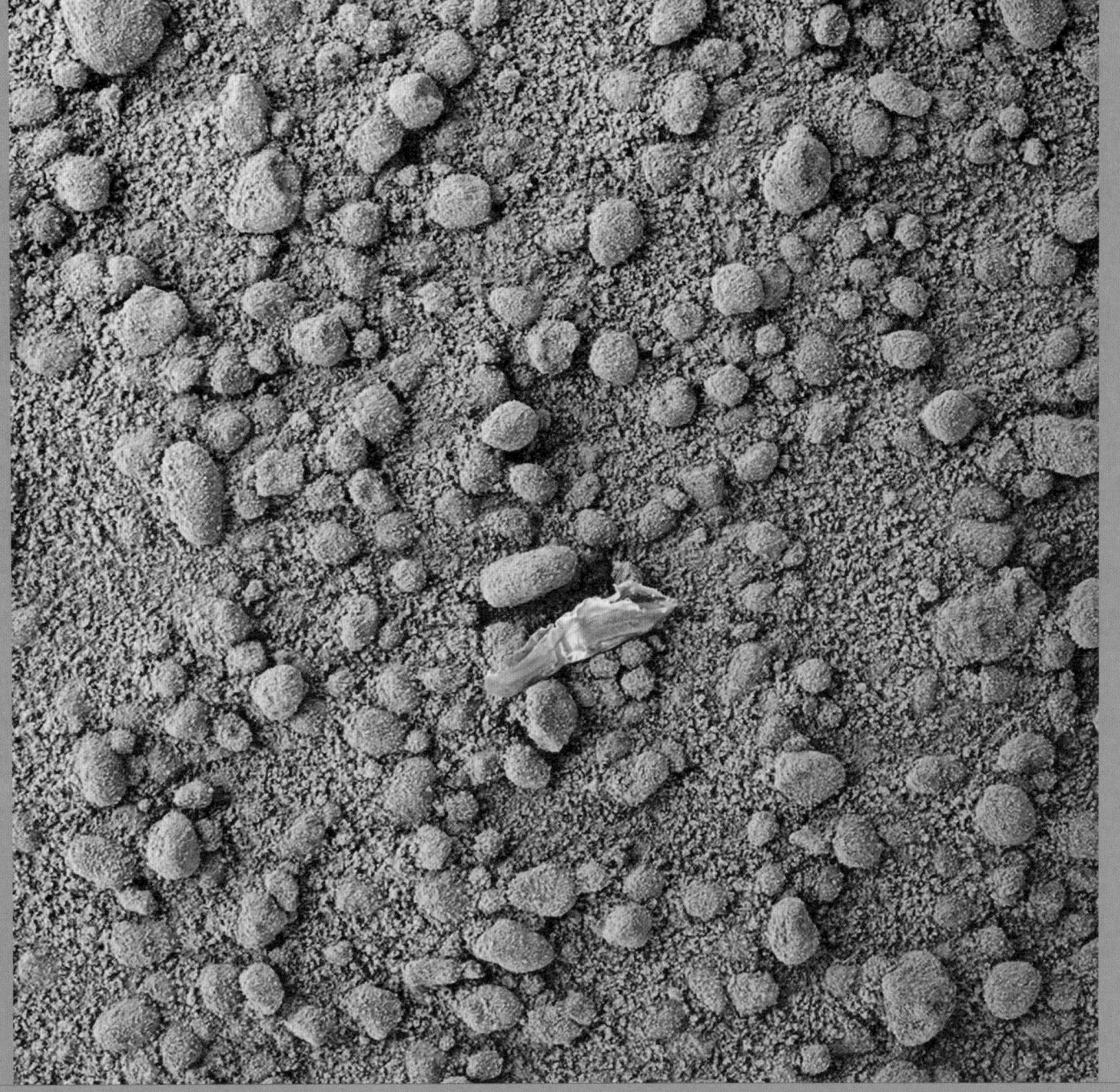

'The Bones of Mars'
Sam Illingworth, 2024

*A weight lingers on red dust –
half-wheels and net-veils,
their shroud sewn
into the stone's hush.
The ash-ghosts of flame wings
mark unbroken passages –
trail-lines pressed
by silent machines.
Each crash-frag
and discarded tether
spreads across
the skin-mirror,
etching our map
of priceless debris.
These fragments,
once tool-spirits,
now rest
as wanderer bones –
an amassed archive
left to unravel
with the winds*

will still be sailing out for an eternity. A puzzle for future alien archaeologists, and a moment for us now to consider how our space technology and wider human culture can coexist. How should we distinguish junk from treasure? How should we interpret Elon Musk's choice of dummy payload on the first Falcon Heavy: his red Tesla Roadster, which is now lapping around the solar system? An inevitable change in the story of space flight? Or a moment to roll our eyes in frustration? Everything we do leaves an imprint, and every imprint is political.

Space archaeology field research can happen at ground level. Launch sites such as Woomera in Australia or the Baikonur Cosmodrome in Kazakhstan provide rich pickings. Baikonur is still the primary Russian launch facility,* and the launch pad known as 'Gagarin's Start' has taken on an almost religious significance, marking the geographical origin of the space age with the launch of Sputnik 1. It is the exact point where we as a species first ascended to the heavens. You feel echoes of the earliest missions everywhere. These days the site is a difficult place to reach, but French photographer Jonathan 'Jonk' Jimenez managed to sneak in and document his clandestine exploration of abandoned Buran Soviet space shuttles. The craft never flew; they have been hidden away for decades in their crumbling tombs, slowly being repainted by rust and entropy. Time machines instead of space machines.

Through various initiatives, Alice has become a vocal advocate for promoting the Moon's cultural heritage as a sensitive archaeological site, at a time when lunar exploration is back on the agenda. As space archaeologists, what might we like to see? The famous flags now bleached white and their flagpole assembly designed by NASA engineer Jack Kinzler. The feather and hammer dropped by Apollo 15 Commander David Scott to test Galileo's theory of falling objects in gravity fields. The surplus food pouches cooked by NASA space chef Rita Rapp, or the golf balls hit by Apollo 14's Alan Shepard. You'd want to see the footprints of course.

What a potent symbol the footprint is for the archaeologist in us all. Gorman reminds us: while the prints were made by the boots of men, the boots themselves were made by the hands of women. A footprint is a sharing of a space through time; it tells us we're not alone. A footprint asks questions: Who? When? Why? On Earth, footprints can fossilize in rock or wash away in hours. Here they will last for aeons, despite being formed in the finest of materials. The Sea of Tranquillity has no tide to erase them. These are imprints that mark a moment when we migrated from our home planet to another, transitioning from *Homo sapiens* (wise) to *Homo spatium* (space).

---

* For how much longer is unclear; the new Vostochny Cosmodrome looks set to take over.

'Every time you look at an image of that first footprint on the Moon, remember it was made by a man's foot and a woman's hand.'

ALICE GORMAN, 'SPACE AGE ARCHAEOLOGY', 2025

33.11 French photographer Jonathan 'Jonk' Jimenez snuck into the Baikonur Cosmodrome, Kazakhstan, to photograph a decaying Soviet Buran orbiter from the 1980s. The Buran never flew in space.

33.12 (OVERLEAF) Footprints found in White Sands National Park, dated to the Last Glacial Maximum by USGS Research Geologists Kathleen Springer and Jeff Pigati. Photo taken in 2023.

33.13 (OVERLEAF) View from Station Apollo Lunar Surface Experiments Package (ALSEP), Heat Flow Probe taken during the third Extravehicular Activity (EVA) of the Apollo 17 mission.

Acknowledgments

The heroes of this book are: Ben Hayes, Phoebe Colley and Aman Phull at Thames & Hudson. Thank you to Pauline Hubner, picture researcher extraordinaire, who has tirelessly put up with my unreasonable picture demands. We shall forever lament the ones that got away. Thank you to Chloe Gott and Anna Muir at KBJ who held my hand getting this book off the ground. Thank you to my family who have helped me tough out the last year with extra kindness and support. Nicola, Lucy and Robbie, Dad and Carina, John, Lenore and Jeremy. Chimp and Chomp of course. Alison Barton for listening to me prattle on in Newcastle. Thank you to Suzie, Cathy and Ian for endless cake, coffee and more. A special thank you to my mystery book-giver who has sent a wealth of inspirational materials over the years, much of which has made its way into these pages. I've no idea who you are, but I'm eternally grateful. Let us continue this entertaining dance. Thank you to Professor Sam Illingworth for his kind permission to publish his poem 'Bones of Mars'. Thank you to George Savona for your friendship and infectious creativity over the years. To Werner Herzog, who I've never met but whose work continues to astound and delight. Most of all, thank you to my mum, to mega-dog and to Reg, who will be eternally missed. Thank you for the Seaton Sluice walks and for making me watch *Breaking Bad* from start to finish. I'm about to watch the final episode ... now.

Picture Credits

Openers
p. 2 MSFC/NASA; p. 14 Fine Art Images/Diomedia; p. 52 detail of 8.9; p. 90 John Frost Newspapers/Mary Evans Picture Library; p. 140 detail of 18.9; p. 210 detail of 27.9; p. 244 NASA

0.1 R. H. Goddard, *A Method of Reaching Extreme Altitudes* (Washington, DC: Smithsonian Institution, 1919); 0.2 Jules Verne, *From the Earth to the Moon* (New York: Scribner, Armstrong, 1874); 0.3 *Popular Mechanics Magazine*, March 1930; 0.4 SpaceX/UPI/Alamy Live News/Alamy; 0.5 Chris Gunn/NASA; 0.6 KSC/NASA; 0.7 Courtesy of The Linda Hall Library of Science, Engineering & Technology, Kansas City, MO; 1.1 Rare Book and Special Collections Division, Library of Congress, Washington, DC; 1.2, 1.3 Kepler/Somnium murals, Anger, Austria, May 2013. © Joshua Ellingson; 1.4 Rare Book and Special Collections Division, Library of Congress, Washington, DC; 1.5 Royal Danish Library, Copenhagen (KBK 2-1); 1.6 Look and Learn/Elgar Collection/Bridgeman Images; 1.7 The Stapleton Collection/Bridgeman Images; 1.8 Houghton Library, Harvard University, Cambridge, MA (STC 11943.5); 1.9, 1.10 © 2025 DACS; 1.11 Wellcome Collection, London; 1.12 Courtesy of The Linda Hall Library of Science, Engineering & Technology, Kansas City, MO; 2.1 Leonard de Selva/Bridgeman Images; 2.2 From Jules Verne, *From the Earth to the Moon* (New York: Scribner, Armstrong, 1874); 2.3 NASA; 2.4 Mary Evans Picture Library; 2.5-2.22 From Jules Verne, *From the Earth to the Moon* (New York: Scribner, Armstrong, 1874); 2.23 Centre d'études verniennes, Fonds Jules Verne, Nantes (MJV 83-14); 2.24 Centre d'études verniennes, Fonds Jules Verne, Nantes (MJV 83-15); 2.25 Science Source/Mary Evans Picture Library; 2.26 Courtesy of The Linda Hall Library of Science, Engineering & Technology, Kansas City, MO; 2.27 Leopoldo Galluzzo, *Altre scoverte fatte nella luna dal Sigr. Herschel* (Naples: L. Gatti e Dura, 1836); 2.28 MSFC/NASA; 2.29 Georges Méliès, *Le voyage dans la lune*, 1902; 3.1 Photo © Ted Streshinsky/Corbis via Getty Images; 3.2 Lucien Rudaux, *Sur les Autres Mondes* (Paris, Larousse, 1937); 3.3 Metropolitan Museum of Art, New York. Gift of the sons of William Paton, 1909 (09.214.1); 3.4 Reproduced courtesy of Bonestell LLC; 3.5 © Bonestell LLC. Reproduced courtesy of Bonestell LLC; 3.6 Courtesy of The Linda Hall Library of Science, Engineering & Technology, Kansas City, MO; 3.7 Courtesy Everett Collection/Mary Evans Picture Library; 3.8 Courtesy George Pal Productions/Ronald Grant/Mary Evans Picture Library; 3.9 Photo courtesy Leo Boudreau; 3.10 B. Christopher/Alamy; 3.11 © Bonestell LLC. Reproduced courtesy of Bonestell LLC; 3.12, 3.13 Reproduced courtesy of Bonestell LLC; 4.1 Mike Wilson and Arthur C. Clarke/Rocket Publishing/SSPL/Getty Images; 4.2 Courtesy pulpcovers.com; 4.3, 4.4 Rocket Publishing/SSPL/Getty Images; 4.5 Arthur C. Clarke/Rocket Publishing/SSPL/Getty Images; 4.6 Cornwall Aviation Company Limited, St Austell/Rocket Publishing/SSPL/Getty Images; 4.7 Courtesy pulpcovers.com; 4.8 *Amazing Stories*, November 1928; 4.9 Photo The University of Liverpool Library (OS/I1/1). Courtesy the Estate of Olaf Stapledon; 4.10 Photo courtesy pulpcovers.com; 4.11 Courtesy MGM/Stanley Kubrick Productions/Ronald Grant/Mary Evans Picture Library; 4.12 Michael Ochs Archives/Getty Images; 4.13 Sunset Boulevard/Corbis Historical/Getty Images; 5.1 Scala, Florence – courtesy of the Ministero Beni e Att. Culturali e del Turismo; 5.2 National Central Library of Florence (Ms. Gal. 88, c. 107v). Courtesy of the Ministry of Culture; 5.3 Scala, Florence; 5.4 Scala, Florence – courtesy of the Ministero Beni e Att. Culturali e del Turismo; 5.5 PHA HMC 241, volume IX, folio 30. Reproduced by kind permission of Lord Egremont and with acknowledgments to the County Archivist, West Sussex Record Office; 5.6-5.11 Galileo Galilei, *Didereus nuncius* (Thomam Baglionum, 1610); 5.12 Marie Claire Eimmart, *Raffigurazione di fenomeni celesti – Aspetto di Giove*, XVII, MdS-124i. Museo della Specola|sistema Museale di Ateneo, Università di Bologna. Photo Marco Pintacorona; 5.13 Scala, Florence; 6.1 Wellcome Collection, London; 6.2 The Royal Society, London; 6.3 Reproduced by kind permission of the Syndics of Cambridge University Library (Adv.b.39.1); 6.4 Reproduced by kind permission of the Syndics of Cambridge University Library (MS Add.3996); 6.5 Reproduced by kind permission of the Syndics of Cambridge University Library (MS Add.3975); 6.6 Reproduced by kind permission of the Syndics of Cambridge University Library (MS Add.3958); 6.7 Isaac Newton, *A treatise of the system of the world* (London: F. Fayram, 1728); 7.1 Courtesy Lowell Observatory Archives, Flagstaff, AZ; 7.2, 7.3 JPL/NASA; 7.4 From *Osservatzioni Astronomiche e fisiche ... Del Planeta Marte ... Memoria del socio G. V. Schiaparelli*. Published by Reale Accademia dei Lincei, Rome, 1878; 7.5-7.10 Courtesy Lowell Observatory Archives, Flagstaff, AZ; 7.11 From W. D. & H. O. Wills 'Romance of The Heavens' series, 1928; 7.12, 7.13 Courtesy Lowell Observatory Archives, Flagstaff, AZ; 7.14 Photo Marriott Library, University of Utah/Library of Congress, Washington, DC; 7.15 Courtesy Lowell Observatory Archives, Flagstaff, AZ; 7.16 Science History Images/Alamy; 7.17 Courtesy Lowell Observatory Archives, Flagstaff, AZ; 8.1, 8.2 Collection Castle Ward, County Down. Photos National Trust Images; 8.3-8.6 The Royal Society, London; 8.7-8.16 Courtesy of The Linda Hall Library of Science, Engineering & Technology, Kansas City, MO; 9.1 Sam Ogden/Science Photo Library; 9.2 David Higginbotham (NASA-MSFC)/NASA; 9.3-9.6 NASA; 9.7 Photo Princeton/NGAS/JPL-Caltech/NASA; 9.8 Photo © Craig Cutler; 10.1 TASS Archive/Diomedia; 10.2 Photo © Gbruev/Dreamstime.com; 10.3 laufer/stock.adobe.com; 10.4 Vic/stock.adobe.com; 10.5 Cameraphoto/akg-images; 10.6 Library of Congress, Washington, DC; 10.7 Sovfoto/Universal Images Group/Getty Images; 10.8 Konstantin Tsiolkovsky, 'Exploration of Space Using Reactive Devices', from *Scientific Review*, no. 5 (St Petersburg, May 1903); 10.9 Konstantin Tsiolkovsky, *The simplest project of a purely metallic aeronaut made of corrugated iron* (Kaluga: Author's edition; S.A. Semenov's printing house, 1914); 10.10-10.13 Archive of the Russian Academy of Sciences, St Petersburg; 10.14 Granger – Historical Picture Archive/Alamy; 10.15, 10.16 Archive of the Russian Academy of Sciences, St Petersburg; 10.17 *Cosmic Voyage* (Mosfilm, 1936). Directed by Vasily Zhuravlyov; 10.18 Photo © Alexander Ovchinnikov/TASS/Mary Evans Picture Library; 10.19-10.28 Photo © Andrii Zhezhera/Dreamstime.com; 10.29, 10.30 Smithsonian National Air and Space Museum, Washington, DC. Gift of Tsiolkovsky Space Museum, Kaluga, Russia (A20110015000); 11.1 NASA; 11.2 Library of Congress, Washington, DC; 11.3 Courtesy of Clark University Archives and Special Collections, Worcester, MA; 11.4-11.6 Smithsonian Institution Libraries, Washington, DC; 11.7 Courtesy of Clark University Archives and Special Collections, Worcester, MA; 11.8 NASA; 11.9, 11.10 Courtesy of Clark University Archives and Special Collections, Worcester, MA; 11.11 NASA; 11.12 MSFC/NASA; 11.13 GSFC/NASA; 11.14-11.16 Smithsonian National Air and Space Museum, Washington, DC. Donated by Dr Robert H. Goddard (A19360022000); 11.17 Photo Dane Penland, Smithsonian National Air and Space Museum, Washington, DC (NASM 2005-35497); 12.1 MSFC/NASA; 12.2-12.7 Private Collection. Photos courtesy Lion Heart Autographs and Gary Cwick; 12.8-12.10 Photos courtesy Iron Library (Eisenbibliothek), Foundation of Georg Fischer Ltd, Schlatt, Switzerland; 12.11 Courtesy Everett Collection/Mary Evans Picture Library; 12.12 Süddeutsche Zeitung/Mary Evans Picture Library; 12.13 Photo Collection Christophel/Alamy; 12.14 Album/IMAGO; 13.1 MSFC/NASA; 13.2 Süddeutsche Zeitung Photo/Alamy; 13.3 Watford/Mirrorpix/Getty Images; 13.4 Photo courtesy Bonhams; 13.5 Photo courtesy Bonhams; 13.6 Smithsonian National Air and Space Museum, Washington, DC (NASM 90-378); 13.7 Photo courtesy Bonhams. Granted by permission of the Family of Wernher von Braun; 13.8 © Barbara Diener 2023; 13.9 Underlying image © Bonestell LLC. Reproduced courtesy of Bonestell LLC; 13.10 Photo Ralph Crane/The LIFE Picture Collection/Shutterstock; 14.1 Süddeutsche Zeitung Photo/Alamy; 14.2 *Die Rakete*, 1927; 14.3 Mary Evans Picture Library; 14.4 Bettmann/Getty Images; 14.5 Science & Society Picture Library/SSPL/Getty Images; 14.8 POST 33/5130, The Royal Mail Archive at The Postal Museum, London;

14.9 Dallas Campbell; **14.10** POST 33/5130, The Royal Mail Archive at The Postal Museum, London; **14.11** POST 33/5130, The Royal Mail Archive at The Postal Museum, London; **14.12** POST 33/5130, The Royal Mail Archive at The Postal Museum, London; **14.13** From the collections of Museum nan Eilean, Stornoway; **14.14** Science & Society Picture Library/SSPL/Getty Images; **14.15** POST 33/5130, The Royal Mail Archive at The Postal Museum, London; **15.1** Historic Images/Alamy; **15.2** Sovfoto/Universal Images Group/Diomedia; **15.3** Photo courtesy Bonhams; **15.4** Sovfoto/Universal Images Group/Diomedia; **15.5** Cosmosphere International SciEd Center and Space Museum, Hutchinson, KS; **15.6** Granger – Historical Picture Archive/Alamy; **15.7, 15.8** John Frost Newspapers/Mary Evans Picture Library; **16.1** Sovfoto/Universal Images Group/Diomedia; **16.2** TASS Archive/Diomedia; **16.3-16.9** Sovfoto/Universal Images Group/Diomedia; **16.10** Onslow Auctions Limited/Mary Evans Picture Library; **16.11** John Frost Newspapers/Mary Evans Picture Library; **16.12** Sovfoto/Universal Images Group/Getty Images; **16.13** Detlev van Ravenswaay/Science Photo Library; **16.14** Universal Images Group/Diomedia; **16.15** Topfoto; **16.16** Ria Novosti/AFP/Getty Images; **16.17** TASS Archive/Diomedia; **16.18** ADN-Bildarchiv/ullstein bild via Getty Images; **16.19** colaimages/Alamy; **17.1** TASS Archive/Diomedia; **17.2** *The Soviet Space Program* (NSA Publication No. 166); **17.3** The Royal Aeronautical Society (National Aerospace Library)/Mary Evans Picture Library; **17.4** cityanimal/stock.adobe.com; **17.5** Vic/stock.adobe.com; **17.6** AlexanderZam/stock.adobe.com; **17.7** Daily Herald Archive/National Science & Media Museum/SSPL/Getty Images; **17.8** The Protected Art Archive/Alamy; **17.9** Fox Photos/Hulton Archive/Getty Images; **17.10** Photo 12/Alamy; **17.11** Photo Patrick Donovan/The Image Bank Unreleased/Getty Images; **18.1** Photo Paul Schutzer/The LIFE Picture Collection/Shutterstock; **18.2, 18.3** Papers of John F. Kennedy. Presidential Papers. President's Office Files. Speech Files. Address at Rice University, Houston, Texas, 12 September 1962. John F. Kennedy Presidential Library and Museum, Boston, MA; **18.4** Photo Robert Knudsen. White House Photographs. John F. Kennedy Presidential Library and Museum, Boston, MA; **18.5, 18.6** Photos Cecil Stoughton. White House Photographs. John F. Kennedy Presidential Library and Museum, Boston, MA; **18.7** KSC/NASA; **18.8** MSFC/NASA; **18.9** Museum of Contemporary Art Chicago, IL (Partial gift of Stefan T. Edlis and H. Gael Neeson, 1998.49). Photo Museum of Contemporary Art Chicago/Art Resource, NY. © 2025 Robert Rauschenberg Foundation/VAGA at ARS, NY and DACS, London; **18.10** Photo Robert Knudsen. White House Photographs. John F. Kennedy Presidential Library and Museum, Boston, MA; **19.1, 19.2** Bettmann/Getty Images; **19.3** Office of the Federal Register. (04/01/1985-). National Archives and Records Administration, College Park, MD; **19.4** NASA Lewis Research Center, *Orbit*, vol. XVI, no. 1, 30 September 1958; **19.5-19.13** NASA; **19.14** SpaceX/NASA; **19.15** JSC/NASA; **20.1** JSC/NASA; **20.2-20.6** T. A. Bergstrahl, Photography from the V-2 rocket at altitudes ranging up to 160 kilometres. Naval Research Laboratory, Washington, DC, NRL Report no. R-3083, 1947. Jeremy Norman Collection on the History of Aerodynamics, Aviation and Aerospace, approximately 1871–1976, The Huntington Library, San Marino, CA; **20.7** US Navy/John Hopkins (APL); **20.8** AP Wirephoto/Sydney Morning Herald/Superstock/Alamy; **20.9, 20.10** US Navy/John Hopkins (APL); **20.11** Photo Karl Mills, Scientific Photo Arts, Berkeley, California/NASA; **20.12** NASA Image Collection/Alamy; **20.13** KSC/NASA; **20.14, 20.15** NASA; **20.16-20.18** JSC/NASA; **21.1, 21.2** Ryan Nagata; **21.3** Keystone-France/Gamma-Keystone/Getty Images; **21.4** JSC/NASA; **21.5** Photo Mark Avino, Smithsonian National Air and Space Museum, Washington, DC (NASM 2008-8610); **21.6** JSC/NASA; **21.7** Smithsonian National Air and Space Museum, Washington, DC (NASM2008-14049); **21.8, 21.9** NASA; **21.10-21.13** Photos Lee Jones/NASA; **21.14** Science & Society Picture Library/SSPL/Getty Images; **21.15** Dallas Campbell; **21.16** Smithsonian National Air and Space Museum, Washington, DC (NASM2008-13657); **21.17** Ryan Nagata; **21.18** Photo Dane Penland. Smithsonian National Air and Space Museum, Washington, DC (NASM2016-00782); **21.19-21.24** Ryan Nagata; **22.1** TRW/PhotoQuest/Getty Images; **22.2** Courtesy MIT Museum, Cambridge, MA; **22.3** Smithsonian National Air and Space Museum, Washington, DC (NASM 9A12593-45506-A); **22.4, 22.5** Photos Ralph Morse/The LIFE Picture Collection/Shutterstock; **22.6** Smithsonian National Air and Space Museum, Washington, DC (NASM NASA-68-H-457); **22.7, 22.8** Dallas Campbell; **22.9** Courtesy ILC Dover Astrospace; **22.10** Mario De Biasi/Mondadori/Getty Images; **22.11, 22.12** Photos Ralph Morse/The LIFE Picture Collection/Shutterstock; **22.13-22.16** Massachusetts Institute of Technology, Magnetic Core Memory records, AC-0337. Box 76, Photos. Department of Distinctive Collections, MIT Libraries, Cambridge, MA; **22.17** From T. E. Ivall, *Electronic Computers: Principles and Applications* (Iliffe, 1956); **22.18, 22.19** Photos Raytheon. Collection David Meerman Scott; **23.1** Courtesy Virgin Galactic LLC; **23.2** Photo John Rensten and Graham Smith. Courtesy Trevor Beattie; **23.7-23.17** Photos John Rensten and Graham Smith. Courtesy Trevor Beattie; **23.18-23.25** Courtesy Virgin Galactic LLC; **24.1** Courtesy Dan Postgate/Smallfilms; **24.2** Gavin Rodgers/Alamy; **24.3, 24.4** Photos courtesy Clive Banks (clivebanks.co.uk); **24.5-24.16** Courtesy Dan Postgate/Smallfilms; **25.1** CBS Photo Archive/Getty Images; **25.2** © 2025 Eames Office, LLC. All rights reserved; **25.3, 25.4** From George Edward Pendray, *The story of the Westinghouse Time Capsule* (East Pittsburgh, PA: Westinghouse Electric & Manufacturing Co., 1939); **25.5** Scherl/Süddeutsche Zeitung/Alamy; **25.6** Library of Congress, Washington, DC; **25.7, 25.8** JPL-Caltech/NASA; **25.10, 25.11** NASA; **25.12** Yutaka Nagata/UN Photo; **25.13** Rice/UN Photo; **25.14** B. Wolff/UN Photo; **25.15** Tsagris/UN Photo; **25.16, 25.17** UN Photo; **25.18** Yutaka Nagata/UN Photo; **25.19** Ray Witlin/UN Photo; **25.20** UN Photo; **25.21, 25.22** Yutaka Nagata/UN Photo; **25.23** UN Photo; **25.24-25.29** NASA; **26.1** © Jocelyn Bell Burnell; **26.2, 26.3** The Papers of Professor Antony Hewish, HWSH 1/1, Churchill Archives Centre, Cambridge. Courtesy the Hewish Estate; **26.4** From Harold D. Craft Jr, *Radio Observations of the Pulse Profiles and Dispersion Measures of Twelve Pulsars* (September 1970). Courtesy Harold D. Craft Jr; **26.5** Landmark Media/Alamy; **27.1** Acey Harper/The Chronicle Collection/Getty Images; **27.2** PictureLux/The Hollywood Archive/Alamy; **27.3** From *Popular Science*, vol. 95 (September 1919); **27.4-27.6** Science Photo Library; **27.7** Courtesy of the Ohio History Connection (AL07146); **27.8** Dr Seth Shostak/Science Photo Library; **27.9-27.11** NASA; **28.1-28.5** Courtesy of Taito Corporation; **28.6, 28.7** From H. G. Wells, *La guerre des mondes* (Brussels: L. Vandamme & Co., 1906); **28.8-28.11** Courtesy of Taito Corporation; **28.12-28.16** Photos Invader. © 2025 ADAGP, Paris and DACS, London; **29.1** Photo Tech. Sgt Doug Gruben/US Air Force; **29.2** Courtesy Ron Jones; **29.3** Ames Research Center/NASA; **29.4, 29.5** Courtesy Ron Jones; **29.6** Ames Research Center/NASA; **30.1** Photo © Estate of Francis Bello/Science Photo Library; **30.2-30.9** Courtesy George Dyson; **30.10** Olaf Stapledon, *Star Maker* (London: Methuen & Co., 1937); **30.11** Courtesy George Dyson; **30.12-30.14** © Jon Lomberg 2025; **31.1, 31.2** Courtesy the Estate of Paul Van Hoeydonck; **31.3** JSC/NASA; **31.4, 31.5** Courtesy the Estate of Paul Van Hoeydonck; **31.6** JSC/NASA; **32.1** Blue Origin; **32.2** Gerard K. O'Neill, *The High Frontier: Human Colonies in Space* (New York: William Morrow and Company, Inc., 1976); **32.3** Ames Research Center/NASA; **32.4** Blue Origin; **32.5-32.8** SpaceX/NASA; **32.9** © Barbara Diener 2023; **32.10** Courtesy Tresco Island; **32.11** Sketch by R. A. Smith. © BIS; **32.12** Blue Origin; **33.1** Photo Ashton Claridge, 2011, Flinders University Archives, Adelaide (APH25 392); **33.2, 33.3** ISSAP/NASA; **33.4-33.6** NASA Orbital Debris Program Office; **33.7** Geoff Tompkinson/Science Photo Library; **33.8** Skylab Protective Helmet: concept and marketing Jeff Hall; design and production William Foard. Photo courtesy Foard Media, Los Angeles; **33.9** Detlev van Ravenswaay/Science Photo Library; **33.10** MSSS/JPL-Caltech/NASA; **33.11** Baikonur III. Photo © Jonk; **33.12** US National Park Service; **33.13** JSC/NASA

Index

Page numbers in *italic* refer to the illustrations

0-9

10 Story Fantasy 49
2001: A Space Odyssey (film) 39, 48–50, *50–1*, 260

A

Abbe, Ernst 13
Adams, Douglas 264
Aelita: Queen of Mars (film) 99
Aerothreads 203
Aldrin, Buzz 8, 24, *27*, *182–3*, 199, *207*, 208, 250
alien life 219, 227–8, 230–4, *230–5*, 256
Allen, Iona 200
Allen Telescope Array (ATA) 230, *234*
Amazing Stories 45
American Civil War 26, 31
Ansari X Prize 204
Ansel, Ruth 169
Apollo 13 (film) 186
Apollo program *27*, 48, 118, 125, 204, 208, 264
  artefacts on Moon 260–1, *260–3*, 279, *281*
  computers 200–3, *202–3*
  Kennedy and 26, 164
  legacy of 174–81, *174–85*
  Lunar Module 39
  spacesuits 187, *190*, 192, *192–3*, *195*, 196–200, *196–201*
  archaeology 272–9, *272–9*
Arecibo Observatory 230, 232, *232*
Aristotle 16, 64
Armstrong, Neil 8, 24, *27*, 112, 174, *174*, 180–1, *181*, 186, 200, *200–1*, *206*, 208

Asimov, Isaac 264
Astro Aerospace/Northrup 88–9
astronauts 204
  cosmonauts 97–8, 99, 137, 142–50, *142–51*
  detritus on the Moon 260–1, *260–3*
  effects of space travel 112, 118
  first men in space 142–50, *142–51*
  Mercury mission *163*
  Moon landings 8, 10, 24, 26, 122, 125, 158, 164, 174–81, *174–85*, 270
  in space stations 64
  spacesuits 186–92, *186–95*, 196–200, *196–201*, 268
  Trevor Beattie 204–8, *204–9*
Atari 236
Atlas ICBM *163*

B

Babbage, Charles 200
Bach, J.S. 224
Baikonur Cosmodrome 137, 142, *278*, 279
ballistic missiles 121, 135–7, 142
Bangor, Henry Ward, 5th Viscount 78
Bassett, Charles A. II *263*
Bayard, Émile *27*
Bayer, Johann *25*
Beattie, Trevor 204–8, *204–9*
Beethoven, Ludwig van 224
Behnken, Bob 172
Belyayev, Pavel 149, *263*
Bergerac, Cyrano de 21, 31
Bergstralh, T.A. *175*
Berry, Chuck 224
Betsill, Morgan 203
Bezos, Jeff 264, *264*, 266, 268–70
Bierstadt, Albert *36*
biosignature gases 86
Birr Castle, Ireland 74, *74–5*

Blue Marble photograph 179
Blue Origin 8, 10, 266, *266*, 268–70, *271*
Boeing 266
Bolton, Scriven 34
Bonestell, Chesley 34–9, *34*, *37*, *38*, 40–3
Boston Post 103
Bowie, David 204
Brahe, Tycho 13, 16, *17*, *18–19*
Branson, Richard 204, 268
Braun, Wernher von 35, 41, 112, 114, 120–4, *120–7*, 160, 249
British Interplanetary Society 44, 270
Brooke Bond *205*
Buran space shuttles *278*, 279
Burchell, Mark *275*
Burnell, Jocelyn Bell 226–8, *226–9*, 232
Bush, George 249
Bykovsky, Valery *142*

C

Callisto 57
Cambridge University 226–8
Capra, Frank 92n.
Carpenter, M. Scott *188*
Carter, Jimmy 224
Cassini, Giovanni 66
Chaffee, Roger B. 180, *263*
Challenger space shuttle 172, *173*, 246
'The Chief Designer' 134–7, *134–9*
The Clangers 212–16, *212–17*
Clark Telescope, Flagstaff 69
Clarke, Arthur C. 8, 13, 44–50, *44*, *46–7*, *49*, *50*, 60, *61*, 64, *65*, 260
Clark University 103
Cocconi, Giuseppe 232, 234
Cold War 26, 137, 158–61, 179, 252
Collier's magazine 35–9, *40*, 122, *124–6*

Collins, Michael 24, *27*
Columbus, Christopher 24
comets 78, *80–1*
Communism 122, 158
computers 200–3, *202–3*
Cooper, L. Gordon Jr *188*
Copernicus, Nicolaus 16, 21, *18–19*, 60
Correa, Henrique Alvim *238*
Cosmic Voyage (film) 97–8, 99
Cosmism 92–5, 150
cosmonauts *see* astronauts
Cosmopolitan 102
Creti, Donna *55*, *59*
Cronkite, Walter 261
Crossfield, Scott 208
Curiosity rover 276

D

da Gama, Vasco 24
Daemon 16, 21
Daily Express *130*, 131
Daily Mail 71
Daily Mirror 207
Daily News (New York) 139
Danne & Blackburn 169–72
Darwin, Charles 71
Davis, Don 249, *251*
Descartes, René *18–19*
Destination Moon (film) *38–9*, 39
DeVoto, Harry 167
Diamandis, Peter 266
Diener, Barbara *125*, 269
Dione 35
Disney, Walt 122
Dobrovolski, Georgi *263*
dogs, Laika 152–4, *152–7*
Dollond telescopes 78
Dombrowski, C.H. 129
Doná, Leonardo 57
Donati's Comet 78, *80*
Dornberger, Major General Walter 120–2
Dostoyevsky, Fyodor 92
Drake, Frank 219, 232, 234

Druyan, Ann 223
Dryden, Hugh 167
Duke, Charlie 181, *192*, 208
Duracotus 16, 21
Dyson, Freeman, 252–6, *252–5*, *257–9*
Dyson spheres 256, *258*

E

Eames, Charles and Ray 218, *218*
Earhart, Amelia 13
Earth, orbit around Sun *18–19*
Earthrise photograph 118, 174–9, *184–5*, 212
Ehman, Jerry 232
Einstein, Albert 256
Eisenhower, Dwight D. 166–7
Eisenstein, Sergei 92
Ellingson, Joshua *17*
Elliott, Vernon 212
Eimmart, Maria Clara *59*
Engle, Joe 208
Enterprise space shuttle *246*, 249
Esnault-Pelterie, Robert 114
Europa (moon) 57
Europa Clipper spacecraft 203
exoplanets 84–7, 234
Explorer satellites 122, 166
'Eyes on Exoplanets' 86, *87*

F

Falcon rockets 172, *172*, *267*, 268, *270*, 279
Fedorov, Nicolai 92, 99
Fellows, Hazel *199*
Fermi, Enrico 232
Fernandez, Jose *192*, 268
Feynman, Richard 256
films 34–5, 39, 48–50, 99, 204, 218, 231, 236–40, 260
Filonov, Pavel *93*
Fiolxhilde 16, 21

Firmin, Peter 212
First Man (film) 186
First World War 128
Flagstaff, Arizona 67, 69
Flammarion, Camille 232
Florida 26
Fly Me to the Moon (film) 186
flying saucers 232
Foard, William 276
Fontenelle, Bernard Le Bovier de 31, 67
footprints 279, *280–1*
For All Moonkind 279
Foraker, Eleanor 200, *200*
Frau im Mond (film) 114–18, *118–19*
Freedom 7 capsule 150
Freeman, Fred *124–5*
Freeman, Theodore C. *263*
fuel, rockets 95
Funk, Wally 268–70

G

Gagarin, Yuri 137, 142–3, *142–5*, *148*, 149–51, *150–1*, 152, 158, *263*, 273, 279
galaxies 57, 60, 84, 232, 256
Galileo Galilei 16, 54–7, *54*, *56*, *58*, *59*
Galluzzo, Leopoldo 32
Ganymede 57
gases, biosignature 86
Gassendi, Pierre *18–19*
Gemini mission 180, *192*
General Atomics 252, *252–5*
George V, King 131
Gernsback, Hugo 45, 50
Giudice, Rick 246–9, *265*
Givens, Edward G. Jr *263*
Glenn, John *161*, *163*, *188*, 270
Glennan, T. Keith 167
Goble, Warwick 240
Goddard, Robert H. 8, *8*, 32, 99, 102–9, *102–11*, 112, 268
Godwin, Francis, Bishop of Hereford 21, *22–3*, 24, 31

285

'Goldilocks' zones 86
Goldin, Dan 172
Gonzales, Domingo 21
Gorman, Alice 272–9, *272*
gravity 60, *63*, 64, *65*
Green, Simon 275
Green Bank Observatory 232
Grissom, Virgil I. (Gus) 180, *188*, *263*
Group for the Study of Reactive Motion (GIRD) 134
Guggenheim, Daniel 107

H

Hale, Edward Everett 32, *32*
Hall, Jeff 276
Hamilton, Margaret H. 196, *196–7*
Harper's Bazaar 169
Harriot, Thomas 56, *57*
Heinlein, Robert 264
Herschel, Sir John 31
Herschel, William 66, 75
Hitler, Adolf 121, *122*
Hoeydonck, Paul von 260–1, *260–3*
Hokusai, Katsushika 240
Holliday, Clyde T. 122, 174, *178–9*
Hollywood 35, *39*, 180, 192
Hubble Space Telescope 84, 275
Hudson River School 35
Hurley, Doug 172
Huygens, Christian 66

I

Iapetus 35
Iceland 16
Illingworth, Sam 280
Industrial Revolution 24, 31
Instagram 172, 192
Integrated Space Plan (ISP) 247, 249, *249–50*

International Air Post Exhibition (APEX), London (1934) 129, *130*
International Geophysical Year (1958) 166
International Latex Corporation 196, *198*
International Space Station (ISS) 60, 64, 137, 172, 264, 270, 273, *273*
Io 57
Ireland 74
Ivanov, Konstantin Konstantinovich 155
Ivanovsky, Oleg 142

J

James Webb Space Telescope (JWST) 8, *10*, 84, *85*
Jarvis, Gregory B. *173*
Jet Propulsion Laboratory (JPL) *88–9*, *203*, 218, 276
Jimenez, Jonathan 'Jonk' *278*, 279
Johnson, Lyndon B. *11*, *160*, 164
Jones, Ron 246–9, *246–51*
Joy Division 229
Jupiter 45, 57, *58*, *83*, *203*, 218, 252

K

Kaluga 95, 99
Kardashev, Nikolai 256
Katzen, Elliot 167
Kennedy, John F. 26, 137, *140*, 158–64, *158–65*, 197
Kennedy Space Center 112
Kepler, Johannes 16–21, *16*, 24
Kepler space telescope 84
Khrushchev, Nikita 152
King, Martin Luther 179
Kinzler, Jack 279
Komarov, Vladimir *142*, *263*
Kopra, Tim 152
Korolev, Sergei Pavlovich 137, 273

Krasser, Friedrich 112
Kubrick, Stanley 48–50, 260
Kuiper belt 256
Kwajalein Atoll 268

L

Laika 152–4, *152–7*
Lake, Simon 32
Lang, Fritz 114–18, *118–19*
Lansberghe, Philippe van *18–19*
Lasser, David 45–8, *50*
Lawrence, T.E. 13
lenses 54–7
Leoni, Ottavio *54*
Leonov, Alexei *139*, 148–9, *149*
Lewis, Cathleen 192, *192*
Ley, Willy 35, *38*, 114
Lick Observatory 34
Life magazine 34, *35*, *127*, 198, *201*
Lindbergh, Charles 107, 111, 158
Lipperhey, Hans 54
Locke, Richard Adams 30, 31, *31*
Lockheed Martin 266
Lomax, Alan 224
Lomberg, Jon *257–9*
Los Alamos, New Mexico 232
Los Angeles Times 138
Lowell, Percival 66–71, *66–71*, 67, *67*, 232
Lucian of Samosata 21
Luminism 35
Luna spacecraft 137, *137*
Lunokhod rovers 260

M

McAuliffe, S. Christa *173*
McLennan, Christina 128
McNair, Ronald E. *173*
Magellan, Ferdinand 24
Malenchenko, Yuri 152
Mallory, George 158
Manifest Destiny 35
Mars
  canals on 66–71, *66–73*, 102

colonization of 10, 60, 249, 252, 270
communication with 231–2, *231*
human artefacts on 276, *277*
Mars Global Surveyor 66
Viking missions 66, *66*, 276
Marshall Space Flight Center 85
Maté, Rudolph 39
Medici family 57
Méliès, Georges 32, *33*, 118
Mendeleev, Dmitri Ivanovich 92
Mercury (planet) 78, 86
Mercury project *161–3*, 169, 187, *188–9*, 194
Meyer-Brandis, Agnes 23
Michel, Jean 125, *127*
microscopes 74–8
Milky Way 57, 84
Mimas 35
missiles, ballistic 121, 135–7, 142
MIT 196, 200
Modarelli, James 'Jim' 167–9, *169*
Monument to the Conquerors of Space *156–7*
Moon 78, 84
  Apollo program 48, 118, 200
  craters *12*, *13*, 21
  early drawings of 56, *57*, *58*
  early stories about 16
  eclipses 82
  Goddard and 103, *105*
  human artefacts on 260–1, *260–3*, 272, 279, *281*
  Luna spacecraft 137, *137*
  Moon landings 8, 10, 24, 26, 122, 125, 158, 164, 174–81, *174–85*, 270
  phases *18–19*
  in science fiction 21–4, 26–32
  travel to 112
moons, Jupiter 57
Morning Oregonian 71
Morrison, Philip 232, 234
Morse, Ralph *198*, *201*
motion, laws of 60, 64, 102

Munnings, Lady 152
Murano 54
Musk, Elon 10, 92n., 264–70, *268*, 279

N

Nagata, Ryan 186–92, *186–95*
National Aeronautics and Space Administration (NASA) 21, 166–72, *166–73*
  Ames Research Center 187
  Ames/Stanford Torus study 246
  computers 200–3, *202–3*
  'Eyes on Exoplanets' 86, *87*
  Goddard Space Flight Center 109
  logos 167–72, *168–73*
  Marshall Space Flight Center 85
  Mercury project *161–3*, 169, 187, *188–9*, 194
  Saturn rockets 122, 137, *160–1*, 252
  seamstresses 196–200, *196–201*
  space junk 276, *277*
  space race 137, 143, 161, 166–7, 179, 216
  starshade *88–9*
  see also Apollo program
Nature 232
Nazi party 120–2, 124, *127*
nebulae 76–7
Nell (rocket) 103–7, *104–7*
Neptune 218
Netherlands 54
New Glenn rocket 10, 270, *271*
New York Times 8
Newton, Isaac 8, 60–4, *60*, *62*, 102, 143
Nikolayev, Andriyan *142*
Nishikado, Tomohiro 236–41, *236–43*
Nixon, Richard 164
Nobel Prize 226, 228
Northcutt, Frances 'Poppy' 196, *196*

nuclear weapons 135–7, 142, 143, 252

O

Oberth, Hermann 32, 99, 103, 112–18, *112–19*, 120
Ohio State University 232
O'Neill, Gerard K. 246, *264*, 266
Onizuka, Ellison S. *173*
Operation Paperclip 122
orbits 64
Orion 57, *58*
oxygen 86

P

Pacific News Service 72
Pacific Ocean 26
Pal, George 39
Patsayev, Viktor *263*
Peake, Tim 152
Peenemünde 120–2
Perry, Katy 8, 270
Perseverance rover 276
Phillips, Frank 45
Phoebe 35
Pickering, William 232
Pioneer spacecraft 219
planets, exoplanets 84–7, 234
  solar system 86
  Voyager spacecraft 218–24, *218–25*
Plato 64
Pluto 256
Poe, Edgar Allen 31–2
Polk, Willis 34
Popovich, Pavel *142*
Popular Mechanics 9
Popular Science 129, 231–2, *231*
Post, Wiley 187, *187*
postage stamps 129–31, *130*, *153*, 155
Postgate, Oliver 212–16, *212–17*
Potočnik, Herman 'Noordung' 45
Prada 192

Praesepe 58
Project Cyclops 234, *235*
Project Mercury 150
Project Orion 252
Project Ozma 232
Proxima Centauri 86, 87
Ptolemy 16, *18–19*
pulsars 227–8, *227–9*, 232

Q

quasars 226–7

R

radio astronomy 226–8, *226–9*, 234
Radkevich, Lyudmila 152–3
Die Rakete 129
Ramsay, Sir William 131
Rapp, Rita 279
Rauschenberg, Robert *164*
Raytheon 200
Red Army 134
Redstone rockets 122
Resnik, Judith A. *173*
reusable spacecraft 266
Rice University 158, 164, *165*
Robins, Denise 152
rockets
  'The Chief Designer' 134–7, *134–9*
  Falcon rockets 172, *172*, 267, 268, 270, 279
  fuel 107, 112, 131
  Goddard and 103–9, *104–11*
  heavy lift rockets 270, *271*
  invention of 32
  multi-stage rockets 112
  N-1 rockets 137
  Oberth and 112–18, *112–19*
  postwar US space programme 122
  Redstone rockets 122
  reusable rockets 8–10, 9
  Rocket Mail 129–31, *130–3*
  Saturn rockets 122, 137, *160–1*, 252

Soyuz rockets 134, 152
Tsiolkovsky and 94–9, *94*, *96–7*, *100–1*
V-2 rockets 35, 109, 114, 120–4, *121*, *123*, 135, 174, *175*, 270
X-15 rocket plane 204–8
Rockwell International *247*, 249, *249–50*
Roeg, Nicolas 204
Rogan, Joe 66
Rogers, Samuel 60
Rosse, William Parsons, 3rd Earl of 74, *74–7*
Rotary Rocket Company 264
Royal Air Force (RAF) 121
Royal Observatory, Greenwich 74
Rudeaux, Lucien 34, *35*
Rutan, Bert 204

S

Sagan, Carl 66, 212, 218–24, *218–25*, 230–1, 232
Salt Lake Tribune 72
Salyut space station 276
Salzman, Linda 219–23
San Francisco 34
Santa Barbara, California *88–9*
satellites 8, *124–5*, 143
  Explorer satellites 122, 166
  first American satellites 166, *166*
  in science fiction 32
  space junk 276
  Sputnik 99, 109, *136*, 137, 152, *153*, 166, 252, *272*, 276, 279
Saturn 34, 35, *83*, 218, 252
Saturn rockets 122, 137, *160–1*, 252
Scarp, Hebrides 128, 131
Schiaparelli, Giovanni 66–7
Schirra, Walter M. Jr *188*
science fiction 16–24, 26–32, 264–6
Science Wonder Stories 45, *45*, 48
Scientific American 102
Scobee, Francis R. 'Dick' *173*

Scott, David 260–1, *262–3*, 279
Scott, Ridley 240
Scott-Heron, Gil 179
Seager, Sara 84–7, *84*
seamstresses 196–200, *196–201*
Second World War 34, 120–1, 131, 134–5, 179, 216, 252
See, Elliot M. Jr *263*
SETI (Search for Extraterrestrial Intelligence) 219, 230–4, *230–5*, 256
Shackleton, Ernest 13
Shatner, William 8, 270
Shepard, Alan B. 150, *188–9*, *193*, 268
Shostakovich, Dmitri 142, 143, 149
Shotwell, Gwynne 268
Shrimpton, Jean 169
Sikorsky, Igor 32
Skylab 276, *276*
Slayton, Donald K. *188*
Smith, Barnabas 64
Smithsonian Museum, Washington 103, 192, 261
solar system 86
Sorensen, Theodore C. 158, *158–9*, 161–4
Soviet Union 99
Soyuz rockets 134, 152
space colonies 264–70
Space Invaders 236–41, *236–43*
space junk 273–9, *274–5*
Space Launch Complex Six *246*
space race 137, 143, 161, 166–7, 179, 216
Space Shuttles 172, 246, *246*, 249, 264
space stations 32, 45, 60, 64, 137, 172, 246–9, *248*, *251*, 264, 270, 273, *273*, 276
space telescopes 8, *10*, 84, *85*, *275*
space travel *93*, 94–9, 252–6, *253–9*
  effects on astronauts 112, 118
  first man in space 137
  Goddard and 102–9
space tourism 268–70

spacesuits 186–92, *186–95*, 196–200, *196–201*, 268
spacewalks *139*, 149, *149*
SpaceX 8–10, 9, *137*, 172, *172*, 192, 266–8, *267*, *268*, 270
Sputnik 99, 109, *136*, 137, 152, *153*, 166, 252, *272*, 276, 279
Stalin, Joseph 99, 134
Stanislavski, Konstantin 92
Stapledon, Olaf 48, *49*, 50, 256, *256*
Star Wars (film) 39
Starshade project 86–7, *88–9*
Starship 8–10, 9, *137*, 270
Sumner, Bernard 229
Sun *18–19*, 78, 86
Sunset magazine 34

T

Taito 236
Tarter, Jill 230–4, *230*
Tchaikovsky, Piotr Il'ich 92
telescopes 54–7, 74–8, *74–5*, 84, 226–8, *275*
Tennyson, Albert, Lord 46
Tereshkova, Valentina *142*, 143–9, *148*
TESS (Transiting Exoplanet Survey Satellite) 84
Thames, River 13
Thomas, Lewis 224
Thornhill, Sir James 60
Thuller, Elisabeth *33*
time capsules 218–24, *218–25*
Time magazine 152
Titan 35
Tsander, Fridrikh Arturovich 135
Tsiolkovsky, Konstantin 32, 92–9, *92–101*, 112, 137, 273
Tupolev, Andrei 134
Tycho Crater, Moon 13

U

United Nations 161
University of Padua 54, 57

Uranus 218
US Army 122
US Navy 107, 166, 187

V

V-2 rockets 35, 109, 114, 120–4, *121*, *123*, 135, 174, *175*, 270
Valier, Max 114
Vandenberg US Air Force base 246
Vanguard rockets 166, *166*, 276
Vanguard Six *147*
Veller, Max 128, *129*
Venera landers 276
Venice 54–7
Venus 276
Verein für Raumschiffahrt (VfR) 114, 120, 128
Verne, Jules 9, 26–32, *27–9*, 48, 94, *94*, 112, 118, 252–6
Versailles, Treaty of (1919) 120, 128
video games 236–41, *236–43*
Vietnam War 179
Viking mission 66, *66*, 276
Virgin Galactic 204–8, *204*, *207–9*, 268
Volkov, Vladislav *263*
Voskhod 2 mission *148–9*, 149
Vostok spacecraft 142, 143–50, *144–7*, 150
Le Voyage Dans La Lune (film) 32, *33*, 118
Voyager Golden Record 64, 218–24, *218–25*, 276–9
Vykukal, Hubert 'Vic' 187, *190–1*

W

Wallace, Alfred Russel 71
Waltham Watch Company 200–3
Wan Hu 21
War of the Worlds (film) 39
Ward, Mary *12*, 74–8, *74–83*
Welles, Orson 236

Wells, H.G. 48, 71, 102, 236, 240, 241
Westinghouse time capsule 218–19, *219*
Westminster Abbey, London 64
Whalen, Grover *219*
When Worlds Collide (film) 39
White, Edward H. II 180, *181*, *263*
White Sands Missile base 122
Wide World 102
Wilkins, John 21–4, *24*
Williams, Clifton C. Jr *263*
Wit, Frederick de *18–19*
women, NASA workers 196–203, *196–203*
World Magazine 72
Wright, Ernie *85*
Wright Brothers 95, 208, *208*

X

X Prize 266

Y

Young, John 279
Young, Lisa 192, *192*

Z

Zubrin, Robert 270
Zucker, Gerhard 128–31, *128–33*
Zwicky, Fritz 13

To Mum and Geordie

Dallas Campbell has presented some of the most ambitious landmark series across the BBC, such as *City in the Sky* with Dr Hannah Fry and *Stargazing Live* with Dara Ó Briain and Brian Cox. He is a regular contributor to the BBC's science magazine *Focus* and *The Times' Eureka* magazine, and also presents the *Patented* podcast, which explores the history of inventions.

Front Cover: Chesley Bonestell, *Man Will Conquer Space Soon: What Are We Waiting For?*, c. 1952. Reproduced courtesy of Bonestell LLC
Spine: Percival Lowell's hand-drawn map of Mars, 1896–97. Courtesy Lowell Observatory Archives, Flagstaff, AZ
Back Cover: Hubert 'Vic' Vykukal modelling the AX-3 spacesuit at the Ames Research Center, 22 July 1977. Photo Lee Jones/NASA

Endpapers:
The Habitat Wheel of the first geostationary space station: view of the side constantly facing the sun, without a concave mirror, partially in cross-section. From Herman Potočnik (Noordung), *Das Problem der Befahrung des Weltraums: der Raketen-Motor* (R. C. Schmidt, 1929); The Habitat Wheel of the first geostationary space station: axial cross section. From Herman Potočnik (Noordung), *Das Problem der Befahrung des Weltraums: der Raketen-Motor* (R. C. Schmidt, 1929); How the diameter of the apparent circular segment of the Earth increases with the height of an orbiting satellite. From A. Sternfeld, *Interplanetary Travel* (Foreign Languages Publishing House, 1958).

Openers:
p. 2 One of twenty-five photos of the Moon's surface captured by the crew of Apollo 12, 14 November 1969.
p. 14 Émile-Antoine Bayard, 'The projectile passing the moon', colour lithograph from an 1877 English edition of Jules Verne's *From the Earth to the Moon*.
p. 52 Mary Ward, 'Miniature Chart of the apparent Path of Donati's Comet' (detail), plate from *Telescope Teachings* (1859).
p. 90 Ed White, the first US astronaut to walk in space. Detail from the front page of *Daily News* (New York), Wednesday, 9 June 1965.
p. 140 Robert Rauschenberg, *Retroactive II* (detail), 1963.
p. 210 Detail from the cover of *Project Cyclops* (1971).
p. 244 Rick Guidice, *Space Colonization Modules* (detail), 1975.

First published in the United Kingdom in 2026 by Thames & Hudson Ltd, 6–24 Britannia Street, London WC1X 9JD

First published in the United States of America in 2026 by Thames & Hudson Inc., 500 Fifth Avenue, New York, New York 10110

Space Journal: Art, Science and Cosmic Exploration
© 2026 Thames & Hudson Ltd, London

Text by Dallas Campbell © 2026 King Prawn Productions Ltd

All Rights Reserved. No part of this publication may be reproduced or transmitted in any form or by any means, electronic or mechanical, including photocopy, recording or any other information storage and retrieval system, without prior permission in writing from the publisher.

EU Authorized Representative: Interart S.A.R.L.
19 rue Charles Auray, 93500 Pantin, Paris, France
productsafety@thameshudson.co.uk
interart.fr

A CIP catalogue record for this book is available from the British Library

Library of Congress Control Number 2025945340

ISBN 978-0-500-02818-6
01

Printed and bound in China by Shenzhen Reliance Printing Co. Ltd

Be the first to know about our new releases, exclusive content and author events by visiting
thamesandhudson.com
thamesandhudsonusa.com
thamesandhudson.com.au

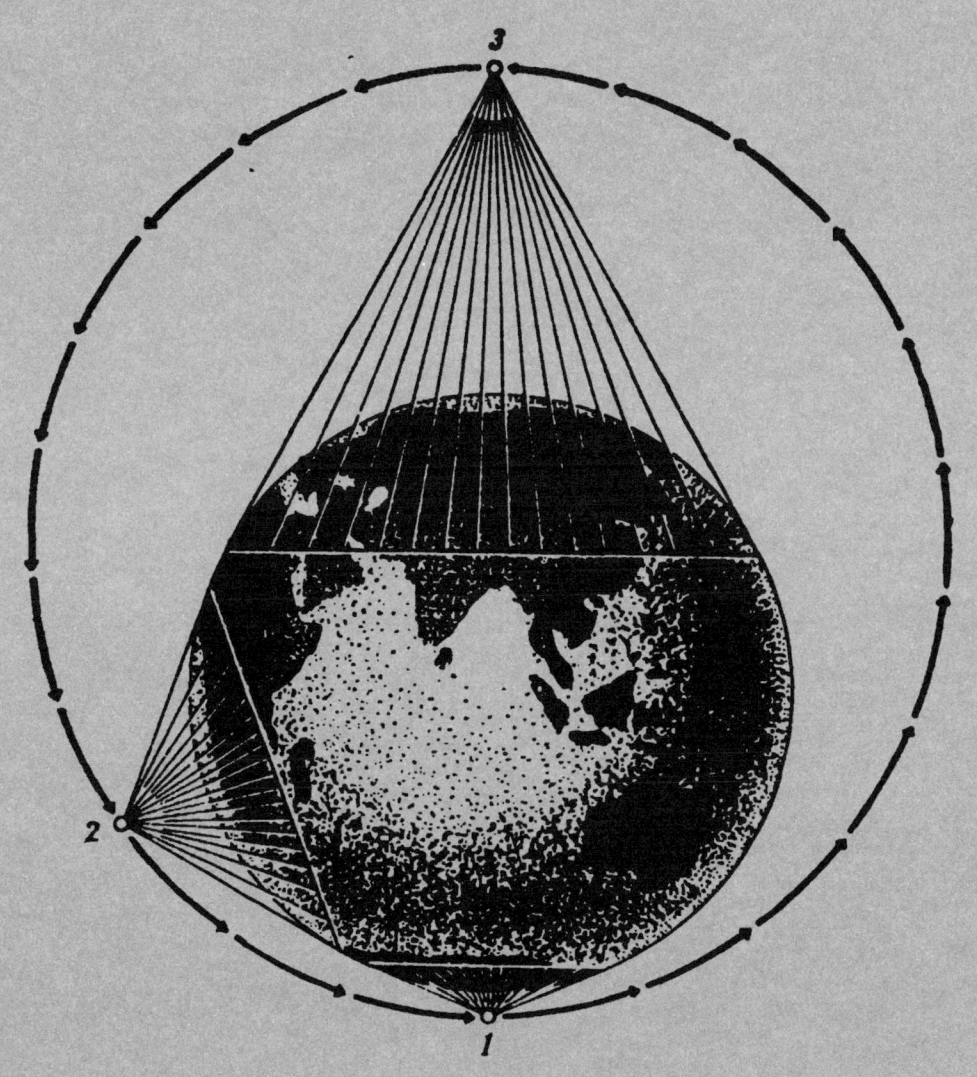

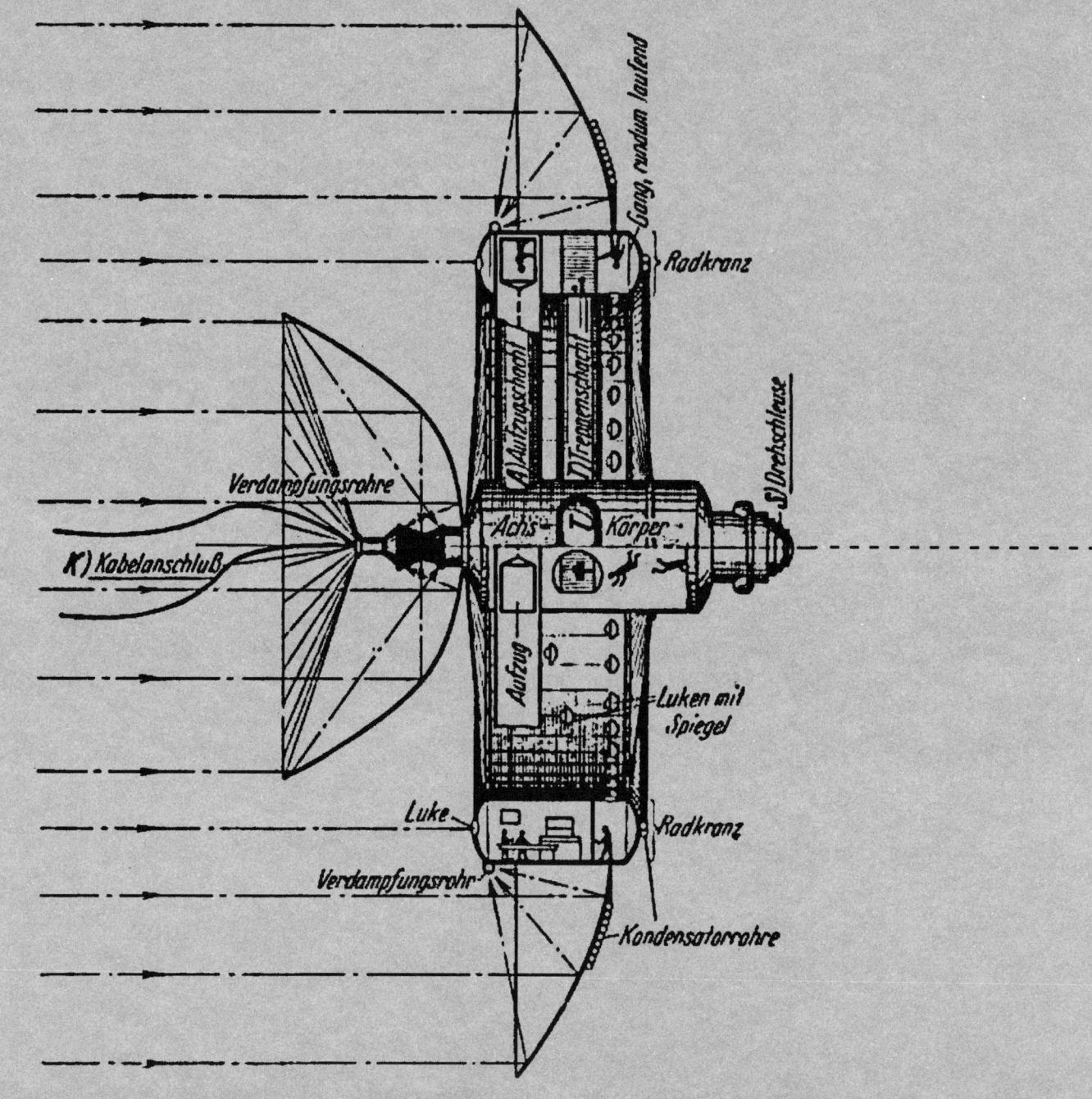